Cover credit: Bendetta Toledo

First edition published 2022
by CRC Press
6000 Broken Sound Parkway NW, Suite 300, Boca Raton, FL 33487-2742

and by CRC Press
2 Park Square, Milton Park, Abingdon, Oxon, OX14 4RN

© 2022 Taylor & Francis Group, LLC

CRC Press is an imprint of Taylor & Francis Group, LLC

Library of Congress Cataloging-in-Publication Data (applied for)

ISBN: 978-0-367-51214-9 (hbk)
ISBN: 978-0-367-51218-7 (pbk)
ISBN: 978-1-003-05286-9 (ebk)

DOI: 10.1201/9781003052869

Typeset in Times New Roman
by Radiant Productions

Preface

Mobile learning is nowadays a frontier trend of the digital world. The aim of this ground breaking book is to solicit teachers, educators, and practitioners to renew their own teaching and learning methodologies narrowing themselves to educational digital models related to mobile technologies. Instructional designers, curriculum developers, and learning professionals can empower their design of innovative educational mobile learning environments. Learning get more and more customized in always-connected and ever-changing educational mobile learning environments, grounded on concepts like space, sound, and bodies.

Mobile devices are indeed to be already considered as extensions of human bodies and brains. Nomadic learners always use them in daily activities. But not all teachers are prepared to use mobile technologies in and outside classroom, and they can find a significative support in this book to manage mobile digital world and to explore the role of technology devices in learning enhancement processes. Innovative digital interactions are studied also deepening Specific Learning Disorders.

In Chapter 1, Flavia Santoianni analyses which can be considered the key aspects of mobile digital education, evaluating how the multiple affordances of mobile devices can be intertwined with curriculum and user interaction, in order to develop the skills of 21st century learners. Technology availability and flexible adaptation to it are all factors involved in effective technology adoption, which is the first step to gain mobile personalized and self-regulated learning. Mobile digital education is nowadays learner-centered, but it represents a frontier pedagogy. As explained in Chapter 2, indeed the history of educational theories of mobile teaching alongside the twentieth century can be related to mobile learning but it has been influenced by different learning metaphors, respectively of acquisition, participation, and knowledge creation. Innovative teaching and learning are rooted on significative educational theories of the twentieth century, but it leverages a huge re-thinking of pedagogical practices. As a consequence, the educational design of mobile learning environments is related in Chapter 3 to collaborative, experiential, personalized, active, authentic, formal and informal approaches to student learning.

In Chapter 4, Corrado Petrucco discusses the role of smartphone as a technological-cultural object and its potential problematic effects in social and cognitive interaction, deepening the concept of smartphone addiction for both teachers and students. Teachers, in particular, should acquire technological skills to improve smartphone usability in teaching and learning. In Chapter 5, it is suggested to overcome the traditional models to adopt instead innovative educational theories, such as situated cognition and experiential learning, in which the social construction of knowledge stands alongside the enhancement of the methodological competencies needed to design integrated learning environments.

In Chapter 6, Alessandro Ciasullo focuses the technological educational approaches on sound, which can be considered as a mediator between individuals, environment, and their own nature. The evolutionary value of sound is analyzed accordingly to the transition from nature to culture, which allows to show in Chapter 7 the interactions between technologies, body, and sound production by highlighting the interchange between natural interfaces and the interpretation of human actions. In line with this direction, in Chapter 8 is deepened the role that sound plays in Specific Learning Disorders and how smartphones, tablets, notebooks, and PCs can develop remote cloud higher-level effective assistive systems.

In Chapter 9, Daniele Agostini explores mixed reality mobile technologies for education and training, explaining the meaning of Augmented Reality (AR) and how handheld devices can be used for virtual, augmented, and mixed reality, outlining the application of projects for the future experiences. Teaching and learning practices, settings, and contexts interact with mobile mixed reality tools in Chapter 10, where the shift from static and place-centred to mobile, active, and co-constructive approaches is studied from the theoretical and methodological point of view of Activity Theory. In Chapter 11, the transition from Mobile Learning to Mixed Reality Mobile Learning, and its affordances, is focused.

This book covers several issues as mobile learning techniques and strategies within teaching settings, highlighting cognition processes and theories, mobile teaching innovation, educational methodologies, and design. The intertwining of teaching methodologies and mobile learning improves teaching and learning methodologies by deepening cognition processes and digital education models related to mobile technologies, and discovers how innovative interactions in teaching/learning, user/device, learning/mixed-reality, and virtual body/sound can affect the educational design of mobile learning environments.

Flavia Santoianni

Contents

Part II: Corrado Petrucco

Part IV: Daniele Agostini

Part I
Flavia Santoianni

CHAPTER 1
Key Aspects of Mobile Digital Education

Flavia Santoianni

||||||||||||||||||||||||||||||||||||||||||||||||||||||||||||||||||||||||||||||||||||||||||||||||||||||||||||||||||||||||||||||||||||||||||||||||||||||||||||

> *Since the dawn of humanity, people have learned outdoors while on the move.*
> —Sharples and Pea, 2014

> *Emergent technologies and emergent pedagogies are interdependent.*
> —Gros, 2016

Introduction

Mobile learning (M-learning) emerged about 20 years ago within the 'Mobile Education Research Project' of 2000 at Berkeley University of California (Ma et al., 2019), but its roots date back to the ground-breaking technological developments of 1970s, when the concept of discovery learning was leveraged through computer-assisted learning, even if information and communication technologies (ICT) were not yet diffused (Crompton, 2013). Handheld computers changed effectively the learning scenario in 1980s, when constructivism enhanced student-centered learning through the spread of computer-assisted instruction, which gradually led to students' opening to socially-shared and contextually-authentic experiences and co-construction of knowledge sustained by constructionism.

After the web revolution of 1990s, mobility became a more and more desired characteristic of multimedia devices, which finally began to be exploited fully worldwide (Ranieri, 2015). Emerging pedagogies shifted towards both socio-constructivist and personalized approaches, moving towards increasingly student-centered educational practices of 2000s. World Wide Web has become more interactive, data have been stored in digital libraries and collections, social networks

Professor of Education, University of Naples Federico II, Italy.
Email: bes@unina.it

have increased communication, and learners have knocked at the doors of virtual learning environments (VLEs) to enter learning platforms. Emergent technologies and pedagogies have become more and more interdependent with a two-fold effect: while technologies without educational design have shifted towards educational aims, education-integrated technologies are challenged to develop according to evolution of educational practices (Gros, 2016).

Nowadays the research field of M-learning is still changing[1] and is on the move. Even if it has been compared to core technological innovations as Gutenberg's printing press, it is still very difficult to define. The main conceptual constructs may be individuated around it, concerning multiple learnings and pedagogies, social and contextual interactions, and various personal approaches to technological devices. Indeed, mobile learning has already been investigated by many fields of research concerning its impact on the student's achievements in learning performance and collaboration, which pedagogies are involved, how they interplay with self-management, student's perception about its adoption versus traditional practices including gender differences, the variables which may influence its contextual teaching and learning, and which applications or mobile learning systems can be implemented (Crompton and Burke, 2018).

Teaching and learning are evolved towards more authentic approaches to introduce new pedagogies which are socially distributed and contextually situated (Gros, 2016). Mobile learning can be mainly evaluated by addressing usability, technology availability, and connectivity; learning effectiveness, curriculum integration, and users' satisfaction since its affordances are ubiquity, flexibility, accessibility, immediacy, interactivity, motivation, engagement, and contextuality to co-create new relationships between users, content, and learning environments (Turner, 2016). Key aspects of mobile learning are identified here in its basic premises, deepening the core issues of seamless learning, skills of 21st-century learners, flexible adaptation, and technology adoption. Emergent concepts of mobile learning are individuated in the theories of personalized adaptive learning, self-regulated learning, and heutagogy. Frontier pedagogies are raised in the intertwining of pedagogy, biological sciences, and neuroscience through experimental models of education related to social-personal interaction and to the affordances of technologies, besides being influenced by a learning interpretation as an individual and collective adaptive process.

According to these trends, basic premises, emergent concepts, and frontier pedagogies may sustain a new learning vision on the move.

Basic Premises of Mobile Digital Education

Mobile education leverages the overcoming of many dichotomies, such as the gap between formal and informal learning, teachers and students, learners and content

---

[1] Mobile learning research passed through three phases of development, shifting the focus from technological devices to informal learning, and to learners' mobility. Nevertheless, mobile learning has been not always linked to Web 2.0 technologies in an explicit manner and there are still some shortcomings due to lack of integration of pedagogical theories, transferable design, and teachers'/ students' support (Cochrane, 2013).

resources at different stages. In its first generation, mobile learning was mainly oriented towards knowledge transfer of contents to learners. Its second generation highlighted knowledge building, and in its third generation, the focus shifted to situated cognition.

Learning characteristics are defined according to mobility, ubiquity, continuity between informal and formal contexts, connectivity to any network (content, resources, learners, etc., collective synergy (regulating communication and interactions), situated contextuality in a broad sense (including cognitions and emotions), multi-modality, and adaptability to foster adaptive materials, services, and support services for specific individuals. In other words, technology-enhanced, contextual, and multi-mode mobile learning can be simply defined by the 3A – *any* content, *any* time, *any* location – linked to the fourth A – *any* mobile device (Wang, 2018).

Basic premises to mobile education are the concepts of seamless learning and which skills of 21st century are required nowadays by mobile learners to manage this kind of innovative learning. Concepts as flexible adaptation and technology adoption are outlined as intertwined since embedded in the environmental interactions through a continuous perceptual and conceptual synergy. According to adaptive learning, environments, format, content, and devices should be adaptive. Adaptive models implement learners' individual peculiarity and cognitive diversity within unpredictable situations. Educational design is indeed dynamic and adaptive within learning environments (Santoianni, 2007).

Mobile learning involves several dimensions like teachers' and students' interactions, design planning, and institutional organisation (Joo et al., 2016). Educational interactions are reciprocal and the continuous intertwined relations between individuals and environments are to be deepened.

Seamless Learning

In general, mobile learning is a term denoting learning which involves the use of a mobile device to support learning of any content at any time and at any place, and may be extendable and interleaved across time and space. Both students and learning resources are mobile, while teaching resources are available anywhere online, thereby broadening the range of educational chances. Its mobility is downloaded across technological devices in physical, conceptual, and social space, in formal and informal learning contexts as universal processing,[2] and over time (Sharples et al., 2009).

Seamless learning is rooted in everyday activities, has distributed management, is always situated in contexts, which can be pre-existent or constructed by learners, and may exploit every experience into a learning opportunity.[3] Connectivity has really changed knowledge building and both space and time where it occurs. During

---

[2] Universal processing means mobile pervasive learning, that is, the incorporation of data innovation into the lives of individuals through formal, informal, and social learning modalities (Shuib et al., 2015).

[3] Mobile technologies and applications powered with social media and cloud computing concern learners on the move and seamless ubiquitous learning experiences (Churchill, 2014) but, to reach learning outcomes, teaching and learning design should consider—apart from mobility resources—other pieces of the interpretative framework, such as active experience, continuous assistance during stand-alone or collaborative tasks, and progress evaluation (Churchill et al., 2016).

seamless learning, learning experiences take place within a synergy of space, time, devices, and collective/individual settings (Gros, 2016).

There is no univocal agreement about which activities have to be considered in mobile learning because the field is in continuous evolution; the inference of mobile technology within learning contexts has to be deepened. Challenges to designing context-dependent mobile learning can be considered under four heads. The first concerns the possibility to interpret learning as an intersectional phenomenon on the move between a nexus of physical locations and interplay of social groups, within which learners can have their own different space. The second involves the influence of mobile technology on formal learning, which is now continuing side by side with informal learning, overcoming any possible gap. As a consequence, the third challenge concurs to blur the boundaries which relate to formal and informal learning, thereby melting them. The fourth acknowledges the complexity of contexts seen as conceptual, physical, and social spaces, as well as technological dimensions, and their role in the design of seamless learning environments, where mobility may concern both physical and social mobility throughout formal and informal contexts, and the transferability of flexible learning content (Jaldemark, 2018).

The idea of learners' mobility is to be re-thought by focusing on the word *mobile* rather than the word *technology*. According to this interpretation, formal and informal learning can be combined and seen as complementing, leveraging on both reflective and authentic learning up to gradually reducing the differences between in-class, academic, curricular, on-campus learning and out-of-class, non-academic, co-curricular, off-campus learning (Sharples and Pea, 2014). Learning occurs effectively within and outside the classroom while offering opportunities for blended learning environments (Karimi, 2016). Mobile learning can be indeed complementary to formal learning environments, which it can support without representing an alternative to them (Haag and Berking, 2019).

Mobile seamless learning (MSL) is indeed considered to encompass both formal and informal learning, personalized and social learning, physical and digital worlds, multiple pedagogical models, learning tasks, and technological devices. Moreover, it develops across time and locations through ubiquitous access to online/in-classroom interactive learning resources co-created by teachers and students, and by multi-disciplinary/multi-level learning (Wong and Looi, 2011). M-learning has narrowed user device since it gives easy access to multimedia fruition and generation; content and learning because it is highly interactive and flexible; and teachers and students inside and beyond the classroom due to its ubiquity.

The dynamic nature of digital change may be accepted by schools' educational approaches if they are interested in curriculum enhancement, have the needed infrastructural affordances, and are ready to undertake and share organizational responsibility (Turner, 2015). Wireless Internet learning devices (WILDs) can sustain innovative learning dynamics by orchestrating access to and management of the learning content through technological devices, which assist teachers in organizing a dynamic lesson plan by regulating the distribution of resources, setting the presentation of materials, supporting individuals in group, and whole-class tasks inside/outside the classroom (Roschelle and Pea, 2002). School mobile devices can increase participation and sharing, facilitate interaction in whole-class settings, and at the same time give more autonomy to individual work in classroom. A school digital portfolio can increase in-school parental access, personal learning with technological devices, and institutional requirements to enhance school digital systems and curriculum (Turner, 2016).

Ubiquity sustains continual learning more than lifelong learning because it modifies the learner's interaction with time. Mobile technologies encourage flexible approaches because the learner begins to be involved in a task only when s/he feels ready, and learning is always possible and available. The access to resources is increased as well as the chances to be in network with co-learners. Learning on the move is authentic and 'just in time, just enough, just for me' but requires accurate management because it implies various learning modes and e-pedagogies (Palkova, 2019).

Skills of 21st-century Learners

Millennials, Generation Y, Digital Natives, or Net Generation—born between 1981 and 1999—are technological, team-oriented, and high-achieving creative learners, interactive and participatory (Vincent-Layton, 2019). Mobile society demands them to continuously adapt to cultural and social transformations due to technology development, and M-learning may represent a chance of joining the evolution of technologies with the rethinking of teaching practises and learners' emerging digital skills (Boude Figueredo and Jimenez Villamizar, 2019). Skills of the 21st century (National Research Council, 2012) are recognized to be relative to cognitive processes and strategies, creativity, critical thinking, reasoning, and innovation; intellectual openness, flexibility, self-evaluation, and metacognition; teamwork, communication, responsibility, leadership, conflict resolution, and more (Dede et al., 2017).

The exchange of ideas in real-time and network connections occurs within diversified environmental scenarios for continuously new student-and-technology partnerships. However, mobile learning is about the learner: technology is what allows him/her to learn. Mobile learning is then a social, and not a technical, phenomenon. It offers flexible access to educational resources and services (Louhab et al., 2018), leading to continuously changing conditions for participation in learning contexts and influencing the relationship between individuals and space (Jaldemark, 2018). This aspect enhances the multi-tasking nature of mobile learning, which empowers both knowledge and learning skills (Hamidi and Chavoshi, 2018) required by the complexity of the actual society and needed to cope with its challenges. Mobile skills concern learning on the move, actively negotiating knowledge in continuous collaborative interaction with technology, and learning inside personal contexts (Cochrane, 2013; Stoerger, 2013), which can be self-generated in one's own learning community.

Learners are required to be highly skilled, able to adapt to different contexts, flexible between the networks, and collaborative within the community, and self-regulated when approaching personal learning. Learners are co-learners and learning leaders at the same time because they share knowledge and visions with others, keeping their minds open to experience and to the possibilities to change their own beliefs according to others' points of view, while fostering personal knowledge management, being entrepreneurial, and resilient. They have to carry on a twofold effort. On the one hand, learners are engaged in disentangling information to be filtered from what they consider to be left out within the huge amount of available knowledge on the web. In other words, they need critical thinking to recognize ambiguous information by themselves, without the need for external control. On the other hand, learners live fully immersed in ambiguity because hyper-complex societies frequently raise ill-structured problems, which need challenging skills to cope with them (Blaschke and Hase, 2016).

Mobile learning can be bounded by different technical barriers, like lack of web access, differences in mobile signals quality, high costs, and even ethical doubts of educators and parents about the opportunity to let young learners use it, but actual learners prefer multimedia, multitasking, and co-learning—all aspects related to mobile learning. At the same time, these characteristics of M-learning have implemented new aptitudes in students, thereby changing their own learning habits and strategies on attention and reflection. Mobile devices allow, for instance, delivery of short messages of information, to which it is possible to dedicate a smaller time, to guarantee a flexible access to knowledge (Yu, 2019). Mobile devices— even if characterized by individuality, portability, availability, connectivity, and interactivity—can generate personal limitations due to distracting effects related to multi-tasking, the overuse of mobile devices, and the consequent lack of interpersonal relationships (Sung et al., 2019).

Learners' skills are continuously changing due to the rapid evolution of digital technologies and devices. The concept of change is nowadays critical to the survival of learners, who have to continuously adapt themselves to the evolving environments. But change is often required quickly, and there is not enough time for learners to be prepared to cope with. This means that flexible adaptation is needed.

Flexible Adaptation

Flexible learning is a term born in the 1990s when began the increased use of new information and communication technologies to indicate first the virtual delivery of education itself and then the variety of students' needs for individual choice of time and space during the learning process or their preferences about the learning contents and strategies. Flexible delivery refers to learners' choice of study mode and to its personal control, enhancing the understanding of students' learning difficulties according to knowledge domains, and improving it through a more accurate design of the learning experience to support a deeper learners' perception of the learning environments (Alexander, 2010).

Evolution of this term has involved self-paced learning experiences. M-learning is flexible and self-paced. Flexible education is given through the freedom to study on the move by choosing learning time, place, and content according to the student's individualized needs (Zhang, 2019). Learning, in general, develops in adaptive environments, as embodied in unpredictable situations, according to the interdependent relation between mind, brain, and organism. The research focus is nowadays on the cognitive, emotional, and organismic (perceptual and behavioral, including the concept of implicit) dynamic complexity, no more detached from the affective, emotional, biological, social, cultural, and contextual evolutionary factors and their interpretations, as space-time discontinuity or adaptive development (Santoianni, 2017).

During a teaching and learning experience, learners are coping with a simulated environment, designed to teach content materials/resources and to implement skills. The effective challenge of mobile learning is nowadays on adapting to learning contents to learners' needs more than to guarantee accessibility to the learning itself. Adaptive learning can improve learners' achievements and educational experiences are empowered through their personalization (Liu et al., 2017). For instance, content personalization can occur, integrating learning styles into adaptive learning through online classification models or adaptive learning systems (Truong, 2016).

In educational hypermedia environments (Santoianni and Ciasullo, 2018a), adaptive systems are indeed designed to sustain flexible student-centered approaches according to individual differences and preferences of learners of different age, gender, and learning experiences (Sadler-Smith and Smith, 2004). Since the earliest 90's, adaptive educational hypermedia systems have overcome the 'one-size-fits-all' tradition focusing on learners' needs and skills to develop hypermedia support web-based instruction relying on learning styles (Brusilovsky, 2003). The research questions here the individual differences which can influence learning interactions (Chen and Paul, 2003), and how to select hypermedia adaptation technologies (Papanikolaou and Grigoriadou, 2004) in adaptive presentation[4] or adaptive navigation support.[5]

Adaptive flexible learning focuses on the concept of self-direction, to be applied both to personalize instructional design for learners and environments. The locus of control of adaptable systems (Wolf, 2002) is balanced between the learner, with the tool of end-user modifiability (Santoianni et al., 2018) and the system, which can manage and control learners' information. Adaptive flexible learning can be also supported by social interaction through communication and collaboration (Paramythis and Loidl-Reisinger, 2004). While learners are actively immersed in real-world situations, they receive adaptive scaffolding.

Mobile technology is indeed considered to be an adaptive and personalized context-aware system to facilitate the interplay between learners and environments inside/outside the classroom through on-demand access to formal and informal learning activities. Mobile learning is recognized to foster learning outcomes in informal learning environments better than in formal ones. The focus is on authentic and situated learning environments, in which learners are embedded in real-life contexts within learning communities. Learning is highly contextual and, to implement authentic learning environments, it relies on learners' awareness about previous knowledge, skill levels, and environment location/setting awareness

---

[4] Adaptive presentation technologies use both adaptive text and adaptive layout to customize learning materials and resources (Chen and Paul, 2003).

[5] Adaptive navigation support is designed to facilitate learners' navigation by prioritizing the appearance of visible links to increase the speed of their finding (Brusilovsky, 2003).

(Kinshuk, 2015). Context can be a learning context, which varies according to learners' preferences, educational strategies, materials, resources, activities, and a mobile context, which is more related to the different adopted mobile devices (Gomez et al., 2014).

The core idea is that mobile learning should be context-aware, and not only adaptive; technological devices and their applications should allow learners to use emergent contextual information in order to select their own preferred contents and to interact with them. Adaptive and context-aware mobile learning provides users with contextual learning resources (Louhab et al., 2018) related to various kinds of mobile devices. Mobile technologies can be adopted in different ways, which are defined alongside their real effective use in everyday learning practices (Sharples et al., 2009).

Ambient learning is a broad definition of the field of research which studies how technology can be adapted to learning environments and how people interact with this embedded technology in everyday life.[6] It has been considered the next generation of mobile learning, in which learning environments provide personalized and contextualized knowledge (Arnone et al., 2011). Ambient learning is an innovative area—near to context-sensitive learning—which uses digital devices to promote learning within enriched environments, and to improve learners' contextual activities (Ranieri, 2015).

Since both individuals and technology are in continuous evolution, they influence one another. Technology varies according to its users' needs and in relation to other technology, while learners' cognitive and emotional development is linked to technology evolution. So ambient learning has been also defined as the ubiquitous meeting point of knowledge development and technology evolution, with the aim of improving this synergy and increasing its levels of co-operation. Learning resources are consequently adapted learning contexts and users' needs (Kölmel, 2004; Lyardet, 2008).

Ambient learning is a system which intertwines innovative knowledge with learning management. Learning resources are enhanced with reliable up-to-date content; all materials are always available on demand within contexts—not only pre-organized as a specific offer, and shared. Ambient learning is developed through multimodal broadband access—broad bandwidth and wireless connections, shared media, etc.; multimedia platforms for asynchronous/synchronous connections, and context management, which allow the fruition of e-learning objects, based on her/his prior knowhow, personal interests, and cognitive identity. Learning resources are integrated with previously existing knowledge and high-quality content through content integration (Paraskakis, 2005).

Technology Adoption

According to the technology acceptance model (TAM) (Davis, 1989), technology adoption depends on the perceived usefulness and perceived ease of use of a

---

[6] Ambient learning aims to provide easy e-learning for personalized approaches, which may concur to co-create learning objects based on users' contexts, their characteristics, training needs, and preferences. Self-directed learners can rely on flexibility of learning contents, teaching approaches, and assessment.

technology device,[7] together with the enjoyment and self-efficacy acknowledgement which can emerge for user satisfaction, while cultural background and learners' prerequisites/skills are now considered as concurring factors antecedent to technology adoption (Balakrishnan and Gan, 2016). Other concurrent factors are attitude, behavior, and usage—all these factors interact with teaching practice and affordances of technological resources (Boude Figueredo and Jimenez Villamizar, 2019).

Models, which analyze acceptance of technologies in daily life, refer to an adoption process subdivided in pre-adoption, adoption, and post-adoption. Technology acceptance model concerns pre-adoption expectation, while the expectation-confirmation model (ECM) (Bhattacherjee, 2001) involves post-adoption expectation, that is, the perception of the congruence between the expectation of a technology use and its effective performance during experience. The technology acceptance model is focused on the adoption of new technologies; the expectation-confirmation model on the adoption of already used technology (Joo et al., 2016).

The unified theory of acceptance and use of technology (UTAUT) is a comprehensive model which involves several technology adoption theories, including the technology acceptance model (Venkatesh et al., 2003). This unified theory implies that performance expectancy—that is how much useful learners consider M-learning to improve their own learning performances and effort expectancy, that is, the degree of ease of use that learners can associate with M-learning, plus perceived playfulness and learning styles, together with personal innovativeness, are all concurrent factors in mobile learning adoption in formal and informal environments (Karimi, 2016).

According to the unified theory of acceptance and use of technology, the choice of a device depends on performance and effort expectancies, social influence, and facilitating conditions—all aspects related to age, gender, personal experience, and individual intentions. This model has been already used to analyze technology acceptance in the interactions of mobile learning through websites, blogs, and mobile-assisted language learning (Hoi, 2020).

The use of a technology relies on its affordances. The term 'affordance', introduced in 1977, means all the latent possibilities of the characteristics of objects/environments to be exploited in order to allow users to activate them to perform a specific skill (Gibson, 1977). For instance, the use of haptic interfaces has an emotional and social evolutionary significance; touchscreen and sensor-based inputs can implement authenticity and engagement within simulated environments. Moreover, technology use can change depending on the device—phones are used more when standing or walking, while tablets are used more when sitting. User interaction preferences[8] and behaviors, as to change the orientation of view of phones

---

[7] Uncertainty avoidance is part of the technology acceptance model, which implies, on the other hand, the absence of predictability. This means that there is a degree of uncertainty with which learners can feel uncomfortable since there are variables and doubts (Hofstede et al., 2010). This aspect has to be put in relation to the feeling of self-efficacy of each learner, as well as with the perceived usefulness and ease of use of technology devices.

[8] A personal approach to technology is BringYourOwnDevice (BYOD) that is based on technological flexibility to let learners choose which technology may really meet their needs in order to access content (Jaldemark, 2018).

and tablets or to switch them from one to two hands, and screen sizes, are all factors which can influence mobile learning design for better readability and uses (Haag and Berking, 2019).

Learners can adopt mobile learning for many reasons: expectations about allowed performances, quality of technology, personal intention to improve self-managed learning, and teachers' vision including mobile devices in classroom. Teachers may indeed first accept the use of mobile devices in the classroom and acknowledge that its characteristics may be useful for teaching and learning in formal contexts; at the same time, they can master their knowledge about mobile devices and transfer prior training for different purposes to the field of mobile learning; then they can play the role of online tutors (Cardoso and Abreu, 2019).

Each case of adoption of mobile technologies has a specific emerging learning design model (Churchill, 2016). Handheld technology adoption has the characteristics of a two-way interaction,[9] in which digital contents allow two-way communication between teachers and students. Each design model has the aim to increase the future use of technology through user's positive satisfaction, to be measured throughout the technology-acceptance process.

Emergent Concepts of Mobile Digital Education

Future research on mobile learning, intended as a research focus, should highlight many issues as objective measures of learning outcomes, mobile instructional methods, theories of learning and motivation, and their experimental support/ evidence-based approaches (Mayer, 2020). A scientific approach on mobile learning is nowadays based on emergent learning concepts relying on individual experiences and environment development as personalized adaptive learning, self-regulated learning, heutagogy, and ambient learning. Mobile learning is indeed both related to the collective/personal interaction and to the affordances of technologies in order to design learning environments and to monitor the network between learners and their contextual resources.

Deeper learning approaches are needed, narrowing learners to real-world situations through case-based learning, providing many different knowledge representations with multi-processing learning, enhancing apprenticeship, self-directed learning, personalized learning, and interdisciplinary, motivational and generalizable learning. On the other hand, learning is intended to be collaborative and connected between formal/informal learning environments (Chris et al., 2017).

---

[9] Students are active in the learning process and improve their self-efficacy through the management of multimedia learning resources. Interactive multimedia technology includes a variety of digital resources, such as streaming videos, video editing, digital audio, social networking, blogs, e-books, podcasts, cloud computing, and other applications. Technological tools, resources, and services are learning facilitators, encouraged by the concept of open access to knowledge, for instance, through digital library services, online courses, or open educational resources (OERs). Open educational resources, especially if they are game-like, allow affordable learning for anyone, but the main shift is represented by the possibility to co-create learner-generated content (Ally and Prieto-Blázquez, 2014).

One of the critical nodes for teaching and learning is represented by the concept of integration of mobile technology, social sharing, and effective learning design (Churchill, 2016). Mobile learning can be indeed interpreted through several aspects: multiple contexts, social interactions, content interactions, and real-time information plus the intertwining of these dimensions. Many questions arise as the challenge for any learner to move across different contexts, to find how instructional design can be useful accordingly, to see how to apply mobile learning to specific knowledge domains, as social studies (Diacopoulos and Crompton, 2020), and learn what is the role of social interaction in learning. Learning on the move is the result of synergy between spaces and content to produce new learning activities, even if the balance between formal and informal levels, the diversity of contexts of implementation, the possible gap between foreseen designs, and their effective use in the real world are still issues to be deepened (Danish and Hmelo-Silver, 2020).

Personalized Adaptive Learning

Personalized teaching and learning overcome the traditional one-size-fits-all didactic, which is top-down, in favor of more dynamic models, which are tailored to meet individuals' needs (Gros, 2016). M-learning can encounter personalized learning, that is, a learning mode near to universal design. Personalized learning means that students can choose the learning content which is more suitable for them, get various learning materials from different sources, and enjoy them when and where they prefer. Mobile devices are seen as effective learning tools and easy/quick ways to access knowledge in synchronous and asynchronous collaboration among learners and teachers (Khan et al., 2016). Mobile learning should enhance the user experience and the personalized management of learning content through easy approaches to simple interfaces of learning units usable within small time-slots.

Mobile learning has been indeed defined as being personalized, situated, and authentic (Traxler, 2007). The idea of personalized learning relys on several pedagogical approaches, acknowledges individual cognitive differences, both in environmental and social interaction, as well as the need to differently design learning contents, and technological interfaces of mobile devices in order to meet individual diversities. Learning is also situated in constantly changing contexts and copes with authentic real-world problems (Cardoso and Abreu, 2019). Mobile learning is no longer interpreted as a mobile delivery of contents but rather as a process which changes daily life into a learning space (Pachler et al., 2010). Seamless learning is related, as a consequence, to the authenticity of the learning experience and to its personalization (Ranieri, 2015) through the learner's agency, that is, individual choice about the physical/virtual learning place and about the time to dedicate to learn within it (Pachler et al., 2013).

Personalized learning,[10] rooted on learner-centered principles, is categorized into the following four factors that influence learning (Watson and Reigeluth, 2018):

- cognitive and metacognitive factors, which concern learners' intention to represent, build, and link knowledge, to develop reasoning and higher-order meta-reflective strategies, and to contextualize them into learning environments;
- motivational and affective factors, which are related to individuals' emotions and personal interests, affecting learners' intrinsic motivation, curiosity, and creativity;
- developmental and social factors, which have to be considered to jigsaw an effective framework of learning chances and their boundaries within interpersonal situations;
- individual-differences factors, which acknowledge learners' different knowledge building as a consequence of their various backgrounds, to be assessed through tailored approaches.

A personal learning-focused educational approach is based on differentiated instructions, in which teachers design how to personalize instructions for each student, according to each's learning readiness, interest, and profile. Differentiated instruction is linked to brain-based education, which is focused on adaptive learning experiences.

Learning is personalized and adaptive. These characteristics are reciprocal and influence each other, giving rise to personalized adaptive learning (PAL), which is regulated by individual differences, personal needs, individual performances, personal development, and adaptive adjustment. Personalized learning objectives and content, as well as educational approaches, are tailored to customize learners' activities. Teaching is learner-centered, inclusive, and universally designed; its focus of research is learners' individual development. At the same time, adaptive teaching monitors student progress and consequentially adjusts instructional design according to learners' ongoing performances, since each individual is in a constant change. Adaptive learning strategies regulate knowledge interactions within learning environments.

Both personalized and adaptive learning lead to differentiated instruction, giving attention to individual differences, characteristics, needs, and to personal development, while adaptive learning is more focused on the implementation strategies which regulate individual performances and strategies of adaptive adjustment. Differentiated instruction is planned as a set of flexible learning environments, which allow students to easily approach knowledge and to reach all the same objectives through customized teaching methodologies and strategies. Adaptive teaching differentiates

---

[10] Since the 1970s, personalized learning was adopted by the National Association of Secondary School Principals (NASSP), the Learning Environments Consortium (LEC) International, and the special education movement (Watson and Reigeluth, 2018). Its core concept was to help learners to be effectively engaged within the learning process and to let them control autonomously the related variables. Personalized learning focuses on the design of learning styles' customization, instructional strategies, teacher-student interaction, small class, parental involvement, and technology adoption.

learning times, as the planning of the learning path, the way of offering learning content, and the resources to be shared are done according to individual interests and development, and change of individual performances (Peng et al., 2019).

Personalized and adaptive learning can be divided in three areas, according to its objectives. Macro-adaptive instruction concerns learning pace in relation to learners' characteristics, aptitude-treatment interaction, which focuses on individual attitudes, abilities, and skills to differentiate instruction, and micro-adaptive instruction, which is a dynamic regulation of how learners process information to adjust instruction accordingly. Instructional design relies on personalization to implement supportive learning environments, overcoming the ideas that adaptive instruction should be tailored only on learners' special needs, and that personalized instruction is foreclosed to large groups of learners.

Intelligent Tutoring Systems (ITSs) are micro-adaptive systems which take into consideration learners' prior knowledge and learning experiences, the intertwining of knowledge domains, instructional strategies within learning environments, and interfaces of learning content presentation. The mission of adaptive learning environments is to link the source of adaptive instruction, i.e., learners' characteristics within contexts, to its source, as the instructional design of a learning content, and its presentation through a modeling process. Stereotype modeling refers to clustering learners into groups according to their cognitive profile; feature-based modeling individuates learners' interests and goals, and tracks their development; new user modeling, at the end, focuses on a new learner profile, about whom no information is still available, which has to be first associated to a group level and then included into a developed feature-based model (Vandewaetere et al., 2011).

Personal learning environments (PLEs)[11] developed within web 2.0, partially overcome learning management systems (LMS) because they focus on each learner rather than all learners, enhancing personalized set of resources more than the tools (Weller, 2018). Personal learning environments are learner-centric approaches which customize the learning experience and support learners in the social building of their own personal knowledge network (PKN). Within a personal learning environment, learners can experience trial and error, inquiry, reflection, networking, and more. Personal learning environments sustain self-regulation, co-regulation, and social share regulation (Gros, 2016). In mobile learning research, learning environments can be both formal and informal, but formal learning has been studied more than informal learning, and more attention is to be given to how learners' individual differences may affect M-learning adoption (Karimi, 2016).

Self-regulated Learning

Self-regulated learning (SRL) is the process of achieving the desired results through adaptive learning strategies and relative adjustments, which allow pursuance of

---

[11] Personal learning environments have been introduced as systems which allow learners to manage their own learning process by setting goals, selecting contents, developing strategies, and sharing efforts. The idea of a personal learning environment means encouraging learners' self-direction through technologies (Gu, 2016).

specific learning goals (Winne and Hadwin, 1998, 2010). Self-regulated learning as a structural function implies activation of cognitive and affective/motivational control systems, concerning knowledge domains and metacognitive knowledge, and cognitive and motivational strategies and beliefs.

Self-regulated learning has been studied more in formal contexts of academic learning and can be intentionally planned with pre-set objectives and identifiable outcomes than in informal contexts of learning, which can be less organized and require other kinds of regulation skills, including resource management, realistic goal setting, effective learning strategies activation, and responsible learning encouraged by teachers, who have to support metacognitive activities guided by learning experiences and self-reflection (Cazan, 2013). Self-regulation skills are particularly significant for online learners rather than classroom learners and have a positive impact on learning achievements in both cases.

The construct of self-regulation has its roots in socio-cultural theory, according to which learners develop regulation of their own cognitions, emotions, and social interactions through monitoring of their caregivers (Whitebread, 2020). Self-regulation has been influenced by social learning theory (Bandura, 1982), according to which motivation—and not only behavioral and emotional regulation—is an activating aspect of self-regulation. Motivational factors also concern evaluation of performances, standards, and activities within learning environments. Self-regulation is indeed the result of interaction between learners and environments, implemented through personal behaviors (Dinsmore et al., 2008).

Self-regulation has been later defined as a goal-oriented process, which individuals activate through a synergy of cognitions, emotions, and behaviors (Schunk and Zimmerman, 1994; Zimmerman and Schunk, 2001). Self-regulation is then considered to be driven by prior knowledge, cognitive, and motivational aspects. Zimmerman's model of self-regulation can be staged in three phases. In the forethought phase, learners design how to perform a task by analyzing their own interest in the task, their expectations for the outcomes, the objectives to be achieved, and the perceived self-efficacy. After this phase, learners monitor the ongoing task through control processes, activating self-instruction, imaging, attention, and specific strategies. Self-observation and self-experimenting can monitor the process variations. In the final phase, learners adopt self-judgment and self-evaluation to check if the desired results have been obtained and to what extent, comparing them to the standards of individual and collective performances.

Applied to academic settings, self-regulation has led to self-regulated learning (SRL), focused between the 1980s and the 1990s and integrating cognitive, motivational, and contextual factors by melting the fields of research of self-regulation and metacognition. The control processes of self-regulation require meta-cognitive and meta-emotional strategies, which monitor thinking, stages of reaching objectives, and cognitive regulation. Learning environments can improve learners' self-regulation through teachers' guidance, encouragement to pursue goals, behavioral engagement, and scaffolding (Pellas, 2014). Instructional scaffolding, to be varied alongside development, can support self-regulation (Boekaerts, 1997; Boekaerts et al., 2000).

In Boekaerts' model of self-regulation, learners first perceive the task environment, which may consist of oral or written learning instructions, the social context, and its situation. Then learners analyze domain-specific knowledge (declarative, procedural, and metacognitive) and its related skills. In the end, learners reflect on their own motivations and goals to cope with and which can be monitored by feedbacks. Winne and Hadwin's model of self-regulation is divided in four recursive phases. In phase one, learners focus internally on task conditions, as to what are the teaching objectives, what is the given time, if the task is individual or collaborative, if any scaffolding is available. In phase two, learners set learning goals, and, in the next phases, learners engage in the task and, if adjustments are needed, metacognitively change their steps according to the emergent issues, eventually going back to the prior phases in a recursive way (Winne and Hadwin, 2010).

There are commonalities between the different models of self-regulation, including the attention to the task environment in which the task can be assigned or self-generated, to the role of individual agency, which allow the learner to make choices based on prior knowledge, individual interests, and environmental aspects. The presented models of self-regulation share some features of the socio-cognitive perspectives of self-regulation, while socio-cognitive perspectives foster individual skills supported by external feedbacks; instead socio-cultural perspectives see self-regulated learning as a social process of scaffolded co-regulation. In this case, self-regulation is first acquired as shared within the learning community and subsequently it is taken up by individual learners.

Self-regulation ranges from the influence of external factors, to which learners have to adapt and give compliance, to the internal control of the self, expressed by monitored processing, critical thinking, problem solving, and attention (Post et al., 2006).

Heutagogy

'Networked individualism' is a new term identified to indicate when individuals find their own networks to go through by self-determined learning (SDL), which allows learners to self-monitor and self-manage their own learning experiences and acquire control over them. According to social cognitive theory, self-efficacy is an intrinsic key factor to determine individuals' behavior in a continuous interaction with environmental factors (Balakrishnan and Gan, 2016). Self-determination theory highlights the role of internal resources and intrinsically-motivated behavior for self-regulation development (Ryan and Deci, 2000), while extrinsically-motivated behavior can evolve in self-determined behavior through internalization and integration of external regulations (Ryan and Deci, 2018).

Self-regulation means that learners use cognitive and metacognitive strategies, as well as resource management reached through self-monitoring, self-judgement, and response to performance outcomes. In formal learning environments, the main self-regulation strategies are recognized to be metacognition, time management, critical thinking, and effort regulation, while in online settings—even if these are anyway active—the main online self-regulation strategies are considered to be rehearsal, organizational, and information processing plus peer learning as a further influencing factor (Broadbent and Poon, 2015).

According to the dual processing self-regulation model, self-regulated learning as a dynamic model acknowledges learners' needs and personal goals, and well-being and developmental pathways influenced by external or internal stimuli (Musso et al., 2019). Cognitive systems need to self-organize during time,[12] and the concept of self-organization is more oriented toward internal forces rather than external pressures (Prokopenko, 2008). To be effectively self-regulated, learners have to self-manage their own strategies and resources, keeping a positive approach towards cognitive tasks, which include the challenge to be able to cope with them (Dermitzaki et al., 2009).

Heutagogy is the holistic study of self-determined learning, which highlights the role of each individual as the main actor of the learning experience, while the teacher provides guidance and resources without taking away from the learner the leadership of the process. Heutagogy is linked to andragogy and self-directed learning, of which it can represent a further extension. Learners create by themselves their own learning path within learner-centered environments. However, since learning environments can be non-linear, ill-structured, and hypercomplex, they require informal, experiential, and collaborative learning to overcome the limits of formal learning in order to find 'just-in-time' solutions, which can be needed to cope with emergent problems.

Heutagogy foresees that teachers and learners work together. The first step is to share the desired learning aims and outcomes in a process called the 'learning contract'. Then the negotiation of the assessment process occurs between teachers and learners, and the curriculum is continuously adapted throughout the teaching and learning activity. Learning process is developed with multimedia, tools, and devices (Blaschke and Hase, 2016).

Heutagogy is learner-centered since the learner is the focus of the teaching and learning world and has three key points. The learner is self-motivated, self-regulated, and self-determined (Hase and Kenyon, 2007, 2003, 2013; Deci and Ryan, 1985, 2002; Deci et al., 1996). The generational change which involves actual digital students implies a shift toward the DIY (do it yourself), which influences creative content construction and knowledge social dissemination. These processes improve learners' skills through the opportunity to narrow change-resistant learners (Santoianni, 2014) to self-regulated learning mediated by technologies, thus offering a tailored support; the challenge to leverage individual and social learning experiences, which can improve self-efficacy, self-confidence, and self-esteem; and the possibility to exploit gained experiences within the learning community to a larger scale.

Mobile learning enhances learning accessibility and promotes the integration of formal and informal learning environments, thus fostering personalized learning experiences (Cardoso and Abreu, 2019). The learner manages her/his own skills, transferring them from one context to another even if in unfamiliar situations, grounding her/his capability on personal self-efficacy (Cairns, 1996). The learner

---

[12] Self-regulated learning is an active process consisting of sequential and simultaneous phases to address a cognitive task as the forethought phase, when learners set goals, plan strategies, and check her/his self-efficacy; the performance phase, in which learners monitor and self-control their own strategies; and the self-reflection phase of self-evaluation, influencing future learning activities (Endedijk et al., 2014).

meta-reflects both on learning contents, as to why it has been acquired, and how it is self-regulated (Metcalfe and Shimamura, 1994; Reder, 1996; Hacker et al., 1998).

Heutagogy leverages the self-determined exploration of online resources and encourages flexible learner-defined curricula. Learners should feel free to create and co-create in many fields, such as writing, drawing, and designing online blogs. Learners are stimulated to work together, aiming at common goals and learning from each other within connected networks; in particular, social networks.

Through information sharing, collaboration can be enhanced, and learners can work together to take care of incoming knowledge, reflect on different topics, and approach issues with critical thinking. The role of the teacher is only a facilitating one. The teacher guides the learner through feedbacks, which are continuously customized according to learner needs. Teacher and learners work together to assess learning outcomes through participative evaluation (Blaschke and Hase, 2016).

Frontier Pedagogies of Mobile Digital Education

Main questions for research on mobile learning concern how to design content for mobile learning curricula according to students' requirements, how to prepare teachers[13] to manage the affordances of digital and mobile-friendly technologies with appropriate teaching and learning approaches, and how to monitor efficacy of self-directed learning of students in the interaction with learning resources and activities (Power, 2019). Mobile learning has made learning more learner-centered, even if individual or social, and fosters experiential learning integrated into daily life. One of the research questions about mobile teaching and learning is to identify which emergent pedagogies can leverage the core shifts related to it, about environments' design, teachers' preparation, learning challenges, technological resources to be used, and the holistic result of all these interactions.

There are three main interpretations of the directions taken by net-aware theories, which have to be considered together with pre-net theories as behaviorism, cognitivism, constructivism, meta-reflection, and early socio-cultural theories because net-aware theories have been developed, incorporating criteria of pre-net theories. These three main interpretations can be identified in network learning theories, social-personal interaction theories, and network affordances/design theories (Gros, 2016).

Network learning theories are sensitive to technological connections between learners, learners and teachers, learning community members, and their content resources. Learning is social, knowledge is experienced together, co-constructed and co-created: the nodes of the network can be represented by humans as well as by artefacts in a continuous cycle of knowledge development which goes from the community to the individuals and back again and again. The 'know what' and 'know how' of knowledge have been overcome by the 'know where' and 'know who'

---

[13] To integrate mobile technologies within the classrooms, it is not enough only to introduce mobile devices, but instead teachers' training is needed about how to integrate these devices into the classroom and to let teachers feel more self-confident about their use (Cochrane, 2014).

(Siemens, 2005). Patterns of connections evolve externally and internally between knowledge sharing, learning outcomes, and learners' minds.

In the next chapter, educational theories of mobile teaching are highlighted, going from pre-net to net-aware theories according to network learning theories. Here are instead focused the educational theories of mobile teaching where core concepts for mobile digital education revolve around the relation between mind, brain, and organism within the actual framework of educational theories of mobile teaching and learning. Mobile learning can be embedded in physical and social contexts as well as embodied through multimodal and interactive learning settings, involving not only the visuo-spatial dimension but also the role of proximity, posture, and gesture; and facial expressions, tone of voice, and touch (Pegrum, 2019).

Experimental models of education are frontier pedagogies grounded into the reciprocal influence of pedagogy, biological sciences, and neuroscience—as in bioeducational sciences (Frauenfelder and Santoianni, 2003). These models interpret learning like an adaptive process and knowledge structures as both individual and collective. The core idea underlying the experimental models of education is the integration between mind, brain, organism, and environment, especially when the learning environments taken into consideration are digital and mobile. Enriched models use multimodal methodologies, focusing on neural networks plasticity and its chances of modifiability, while organismic models highlight the interaction between the affective/emotional and bodily/organismic levels within the evolutive knowledge management processes, focusing, in particular, on learning diversities. Adaptive models interpret learning environments as continuously changing according to the emergent and long-lasting learners' adaptive needs. Learning environments are then studied in their evolution, to design them in function of cognitive development, modifiability, and cognitive-contextual learning interaction. Educability is the process which regulates adaptive development (Santoianni, 2017).

Correlation, change, and adaptive modifiability can be considered the keywords of educability (Santoianni, 2006). Correlation means the study of multiple synergistic variables of developing systems where complex interaction involves the activation of various co-acting factors. Individuals are constantly subject to paths of change that rely on experiences as well as on processes of interaction. At the same time, experiential dynamics and conceptual change can be influenced by processes of resistance to change that slow down individuals' development.

Learning environments are evolutive and their design should be flexible and adaptive in order to respect their inherent unpredictability and dynamism. Adaptive learning environments are modifiable and open to on-demand users' requests, thus leaving to learners the possibility of choice between several alternatives to customize their own experiences and be the leading actors of the learning process (Santoianni, 2007).

Basic Assumptions of Experimental Models of Education

Bioeducational sciences are part of the brain-based education, which promotes a multivariate approach to educational frontier research rethinking mind as linked to the brain. Experimental scientific research in bioeducational field ranges between many emerging fields, as spatiality (Santoianni, 2016), and can be applied to digital research (Santoianni and Ciasullo, 2018a, 2018b), highlighting five core principles (Santoianni, 2019, 2020), which can be useful to interpret mobile digital education:

- Principle of *differentiation*. Each individual is a specific biodynamic adaptive system. Personal learning approaches have to be acknowledged in their qualitative variability and represented by the specific modularity of each cognitive identity, which is to be underlined as well as her/his cognitive, emotional, and organismic developmental potential;
- Principle of *modifiability*. Cognitive, emotional, and organismic developmental potential of each adaptive system depends on individual experiences embedded in interactive learning environments. Levels of educability have to be managed by teachers through analysis of openness and resistance to change alongside personal self-efficacy development;
- Principle of *discontinuity*. Every adaptive system is always continuously educable but her/his development is to be considered variable. Personal discontinuity varies according to paces and ways specific to individual development, and teachers have to pay attention to the evolutive response feedbacks to learning environments given by each adaptive system;
- Principle of *integration/interaction*. Cognitive systems implement developmental synergistic integration of individual functional processes and evolutionary compatibility of social dynamic interactions. The personal history of learning modulates the intertwining between internal and external stimulations, which ongoing self-regulated balance organizes as cognitive structuring. The teacher has to identify which aspects of cognitive interactions with the social learning environment tend to be constant and which are instead modifiable, in order to address adaptive behavior towards 'active' imitative, not homologated, responses;
- Principle of *mutual support*. In a cognitive adaptive system, explicit and implicit levels are related. Implicit level may represent a prototypical base of cognitive processing, and a continuous on-demand support. Teachers should foster collaboration between explicit and implicit levels in cognitive development. Spatial dimension is a possible sphere of collaboration between explicit and implicit.

To learn effectively, learners have to adapt to environments. This challenge can be achieved only if learners manage to change the learning environment because the action itself of modifying the context within which individuals interact is adaptive. When knowledge is adaptive, it implies both explicit and implicit levels of cognition and can be interpreted as a distributed and situated process.

Reciprocal and interdependent on-demand relationships between individuals and environments rely on the interactions regulated by structural coupling (Riegler, 2002), which is a dynamic situation of adaptation. In this situation, synergic interactions within contexts are two-way: coping with any environment requires each individual to adaptively reply to stimulations, while educational environments should foresee, vice versa, modification according to heterodirect change needed by individuals (Santoianni, 2006).

Within learning environments design, aspects of variability and randomness have to be considered which can qualify them as adaptive to the extent that they require dedicated learners' efforts to actively integrate with them. Situatedness also has an adaptive role since it is linked to the contextual contingency in which cognition evolves. Experience is the main human adaptive tool.

Knowledge forms are continuously modified by experiential and adaptive interaction with other individuals and their cultural productions. If the learner internalizes socio-cultural community models and their practices as a customized experience, personal histories of learning are constructed over time, according to which cognitive identities develop differently for each learner and are rooted in environmental synergies, continually re-defined by culture and cognition.

Learners continuously activate adaptative processes that stimulate and enhance their own personal learning potential, regulated by cognitive modifiability, which leverages on educability of adaptive systems during the personal history of learning (Santoianni, 2007). Modifiability is related to the unpredictability of educational contexts because mobile contexts are not well defined, since they can continuously change according to social environment, learners' attention and posture, background noise, lighting, and so on. Consequently, learning processes can be considered unpredictable because mobile learning occurs in both formal and informal learning contexts, and traditional assessment is considered to be updated.

Mobile Internet associated technologies (MIATs) focus nowadays on relational interchange, user generated content, and engagement in the physical world (Davie, 2016). *Technobiophilia* (Thomas, 2013) is a field of research which notes that some educational concepts as classrooms or seminars can be implemented in physical learning environments as well as in virtual learning environments. Learning environment design is the result of comparison between physical reality and its virtual counterpart. Users' activities—online and face-to-face—are organized in a way which reflects experiences of life within the natural world, with the final result of humanizing the web.

Even if face-to-face learning fosters emotional transfer through bodily expressions and leverage active experiences, mobile learning can enhance interpersonal communication and exchange (Yu, 2019). Blended face-to-face and online learning is regulated by flexibility of time and place, by two-way interaction—within the transactional distance, which has to be narrow to facilitate social interaction—and by self-regulation strategies, as orienting, planning, monitoring, adjusting, and evaluating.

If the transactional distance increases, learners tend to feel isolation and experience a reduced motivation to learn. For this reason, teachers are suggested to foster a positive affective learning climate to motivate learners through empathy, encouragement, and appraising individual differences (Boelens et al., 2017). To foster skills of 21st-century learners, learning processes are considered as ongoing development addressing seamless learning, and cognitive systems are seen as embodied within the entanglement of both explicit and implicit (Santoianni, 2006, 2010).

The concept of adaptive knowledge implies a holistic interpretation of the mind in which the sphere of explicit/implicit cognitive processing and the dimension of emotionality/corporeality are intertwined as well as perception and higher processing levels. Knowledge is no longer abstract and rational, detached from the ground level of cognition, but is instead considered as a personal/social and embodied/ situated process. Adaptive functions of knowledge are relative to individual and social co-construction, including personal interiorization of meanings and social representations of them.

References

Alexander, S. (2010). Flexible learning in higher education. pp. 441–447. *In*: Peterson, P., Baker, E. and McGaw, B. (eds.). *International Encyclopedia of Education*. Elsevier, Amsterdam.

Ally, M. and Prieto-Blázquez, J. (2014). What is the future of in education? Mobile learning applications in higher education. *Revista de Universidad y Sociedad del Conocimiento*, 11(1): 142–151.

Arnone, M.P., Small, R.V., Chauncey, S.A. and McKenna, H.P. (2011). Curiosity, interest and engagement in technology-pervasive learning environments: A new research agenda. *Educational Technology Research and Development*, 59: 181–198.

Balakrishnan, V. and Gan, C.L. (2016). Mobile technology and interactive lectures: the key adoption factors. pp. 111–126. *In*: Churchill, D., Lu, J., Chiu, T. and Fox, B. (eds.). *Mobile Learning Design*. Springer, Singapore.

Bandura, A. (1982). Self-efficacy mechanism in human agency. *American Psychologist*, 37: 122–147.

Bhattacherjee, A. (2001). Understanding information systems continuance: an expectation-confirmation model. *MIS Quarterly*, 25(3): 351–370.

Blaschke, L.M. and Hase, S. (2016). Heutagogy: A holistic framework for creating twenty-first-century self-determined learners. pp. 25–40. *In*: Gros, Kinshuk, B. and Maina, M. (eds.). *The Future of Ubiquitous Learning. Learning Designs for Emerging Pedagogies*. Springer, Berlin.

Boekaerts, M. (1997). Self-regulated learning: A new concept embraced by researchers, policy makers, educators, teachers, and students. *Learning and Instruction*, 7(2): 161–186.

Boekaerts, M., Pintrich, P.R. and Zeidner, M. (2000). *Handbook of Self-regulation*. Academic Press, San Diego.

Boelens, R., De Wever, B. and Voet, M. (2017). Four key challenges to the design of blended learning: a systematic literature review. *Educational Research Review*, 22: 1–18.

Boude Figueredo, O.R. and Jimenez Villamizar, J.A. (2019). Framework for design of mobile learning strategies. pp. 257–272. *In*: Zhang, Y. and Cristol, D. (eds.). *Handbook of Mobile Teaching and Learning*. Springer, Singapore.

Broadbent, J. and Poon, W.L. (2015). Self-regulated learning strategies academic achievement in online higher education learning environments: A systematic review. *Internet and Higher Education*, 27: 1–13.

Brusilovsky, P. (2003). Adaptive navigation support in educational hypermedia: the role of student knowledge level and the case for meta-adaptation. *British Journal of Educational Technology*, 34(4): 487–497.

Cairns, L. (1996). Capability: Going beyond competence. *Capability*, 2: 80.

Cardoso, T. and Abreu, R. (2019). Mobile learning and education: synthesis of open-access research. pp. 313–332. *In*: Zhang, Y. and Cristol, D. (eds.). *Handbook of Mobile Teaching and Learning*. Springer, Singapore.

Cazan, A.M. (2013). Teaching self-regulated learning strategies for psychology students. *Procedia - Social and Behavioral Sciences*, 78: 743–747.

Chen, S.Y. and Paul, R.J. (2003). Individual differences in web-based instruction. An overview. *British Journal of Educational Technology*, 34(4): 385–392.

Churchill, D. (2014). Presentation design for "conceptual model" learning objects. *British Journal of Education Technology*, 45(1): 136–148.

Churchill, D. (2016). Preface. pp. VII–X. *In*: Churchill, D., Lu, J., Chiu, T. and Fox, B. (eds.). *Mobile Learning Design*. Springer, Singapore.

Churchill, D., Fox, B. and King, M. (2016). Framework for designing mobile learning environments. pp. 3–26. *In*: Churchill, D., Lu, J., Chiu, T. and Fox, B. (eds.). *Mobile Learning Design*. Springer, Singapore.

Cochrane, T. (2013). M-learning as a catalyst for pedagogical change. pp 24–34. *In*: Berge, Z. and Muilenburg, L. (eds.). *Handbook of Mobile Learning*. Routledge, New York.

Cochrane, T. (2014). Mobile social media as a catalyst for pedagogical change. pp. 2187–2200. *In*: Viteli, J. and Leikomaa, M. (eds.). *Proceedings of Ed Media: World Conference on Educational Media and Technology*. Association for the Advancement of Computing in Education, Waynesville.

Crompton, H. (2013). A historical overview of mobile learning: toward learner-centered education. pp. 3–14. *In*: Berge, Z. and Muilenburg, L. (eds.). *Handbook of Mobile Learning*. Routledge, New York.

Crompton, H. and Burke, D. (2018). The use of mobile learning in higher education: a systematic review. *Computers & Education*, 123: 53–64.

Danish, J. and Hmelo-Silver, C.E. (2020). On activities and affordances for mobile learning. *Contemporary Educational Psychology*, 60: 101829.

Davie, R. (2016). Ceaselessly exploring, arriving where we started and knowing it for the first time. *Studies in Philosophy and Education*, 35: 293–303.

Davis, F.D. (1989). Perceived usefulness, perceived ease of use, and user acceptance of information technology. *MIS Quarterly*, 13(3): 319–340.

Deci, E.L. and Ryan, R.M. (1985). *Intrinsic Motivation and Self-determination in Human Behavior*. Plenum, New York.

Deci, E.L., Ryan, R.M. and Williams, G.C. (1996). Need satisfaction and the self-regulation of learning. *Learning and Individual Differences*, 8(3): 165–183.

Deci, E.L. and Ryan, R.M. (2002). *The Handbook of Self-determination Research*. University of Rochester Press, Rochester.

Dede, C., Grotzer, T.A., Kamarainen, A. and Metcalf, S. (2017). EcoXPT: Designing for deeper learning through experimentation in an immersive virtual ecosystem. *Journal of Educational Technology & Society*, 20(4): 166–78.

Dermitzaki, I., Leondari, A. and Goudas, M. (2009). Relations between young students' strategic behaviors, domain-specific self-concept, and performance in a problem-solving situation. *Learning and Instruction*, 19: 144–157.

Diacopoulos, M.M. and Crompton, H. (2020). A systematic review of mobile learning in social studies. *Computers & Education*, 154: 103911.

Dinsmore, D.L., Alexander, P.A. and Loughlin, S.M. (2008). Focusing the conceptual lens on metacognition, self-regulation, and self-regulated learning. *Educational Psychology Review*, 20: 391–409.

Endedijk, M.D., Brekelmans, M., Verloop, N., Sleegers, P.J.C. and Vermunt, J.D. (2014). Individual differences in student teachers' self-regulated learning: an examination of regulation configurations in relation to conceptions of learning to teach. *Learning and Individual Differences*, 30: 155–162.

Frauenfelder, E. and Santoianni, F. (eds.). (2003). *Mind, Learning and Knowledge in Educational Contexts*. Cambridge Scholars Press, Cambridge.

Gibson, J.J. (1977). The theory of affordances. pp. 67–82. *In*: Shaw, R. and Bransford, J. (eds.). *Perceiving, Acting, and Knowing: Toward an Ecological Psychology*. Lawrence Erlbaum, Hillsdale.

Gomez, S., Zervas, P., Sampson, D.G. and Fabregat, R. (2014). Context-aware adaptive and personalized mobile learning delivery supported by UoLmP. *Journal of King Saud University – Computer and Information Sciences*, 26: 47–61.

Gros, B. (2016). The dialogue between emerging pedagogies and emerging technologies. pp. 3–24. *In*: Gros, Kinshuk, B. and Maina, M. (eds.). *The Future of Ubiquitous Learning. Learning Designs for Emerging Pedagogies*. Springer, Berlin.

Gu, N. (2016). Implementing a mobile app as a personal learning environment for workplace learners. pp. 285–300. *In*: Churchill, D., Lu, J., Chiu, T. and Fox, B. (eds.). *Mobile Learning Design*. Springer, Singapore.

Haag, J. and Berking, P. (2019). Design considerations for mobile learning. pp. 221–240. *In*: Zhang, Y. and Cristol, D. (eds.). *Handbook of Mobile Teaching and Learning*. Springer, Singapore.

Hacker, D.J., Dunlosky, J. and Graesser, A. (eds.). (1998). *Metacognition in Educational Theory and Practice*. Lawrence Erlbaum, Mahwah.

Hamidi, H. and Chavoshi, A. (2018). Analysis of the essential factors for the adoption of mobile learning in higher education: a case study of students of the university of technology. *Telematics and Informatics*, 35: 1053–1070.

Hase, S. and Kenyon, C. (2003). Heutagogy and developing capable people and capable workplaces: strategies for dealing with complexity. pp. 25–27. *In*: *The Proceedings of the Changing Face of Work and Learning Conference*. University of Alberta, Alberta.

Hase, S. and Kenyon, C. (2007). Heutagogy: A child of complexity theory. *Complicity: An International Journal of Complexity and Education*, 4(1): 111–119.

Hase, S. and Kenyon, C. (2013). The nature of learning. pp. 19–38. *In*: Hase, S. and Kenyon, C. (eds.). *Self-determined Learning: Heutagogy in Action*. Bloomsbury Academic, London.

Hofstede, G., Hofstede, G.J. and Minkov, M. (2010). *Cultures and Organizations: Software of the Mind*. McGraw-Hill, New York.

Hoi, V.N. (2020). Understanding higher education learners' acceptance and use of mobile devices for language learning: a research-based path modeling approach. *Computers & Education*, 146: 103761.

Jaldemark, J. (2018). Contexts of learning and challenges of mobility: designing for a blur between formal and informal learning. pp. 141–155. *In*: Yu, S., Ally, M. and Tsinakos, A. (eds.). *Mobile and Ubiquitous Learning: An International Handbook*. Springer, Singapore.

Joo, Y.J., Kim, N. and Kim, N.H. (2016). Factors predicting online university students' use of a mobile learning management system (m-LMS). *Educational Technology Research and Development*, 64: 611–630.

Karimi, S. (2016). Do learners' characteristics matter? An exploration of mobile-learning adoption in self-directed learning. *Computers in Human Behavior*, 63: 769–776.

Khan, S.H., Abdou, B.O., Clement, C.K. and Kum, C. (2016). University student conceptions of M-learning in Bangladesh. pp. 127–138. *In*: Churchill, D., Lu, J., Chiu, T. and Fox, B. (eds.). *Mobile Learning Design*. Springer, Singapore.

Kinshuk. (2015). Roadmap for adaptive and personalized learning in ubiquitous environments. pp. 1–13. *In*: Kinshuk and Huang, R. (eds.). *Ubiquitous Learning Environments and Technologies*. Springer, Berlin.

Kölmel, B. and Kicin, B.S. (2004). Ambient learning. The experience of ambient technologies in eLearning. pp. 179–180. *In*: Courtiat, J.P., Davarakis, C. and Villemur, T. (eds.). *Technology Enhanced Learning*. Springer, Switzerland.

Liu, M., McKelroy, E., Corliss, S.B. and Carrigan, J. (2017). Investigating the effect of an adaptive learning intervention on students' learning. *Educational Technology Research Development*, 65: 1605–1625.

Louhab, F.E., Bahnasse, A. and Talea, M. (2018). Towards an adaptive formative assessment in context-aware mobile learning. *Procedia Computer Science*, 135: 441–448.

Lyardet, F. (2008). Ambient learning. pp. 530–549. *In*: Mühlhäuser, M. and Gurevych, I. (eds.). *Handbook of Research on Ubiquitous Computing Technology for Real Time Enterprises*. IGI Global, Hershey.

Ma, N., Zhang, X. and Zhang, Y.A. (2019). The development of mobile learning in China's universities. pp. 521–548. *In*: Zhang, Y. and Cristol, D. (eds.). *Handbook of Mobile Teaching and Learning*. Springer, Singapore.

Mayer, R.E. (2020). Where is the learning in mobile technologies for learning? *Contemporary Educational Psychology*, 60: 101824.

Metcalfe, J. and Shimamura, A. (eds.). (1994). *Metacognition: Knowing about Knowing*. MIT Press, Cambridge.

Musso, M.F., Boekaerts, M., Segers, M. and Cascallar, E.C. (2019). Individual differences in basic cognitive processes and self-regulated learning: their interaction effects on math performance. *Learning and Individual Differences*, 71: 58–70.

National Research Council. (2012). *Education for Life and Work: Developing Transferable Knowledge and Skills in the 21st Century*. National Academies Press, Washington.

Pachler, N., Bachmair, B. and Cook, J. (2010). *Mobile Learning: Structures, Agency, Practices*. Springer, New York.

Pachler, N., Bachmair, B. and Cook, J.A. (2013). Socio-cultural frame for mobile learning. pp. 35–46. *In*: Berge, Z. and Muilenburg, L. (eds.). *Handbook of Mobile Learning*. Routledge, New York.

Palkova, Z. (2019). Mobile Web 2.0 tools and applications in online training and tutoring. pp. 609–634. *In*: Zhang, Y. and Cristol, D. (eds.). *Handbook of Mobile Teaching and Learning*. Springer, Singapore.

Papanikolaou, K.A. and Grigoriadou, M. (2004). Accommodating learning style characteristics in adaptive educational hypermedia. pp. 77–86. *In*: Magoulas, G. and Chen, S. (eds.). *The Proceedings of the Workshop on Individual Differences in Adaptive Hypermedia in AH2004.* Eidhoven, The Netherlands.

Paramythis, A. and Loidl-Reisinger, S. (2004). Adaptive learning environments and e-Learning standards. *EJEL Electronic Journal of e-Learning,* 2(2): 181–194.

Paraskakis, I. (2005). Ambient learning: a new paradigm for e-Learning. pp. 26–30. *In: The Proceedings of the 3rd International Conference on Multimedia and ICT in Education.* Cáceres, Spain.

Pegrum, M. (2019). Mobile lenses on learning. *Languages and Literacies on the Move.* Springer, Singapore.

Pellas, N. (2014). The influence of computer self-efficacy, metacognitive self-regulation and self-esteem on student engagement in online learning programs: evidence from the virtual world of second life. *Computers in Human Behavior,* 35: 157–170.

Peng, H., Ma, S. and Spector, J.M. (2019). Personalized adaptive learning: an emerging pedagogical approach enabled by a smart learning environment. pp. 171–176. *In:* Chang, M., Popescu, E., Kinshuk, Chen, N.S., Jemni, M., Huang, R., Spector, J.M. and Sampson, D.G. (eds.). *Foundations and Trends in Smart Learning.* Springer, Singapore.

Post, Y., Boyer, W. and Brett, L. (2006). A historical examination of self-regulation: helping children now and in the future. *Early Childhood Education Journal,* 34(1): 5–14.

Power, R. (2019). Design of mobile teaching and learning in higher education: an introduction. pp. 3–11. *In:* Zhang, Y. and Cristol, D. (eds.). *Handbook of Mobile Teaching and Learning.* Springer, Singapore.

Prokopenko, M. (2008). Design vs. self-organization. pp. 3–17. *In:* Prokopenko, M. (ed.). *Advances in Applied Self-organizing Systems. Advanced Information and Knowledge Processing.* Springer, London.

Ranieri, M. (2015). *Linee di ricercaemergentinell'educational technology, Form@re - Open Journal per la formazione in rete,* 3(15): 67–83.

Reder, L.M. (ed.). (1996). *Implicit Learning and Metacognition.* Lawrence Erlbaum, Mahwah.

Riegler, A. (2002). When is a cognitive system embodied? *Cognitive Systems Research,* 3(3): 339–348.

Roschelle, J. and Pea, R.D. (2002). A walk on the WILD side: How wireless handhelds may change computer-supported collaborative learning (CSCL). *The International Journal of Cognition and Technology,* 1(1): 145–168.

Ryan, R.M. and Deci, E.L. (2000). Self-determination theory and the facilitation of intrinsic motivation, social development, and well-being. *American Psychologist,* 55(1): 68–78.

Ryan, R.M. and Deci, E.L. (2018). Self-determination theory. *Basic Psychological Needs in Motivation, Development, and Wellness.* Guilford Press, New York.

Sadler-Smith, E. and Smith, P.J. (2004). Strategies for accommodating individuals' styles and preferences in flexible learning programmes. *British Journal of Educational Technology,* 35(4): 395–412.

Santoianni, F. (2006). *Educabilità cognitiva, Apprendere al singolare, insegnare al plurale.* Carocci, Roma.

Santoianni, F. (2007). Bioeducational perspectives on adaptive learning environments. pp. 83–96. *In:* Santoianni, F. and Sabatano, C. (eds.). *Brain Development in Learning Environments. Embodied and Perceptual Advancements.* Cambridge Scholars Publishing, Cambridge.

Santoianni, F. (2010). *Modelli e strumenti di insegnamento, Approcci per migliorare l'esperienza didattica.* Carocci, Roma.

Santoianni, F. (2014). *Modelli di studio, Apprendere con la teoria delle logiche elementari.* Erickson, Trento.

Santoianni, F. (2016). Spaces of thinking. pp. 5–14. *In:* Santoianni, F. (ed.). *The Concept of Time in Early Twentieth-century Philosophy. A Philosophical Thematic Atlas.* Springer, Switzerland.

Santoianni, F. (2017). Models in pedagogy and education. pp. 1033–1049. *In:* Magnani, L. and Bertolotti, T. (eds.). *Springer Handbook of Model-based Science.* Springer, Cham.

Santoianni, F. and Ciasullo, A. (2018a). Adaptive educational environments. Adaptive design for educational hypermedia environments and bio-educational adaptive design for 3d virtual learning environments. *REM Research on Education and Media,* 10(1): 30–41.

Santoianni, F. and Ciasullo, A. (2018b). Digital and spatial education intertwining in the evolution of technology resources for educational curriculum reshaping and skills enhancement. *International Journal of Digital Literacy and Digital Competence,* 9(2): 34–49.

Santoianni, F., Ciasullo, A., De Paolis, F., Nunziante, P. and Romano S.P. (2018). Federico 3DSU. Adaptive educational criteria for a multi-user virtual learning environment. *Journal of Virtual Studies, Special Proceedings of the Immersive Learning Education Conference*, 9(1): 9–16.

Santoianni, F. (2019). Brain education cognition. *La ricerca pedagogica italiana. RTH Research Trends in Humanities*, 6: 44–52.

Santoianni, F. (2020). Brain-based education. *La ricerca bioeducativ asperimentale, RTH Research Trends in Humanities*, 7: 28–33.

Schunk, D.H. and Zimmerman, B.J. (1994). *Self-regulation of Learning and Performance: Issues and Educational Applications*. Lawrence Erlbaum, Hillsdale.

Sharples, M., Arnedillo-Sanchez, I., Milrad, M. and Vavoula, G. (2009). Mobile learning small devices, big issues. pp. 233–249. *In*: Balacheff, N., Ludvigsen, S., de Jong, T., Lazonder, A. and Barnes, S. (eds.). *Technology-enhanced Learning*. Springer, Dordrecht.

Sharples, M. and Pea, R. (2014). Mobile learning. pp. 501–521. *In*: Sawyer, K. (ed.). *The Cambridge Handbook of the Learning Sciences*. Cambridge University Press, New York.

Shuib, L., Shamshirband, S. and Ismail, M.H. (2015). A review of mobile pervasive learning: applications and issues. *Computers in Human Behavior*, 46: 239–244.

Siemens, G. (2005). Connectivism: A learning theory for the digital age. *International Journal of Instructional Technology and Distance Learning*, 2(1): 3–10.

Stoerger, S. (2013). Becoming a digital nomad: transforming learning through mobile devices. pp. 473–482. *In*: Berge, Z. and Muilenburg, L. (eds.). *Handbook of Mobile Learning*. Routledge, New York.

Sung, Y.T., Lee, H.Y., Yang, J.M. and Chang, K.E. (2019). The quality of experimental designs in mobile learning research: a systemic review and self-improvement tool. *Educational Research Review*, 28: 100279.

Thomas, S. (2013). *Technobiophilia: Nature and Cyberspace*. Bloomsbury Academic, London.

Traxler, J. (2007). Defining, discussing and evaluating mobile learning: the moving finger writes and having writ. *The International Review of Research in Open and Distance Learning*, 8(2): 1–12.

Truong, H.M. (2016). Integrating learning styles and adaptive e-learning system: current developments, problems and opportunities. *Computers in Human Behavior*, 55: 1185–1193.

Turner, J. (2015). Mobile learning in K-12 education: Personal meets systemic. pp. 630–658. *In: The Proceedings of the International Mobile Learning Festival 2015: Mobile Learning, MOOCs and 21st Century Learning*. Hong Kong, China.

Turner, J. (2016). Mobile learning in K-12 education: personal meets systemic. pp. 222–238. *In*: Churchill, D., Lu, J., Chiu, T. and Fox, B. (eds.). *Mobile Learning Design*. Springer, Singapore.

Vandewaetere, M., Desmet, P. and Clarebout, G. (2011). The contribution of learner characteristics in the development of computer-based adaptive learning environments. *Computers in Human Behavior*, 27: 118–130.

Venkatesh, V., Morris, M.G., Davis, G.B. and Davis, F.D. (2003). User acceptance of information technology: toward a unified view. *MIS Quarterly*, 425–478.

Vincent-Layton, K. (2019). Mobile learning and engagement: designing effective mobile lessons. pp. 241–256. *In*: Zhang, Y. and Cristol, D. (eds.). *Handbook of Mobile Teaching and Learning*. Springer, Singapore.

Wang, Q. (2018). Core technologies in mobile learning. pp.127–139. *In*: Yu, S., Ally, M. and Tsinakos, A. (eds.). *Mobile and Ubiquitous Learning. Perspectives on Rethinking and Reforming Education*. Springer, Singapore.

Watson, S.L. and Reigeluth, C.M. (2018). The learner-centered paradigm of education. pp. 758–783. *In*: West, R.E. (ed.). *Foundations of Learning and Instructional Design Technology. The Past, Present, and Future of Learning and Instructional Design Technology*. Ed Tech Books.

Weller, M. (2018). Twenty years of ed tech. pp. 88–116. *In*: West, R.E. (ed.). *Foundations of Learning and Instructional Design Technology. The Past, Present, and Future of Learning and Instructional Design Technology*. Ed Tech Books.

Whitebread, D. (2020). Self-regulation processes in early childhood. pp. 107–115. *In*: Benson, J.B. (ed.). *Encyclopedia of Infant and Early Childhood Development*. Elsevier, Amsterdam.

Winne, P.H. and Hadwin, A.F. (1998). Studying as self-regulated learning. pp. 277–304. *In*: Hacker, D.J., Dunlosky, J. and Graesser, A.C. (eds.). *Metacognition in Educational Theory and Practice*. Lawrence Erlbaum, Mahwah.

Winne, P.H. and Hadwin, A.F. (2010). *Self-regulated Learning and Socio-cognitive Theory*. Elsevier, Amsterdam.

Wolf, C. (2002). iWeaver: Towards an interactive web-based adaptive learning environment to address individual learning styles. *The European Journal of Open, Distance and E-Learning*, 5: 1–14.

Wong, L.H. and Looi, C.K. (2011). What seams do we remove in mobile-assisted seamless learning? a critical review of the literature. *Computers & Education*, 57(4): 2364–2381.

Yu, C., Lee, S.J. and Ewing, C. (2015). *Mobile Learning: Trends, Issues, and Challenges in Teaching and Learning*. IGI Global, Pennsylvania.

Zhang, F.J. (2019). A novel education pattern applied to global crowd of all ages: mobile education. pp. 341–358. *In*: Zhang, Y. and Cristol, D. (eds.). *Handbook of Mobile Teaching and Learning*. Springer, Singapore.

Zimmerman, B.J. and Schunk, D.H. (2001). *Self-regulated Learning and Academic Achievement: Theoretical Perspectives*. Lawrence Erlbaum, Mahwah.

CHAPTER 2

Educational Theories of Mobile Teaching

Flavia Santoianni

People learn in different ways.
—Sun et al. (2007)

What is needed is cognitive diversity.
—Pegrum, 2019

Introduction

Learning theories of the twentieth century are related to mobile learning, even if they are diversified and attributable to different years. Each of them has specific characteristics influencing the mobile approach—from classic and contemporary to innovative mobile teaching. All these theories are addressed to design teaching and learning in formal and informal contexts, which are narrowly related to mobile technology and its potential factors, as educational learning-environments design and user-acceptance/adoption models in educational technology (Hamidi and Chavoshi, 2018).

Mobile learning has been defined as a 'catalyst' for educational change since it promotes overcoming of traditional teacher-directed pedagogies in order to implement new teaching strategies (Cochrane, 2014) according to the issues that current pedagogical approaches are not adequately advanced for the new generation of learners, and where an educational paradigm shift is needed (UNESCO, 2012). Reproducing already existing educational activities seems to be a palliative, while a re-thinking of pedagogical practices for mobile learning should be based on emerging ground-breaking ideas about learner-centered environments.

Professor of Education, University of Naples Federico II, Italy.
Email: bes@unina.it

Mobile learning has been defined as learning across contexts to extend the reach of teaching and learning, to enhance knowledge co-construction and exchange, and to foster both lifelong collaborative and independent learning (Yu et al., 2015). Educational theories of mobile learning converge around key features concerning the learning processes and environments, as well as teachers' professional behavior, students' developing skills, and mobile technological affordances. Since mobile learning is nowadays seamless and ubiquitous in formal and informal spaces, interactive relations with educational materials, online resources, other members of the learning community and their activities are encouraged and mediated by mobile devices (Bernacki et al., 2020).

Technology-mediated learning involves learning management systems (LMS), computer-based learning (CBL), computer-assisted instruction (CAI), computer-based education (CBE), and so on. Common aspects of technology-mediated learning are learners' engagement and feedbacks from programmed learning applications; structured learning with which learners can actively interact; and content storage, management, and availability through different learning theories (Walling, 2014).

During the twentieth century, models of learning were characterized by the following three main metaphors. The metaphor of acquisition has influenced the classic teaching of behaviorism and cognitivism, in which the research focus is on the amount and quality of information transmitted from teachers to students. The metaphor of participation viewed learning as a process rooted in cultural practices and shared activities and has been adopted by contemporary teaching. The metaphor of knowledge creation is the last one and emphasizes not only situativity and social practices, but also their development through mediated activities of co-creation which leverage innovative teaching (Paavola et al., 2004).

These three metaphors still co-exist and for this reason, the re-thinking of pedagogical practices for mobile learning cannot be based only on emerging ground-breaking ideas, but also need recognition of significative educational theories of the twentieth century to understand which of their characteristics are still surviving to address mobile teaching. Educational theories of learning are to be considered complementary, rather than oppositional. Traditional approaches should not be abandoned but instead integrated within a palette of contextualized pedagogies, which can highlight the possible overlapping between them (Pegrum, 2019). Innovative pedagogies have to rethink traditional ones in order to let new meanings emerge (Gros, 2016).

In this chapter, educational theories will be highlighted to understand how they evolved during the 1900s and why they can be still considered of some interest when looking for innovative pedagogical approaches to manage mobile teaching. At the same time, these educational theories will be updated because current ideas in the teaching and learning field of 2000–2020 have already renewed their own matrices developed during the last century and have to be taken into consideration in order to have a whole framework of interpretation of which issues can be more useful to analyze mobile learning nowadays. This recognition will introduce the next chapter, where emerging educational indications will be outlined to re-think the developing field of mobile learning design and co-create learner-centered environments in relation to new teaching strategies.

Classic Mobile Teaching and the Metaphor of Acquisition

The several theoretical and methodological teaching models that have followed each other during the last century, highlighting different educational implications, can be then revisited or better replaced, as suggested nowadays to deal with mobile learning management. However, attention has to be paid to their own evolution because mobile innovation is to be linked to ongoing research on teaching and learning. Also, dated educational models are still used today and have effective characteristics that persist over time without being necessarily opposed; on the contrary, the manifold panorama of educational theories often presents an intertwined overlapping to current research acquisitions which are rather complicated to untangle (Santoianni, 2010).

Even if any schematization of educational models may seem reductive in the attempt to systematize the huge number of formalized learning theories, it is anyway possible to sketch the criteria guidelines and the interpretative categories which characterize the currents of thought that have contributed to outline and influence the longer-lasting models of teaching of the twentieth century. These categories may still encounter the emerging characteristics of the new trends in the teaching and learning field, since they can be longlasting because the acknowledgement of the role played by previous theoretical models in influencing methodological and applicative addresses for teaching practices may encourage and sustain the incoming growth of educational approaches, which may find their roots in their own history to better address how to further develop.

Behaviorist Mobile Teaching

Behaviorist education, just to begin, is still alive. It belongs to the group of educational *asymmetrical* models (Santoianni, 2010) in which teacher and students are involved in an asymmetric relation, and knowledge is transmitted by the teacher to students. The effectiveness of this transfer is based on the amount of information that can be transmitted. The behaviorist model highlights the figure of the teacher and her/his role in the learning environment design (Santoianni, 2003a). The behaviorist instructor can be authoritative and directive, while students are scarcely involved within the learning process (Kraglund-Gauthier, 2019).

The behaviorist model—the conduit model (Fischer, 2009)—arose in the early twentieth century with behavioral studies and developed until the middle of the last century. Learning is associative. It is a process that consists of a series of related steps, which associate new connections between stimuli and responses to develop conditioned habits (Block and Anderson, 1977; Lieberman, 1990). Learners' behavior can be studied because it is directly observable.

Learning by conditioning can be *respondent* (the learner responds to the environmental stimulus) or *operant* (the learner operates in the environment, modifying it). In respondent conditioning, individual behavior changes in relation to the laws of intensity, frequency, and continuity of stimulation. The reinforcement precedes and elicits the conditional response. In operant conditioning, individual behavior changes in relation to the laws of exercise (use and disuse) and of effect. The reinforcement consists of the state of satisfaction that follows the response, which tends to be repeated by trial and error to achieve it. Both types of conditioning are based on implicit and explicit adaptive mechanisms for the development of behavioral patterns. Learner's activity depends on natural, social, and cultural/environmental conditions.

Respondent and operant conditionings are parallel processes in teaching and learning, but they are distinct. In respondent conditioning, learning is activated by responsiveness to the signal: each student learns the answer by repeating it and the teacher exercises control (including emotional control) over the class through signals, rules, and behavioral sequences. In operant conditioning, learning is motivated by orientation to the goal: the learner gives more than one answer and chooses between them the one which seems to be more adaptive while the teacher exercises control through reinforcements, subsequent to the learner's response.

As learning is sequential, the teaching methodology too foresees a sequence of procedures on the basis of the review of learner's prior knowledge. This means deepening of the prerequisites s/he has, and the explicit design of planned goals to be reached through learning steps linked to specific learning units. The quantification of teaching content is valued in relation to common learning standards. If the teacher explains well how to implement a given cognitive task through a continuous process of apprenticeship and imitation, every student can reach, as well as the others, the foreseen standard of mastery performance. The organization of regular evaluation is divided in ongoing as well as final sessions in order to leverage behavioral changes and to calculate the optimized time needed to perform cognitive tasks. Any learner's cognitive development is due to the learning opportunities given by the teacher, who encourages students to gain self-control by modeling their behavior through auto-shaping or disciplined response to the teacher's transmission of information, which they passively receive and exercise through repetition. Learning is then an association between a stimulus and a response, conditioned by the environment.

Transferring these concepts in mobile learning research, any kind of questions, listening and speaking exercises providing feedback, exams, and any mobile response system through content transfer via text messages may represent an expression of the behaviorist model, as language-learning applications, voice-recording software, short message system (SMS), and multimedia messaging (MMS) (Pimmer et al., 2016). Behaviorist mobile teaching allows the students to learn habits, while conditioning them through reinforcements, as exercises. Learning—but also teaching—is monitored and frequently verified by ongoing checks (Santoianni, 2017). Educational action is planned through programmed instruction, methodological and content explicit objectives, and students/users are conditioned by audio and video signals.

Cognitivist Mobile Teaching

The cognitive model[14] emerged in the mid-twentieth century with the birth of cognitive science and developed from the second post-war period (Gardner, 1985).

---

[14] Cognitivist and meta-reflective teaching may also be considered as belonging to *asymmetrical* models, since all of them involve processes of knowledge transmission—even if cognitivism focuses on learners' personal strategies of content management; and meta-reflection, linked to post-cognitivism, is mainly concerned with the tutoring control of the processes of acquisition of information by the individual mind (Santoianni, 2010).

Learning is processing of information. The learner uses individualized strategies of acquisition and organization of information. Cognitive models of learning research on the functioning of mental processes than on the manifest behavior of individuals. The basic concept of behaviorism, that is, the stimulus-response association, is replaced by the concept of hierarchical organization of procedures which regulate informational coding and processing (Santoianni, 2003b).

The emerging cognitive science was studying the computing machines, assimilable to the human mind as computers, while cybernetics was exploring the nervous system and its adaptive processes. Their research paths branched out into two distinct approaches: the computational approach, characterized by interest in quantitative, logical-symbolic, sequential, and programmable calculating machines, and the connectionist approach, according to which cognitive organization is biological, adaptive, and evolutionary, and is regulated by parallel self-regulatory processes, followed by feedback mechanisms.

Comparing the mind to a computer, computational learning is based on information processing, which takes place sequentially: from perceptual stimulation to symbolic data coding that occurs when the cognitive system receives the stimulation of the environmental information (input). During information processing, the hierarchical functions of the cognitive system shape the incoming information as mental representation, i.e., linguistic symbols and numbers, and plans the response (output). Again, as a computer, the cognitive system is seen as a box with limited capacity to be filled in with a defined amount of information to be stored in the short- or long-term archives of memories, where learners may find them through their own retrieval methods. Since the mind is seen as a limited capacity container, as the hard disk of a personal computer, teachers have to calibrate the educational offer to meet students' needs, in order to give them a balanced amount of information.

Knowledge transmission is significant because it does not only consist of transferring the contents, but there are issues concerning how such contents can be codified, elaborated, and memorized. Learning is based on understanding and no more on mnemonic retention. The learner's experience strategies of procedural knowledge (know how) and declarative propositional knowledge (know that) management are based on qualitatively differentiated skills. To be stored, information has to be then significant that is individually reworked in a reflective way. The cognitive system processes information through sequences of procedures, guided by predefined actions as the if/then preliminary rules.[15]

Teaching methodology promotes the learners' possibility to autonomously organize strategies of knowledge management. Learning contents are exemplified and explained in a multimodal way to be well understood and joined into units of meaning which are stored in the long- or short-term memories. The recall of knowledge depends on how the previously acquired information has been memorized. Cognitive

---

[15] The sequential nature of learning represents one of the focal points of the criticism addressed to the computational approach by connectionism because in this other approach, learning is seen as a parallel process which regulates itself in continuous interaction with the environment. Moreover, in the computational model structured data (frames, schemes, scripts) are symbolized within the representative level, while the connectionist model is sub-symbolic, meaning that data are considered quantitative values (Santoianni, 1997).

acquisition develops through increasingly complex sequences of instructions, which need to be calibrated to prevent generation or demotivation or confusion during the learning process. This process has indeed to be monitored in order to continuously modify the teaching practice according to the learners' response feedback.

Cognitivism leads to student-centered, self-paced, multimedia interactive teaching and learning which can be implemented in mobile technology. Multimedia mobile learning—including image, audio or video, text, and animation—short and multimedia messaging, e-mail, podcasts, and mobile TV are all examples of the multiple possibilities to express any educational offer in a really stimulating and more personalized way. At the same time, programmed learning pathways in which contents are organized through questions and samples useful for problem solving and decision making can enhance participation, interaction, and engagement in mobile learning, through handheld games, simulation, and virtual reality (Shin and Kang, 2015; Koç et al., 2016).

Cognitivist mobile teaching gives sequences of instructions to prepare students/ users for specific tasks, since it is not contextually oriented but aims instead at leveraging individual learning *in vitro* and meaningful understanding through calibrated explanations, organization of learning strategies, and repetitive practice. Around the 1980s, the perception that the educational offer should be standardized was overcome by the idea to customize education through tailored teaching (Gardner, 1983; Sternberg, 1985). Cognitive variability has been consequently studied inside the cognitive prism (Santoianni and Ciasullo, 2021). The cognitive prism is a dynamic set of structural and functional elements intertwined by synergistic relationships and concerns a plurality of aspects at perceptual, processing, and meta-reflective level (Santoianni, 2014).

The perceptual level—auditory, visual, or tactile—has been mainly studied as a style level, even if discussed in literature as a stand-alone theoretical construct (Sternberg and Zhang, 2001). Learning, teaching, and thinking styles (Dunn and Dunn, 1978; Fischer and Fischer, 1979; Sternberg, 1997; Zhang and Sternberg, 2005) have been individuated, and more. The processing level has been mainly explored through analysis of the multiplicity of intelligence (Gardner, 1995, 1996, 1997; Gardner and Boix Mansilla, 1997) and its triarchic architecture (Sternberg, 1990).

Mobile educational systems should be then individualized to provide students personalized, dynamic, and adaptive experiences. The issue of adaptivity has been studied according to learning style theories to individual learning behaviors, and in relation to the systematization of the internal division of learning objects to meet the requirement of reusability, since a learning object is an autonomous discrete part of content with an educational objective—reusable as a learning material in different instructional contexts (Sun et al., 2007). Personalization of mobile learning can involve several learning applications for mobile devices as the ones concerning data collection and cloud computing; hyper-documents; first-person assessments, and performance support and coaching; mobile gaming, and multimedia production and playback; virtual worlds and augmented reality; contextual learning, environmental control, and haptic feedback; portable collaborative communities and self-organized collective behavior.

Problem-solving and Problem-based Learning

Problem solving is a cognitivist active process by which the learner aims to transform the initial state of a problem into the one to be achieved. Each problem has data about the initial condition of the problematic situation, objectives as forthcoming answers and solutions, and obstacles which can complicate the path of problem solving. If a learner tries to solve a situation which is in a given state by moving it towards a goal state and by using a set of rules, but some obstacles occur during this transition, a problem may arise (Mayer, 2010).

Problem solving is considered a cognitive process because its development is internal to the learner's cognitive system, involves cognitive representations, and is monitored and controlled by the learner. To solve a problem, it is needed to recognize the problem to be solved, to understand what it is about, and how to reach the possible solution. This process involves some steps, as to identify and define the problem, to mentally represent it—i.e., to build a situation model—to plan resolution strategies, through the cognitive processes of planning, executing, and monitoring, and to evaluate the learner's skills to face and solve problems. All these phases of the process may vary according to the problem characteristics, the cognitive qualities of the solver, and the context in which the problem occurs (Davidson et al., 1994).

Problems can be well defined with a clear given/goal state, or ill defined, when the given/goal states are poorly specified. In well-structured problems, the sequences of steps leading to the solution are clearly identifiable by the solver and can be heuristically arranged. In ill-structured problems, objectives are not clearly defined and the less experienced solver may not grasp the entirety of the problem, thus risking the consolidation of incomplete/incorrect solutions or behavioral patterns. Everyday life is dotted with ill-defined problems, while school problems are often well-defined. Learning to solve work-related, home-related, hobby-related problems is a daily-life skill, while research has mainly focused on problem solving in school-based and university-based settings.

When coping with authentic ill-structured problems, learners should be provided with scaffolds and tools to manage their complexity. One strategy for reducing this complexity can be to scaffold learning in real-world settings and to integrate knowledge and everyday practices, in order to allow learners to engage in deep learning. Learners can consequently analyze new information, connect incoming data, internalize concepts, and transfer them in different settings. In a socio-cultural perspective, problem solving is no more only a cognitive process but includes emotional involvement and environmental embeddedness. In informal settings, problems are dynamic and domain specific, while learners can have personal connections with them, and then need to be scaffolded to develop strategic performances (Choi et al., 2018).

Learning is considered to be more effective if it concerns ill-structured, authentic, complex, and dynamic problems with plus than one solution as case studies, strategic decision making and planning, which require skills of analysis and reflection rather than ready-made knowledge (Jonassen, 2000). If individuals constantly face and attempt to solve daily-life work/home-related problems, while research on problem-solving mainly focuses on school/university settings, they usually cope with nonroutine problems, which may be very different from any previous problem that learners already faced.

Productive thinking emerges, which is divergent and leads to many solutions, therby enhancing deep learning. A strategy for supporting deep learning is to situate it in authentic real-world settings through technological interface utilities, which allow scaffold learning in natural settings in order to integrate scientific knowledge with everyday experiences. Informal education can address the design of experiences within learning environments beyond classrooms through mobile devices, since they are ubiquitous and can easily support skills to acquire knowledge in the outdoors. Problem solving is no more seen as an 'in-the-head' cognitive-only phenomenon, but implies both cognitive and emotional dimensions, involvement in authentic activities, inquiry related to the relevance of problems to the learner, and teaching or scaffolding provided by the learning environments (Choi et al., 2018).

The openness of problem solving to knowledge social co-construction has led to problem-based learning, where learners actively work together on authentic tasks in reassembled contexts, while the teacher pays attention to learners' meta-reflective processes (Savery and Duffy, 1995). To focus on contexts also means to implement reflective and collaborative behaviors of learners as lifelong learning skills to enhance generative learning (Grabinger and Dunlap, 1997).

Contemporary Mobile Teaching and the Metaphor of Participation

Since the last decades of 1900, a convergence of learning theories occurred, addressing research on both learners and social knowledge within contexts. Conceptual matrices of socio-constructivism and socio-cultural theory, distributed and situated informal cognition, lifelong learning and embodied bio-educational sciences—because they are continuously in development, often overlapping, and hardly distinguishable from each other—are intertwined and rely on three main ideas: active learning is individual and collaborative, knowledge is rooted in situated socio-cultural environments, and cognition—even if distributed and co-created—is a personal customized interactive process (Santoianni, 2017). The focus of investigation on teaching and learning as had a gradual shift from one-sided classic didactic (teacher- or student-centered) towards various educational models outlining the dynamic relation between teachers and learners.

Knowledge is distributed, situated, and embodied (Kirshner and Whitson, 1997; Cantwell and Smith, 1999; Bereiter, 2002). The current wide and complex interpretative framework of learning management has been reconsidered, focusing on knowledge as distributed among individuals in learning communities, between cognitive artefacts and cultural tools, peripheral devices and technological innovations; as situated in space and time, in contingent contexts, and in specific domains of knowledge. Knowledge, embedded within cultural practices and attention, is given more to activities related to knowing than to the products of the knowledge previously intended as a world of its own or inside individual minds, while knowing is now considered as distributed over individuals and environments.

Learning is situated in a network of shared activities and cannot be interpreted if detached from where it occurs.

Teaching and learning are rethought as a symmetric relation in which knowledge is no longer transmitted but co-constructed in specific space and time. Post-cognitivist models leverage a renewed consideration of individuals working together to negotiate inter-subjective knowledge, to manage learning, and to share practises within communities of learners, rework meanings in the contextual interaction through collaborative discourses, and contribute to fully exploit and preserve the cultural heritage of the historical continuity of rooted tools, cognitive artifacts, and peripheral devices. Teaching and learning are reciprocally influencing students, teachers, and contents, since learning is no more a task-oriented process but an individual and collective making-meaning process which is only facilitated by teachers.

Systemic Computer-supported Collaborative Mobile Teaching

Computer support (CS) and collaborative learning (CL) can be systematized into two groups of *systemic* and *dialogical* approaches. Specific technological affordances of the systemic approach implement cognitive collaboration, reasoning, and inquiry in knowledge domains. Thinking and acting are analyzed through internalization—when the learner is within the phase of knowledge acquisition or improving—and transfer—when the learner applies knowledge gained in one learning context to another learning context (Ludvigsen and Mørch, 2010). Knowledge Forum (Scardamalia and Bereiter, 2006) is an example of computer-supported collaborative learning (CSCL) which enhances group learning mediated by technologies.

Collaborative knowledge building is a theoretical construct of learning activity which is shared as public knowledge available to students within a community of learners to be negotiated, re-worked, and synthesized (Bereiter and Scardamalia, 2003). Collective activities leverage both the development of cognitive identities and cultural responsibilities to support the community practises (Kruger and Tomasello, 1996). Knowledge is systematically produced together by the members of the learning community through knowledge building, which is a collective work for the implementation of conceptual artifacts as theories, ideas, and models (Bereiter, 2002). Knowledge building is like a scientific-inquiry process, whose phases generate personal hypotheses, experimenting and/or critically evaluating them, followed by scientific explanation, and data interpretation.

Knowledge structures development overcomes the previous idea of learning processes and is related to a multiplicity of factors as contextual contingency, cultural roots, activities of sharing and co-creation, domain-specific interaction, and individual's role within the learning community. The meaning of individuality is reconsidered in relation to the co-construction of learning communities and learning environments which influence individuals and, at the same time, are produced by them. Knowledge becomes an activity of mutual construction—a transformative activity where individuals and contexts constitute each other in dynamic ways within the social relations of the community of practices, and the acknowledgement of the idiosyncrasy of each individual.

Learning is a process of interaction, activity, enculturation, and legitimate peripheral participation, which consists of participation in the community activities of practices. During the process of participation, newcomers are still peripheral to the community; when they become more experienced, their peripheral participation becomes more and more legitimate (Brown et al., 1989; Lave and Wenger, 1991). Participation in the learning community can be disciplinary or conceptual which are different kinds of agencies of the learners. If disciplinary, the members of the community receive and reproduce meanings, but if conceptual, learners co-construct knowledge alternatives and develop critical thinking (Collins and Greeno, 2010).

The objective of knowledge sharing by community learners is not to learn but to develop new knowledge in shared spaces. The learning community is to be considered as the belonging group, in which learners begin to build culture according to their own cultural heritage (Bruner, 1996). After the identification of cognitive responsibilities of learners through apprenticeship and scaffolding of experienced mediators considered as adult models, this process leads to the acknowledgement of personal heritage, which is the personal cognitive identity, to effectively become a member of the community. Learning is supported by scaffolding, which is an instructional technique by which teachers sustain students during a learning task, gradually leaving them with the responsibility of knowledge acquisition. Technological scaffolding relies on educational software whose features are designed to perform virtual guidance functions.

Dialogical Computer-supported Collaborative Mobile Teaching

The dialogic approach analyzes learning as a process of externalization mediated by conceptual artifacts and technological tools within social practice. It is influenced by socio-cultural and situated cognition theories. Mobile learning is rooted in social and cultural theories on learning as mediated by tools. Technology is a mediating artifact in itself. The core level of analysis is the participation in an activity system and the synergic interactions within it.

Born at the beginning of 1900 and diffused in 1960s (Vygotsky, 1986), socio-cultural or cultural-historical approach emphasizes how social and cultural environments can influence development through mediational symbolic and material means, cultural tools, and signs. Cognitive development is socially constructed; it is an emergent property of social interaction and cultural experiences. Basic cognitive processes can be transformed in (and supported by) higher mental functions, which are not only the complex version of elementary functions, but are related to socio-cultural phenomena mediated by signs, tools, and community participants (Gauvain, 2008). The learning community is a shared place, real and/or virtual, where learners can meet and participate in knowledge co-construction (Retallick et al., 1999; Rogoff et al., 2003). Learning is linked to cultural frameworks and to their evolution; cognitive artefacts and peripheral devices are located in contingent situations and belong to specific cultures (Olson and Torrance, 1996; Cole, 1996; Egan, 1997).

Cognition is a mediated activity and a complex social phenomenon distributed between minds, social activities, contexts, and cultures (Lave and Wenger, 1991). Knowledge is historically and culturally rooted in social contexts, downloaded in individual minds, in cognitive artefacts (signs and symbols), and peripheral devices (technology and multimedia tools). In order to promote situated co-construction of learning within community practices, everyone can give her/his contribution to it through shared meanings, thus shifting from learning processes to knowledge structures. Despite distributed knowledge being in the foreground, situated cognition, which promotes learning within authentic contexts and cultures, recovers the role of individual subjectivity (Kirshner and Whitson, 1997) because embeddedness and embodiment are closely related. Interactions with the social environment are firstly interpersonal and then individual.

Situated cognition studies knowledge in the contextual areas in which it is synergically produced and co-created by structuring minds for social, educational, ecological, and more, purposes through interactions mediated by technology and multimedia. Learning is a socially organized activity, mediated by physical and abstract tools within which there is language. According to the dialogic approach, learning tasks can be often open-ended because they are continuously co-constructed by learners, thus becoming a learning activity in themselves. The dialogic approach develops skills for collaboration, negotiation, knowledge sharing, critical thinking, and decision making. Technological automated feedback can support users' interaction through guidance, awareness, and meta-reflection (Ludvigsen and Mørch, 2010).

The multidirectional network of connections across actors, objects, and tools provides a way to understand how learners can be shaped by their environment and how they can shape it through activities (Bernacki et al., 2020). Cultural-historical activity theory (CHAT) analyzes the specific components of an human activity system (Engeström, 1987); for instance, which community is involved in the process, how the working flow has been divided, which are the roles of all, which tools have been used, if there are related activities, and what is the level of interactivity. Interactivity is regulated by purpose-specific uses of resources in relation to the designed complexity of the task (Churchill et al., 2016). The activity theory emphasizes the role of authentic activities and conceptual tools—within which concepts are situated and developed—embedded in cultures of practice, where emerging meanings are synergically shaped by social engagement and culture (Brown et al., 1989). Its basic unit of analysis is human activity rooted in socio-cultural context and mediated by cultural and conceptual artifacts, both abstract and concrete, through expansive learning. Activities are intertwined with cultures and their related concepts, while knowledge becomes dynamically representative of the interactive domains in which situated cognition is activated and contextualized.

Constructivist and Socio-constructivist Mobile Teaching

Constructivism is a learning theoretical framework in which learners build ideas and develop knowledge concepts by giving individualized significant meaning to the environment within which they have experiences. Individuals have a specific autonomous role in the critical and interpretive process of knowledge construction. The mind filters the incoming information through her/his knowledge structures, regulated by previous and current knowledge and experiences. Learning is a process of guided discovery, which can be shared in situations of collective co-construction.

Learners autonomously develop personal concepts about phenomena—individually or collaboratively—through interaction with the external world and organize them as knowledge structures (Sjøberg, 2010).

Constructivism implements a multiple representation of real-world complexity without any simplification, and encourages construction and co-construction, learning by doing, reflection, and environmental feedbacks (Pandey, 2019). Constructivist approach is particularly significant for M-learning because—when mobile—learning may be informal since embedded in real-world contexts outside the classroom.[16] Context is a core concept of mobile learning. While a classroom context can seem a common ground for education, informal contexts are instead less stable learning environments, which require mobile construction of a temporary stability (Sharples et al., 2009). By visiting the worlds of knowledge, learners can autonomously and personally interpret the incoming information.

Mobile activities can promote learning through informal social interaction and learning can occur outside formal curriculum or designed learning environments. Learning is indeed an adaptive process in which knowledge is not acquired or possessed, but produced and constructed through experiential interactions with learning environments. This means that appropriate learning environments are needed, which focus on authentic representations of interactive challenges (Haag and Berking, 2019). Teachers collaborate as guides to the learning process through on-demand interventions, that is, when students request for them. Learning environments are structured as semi-formal contexts in which the processes of change development can be self-driven or guided by teachers.

Constructivist learning environments (CLEs) are constrained by a range of choices between which learners discover the searched solution analyzing the given options and relying on data capture and communication features of mobile devices. Mobile constructivist-learning environments can be structured for analyzing students' characteristics, different kinds of learnings, and educational mobile technologies (Göksu and Atici, 2013). These environments support knowledge building—a process during which learners share and co-construct their own understanding of the contents of learning materials (Jonassen, 1999). Within learning communities, implicit (naïve, common sense) theories about the surrounding minds or environments (Wellman, 1990; Carruthers and Smith, 1995) are confirmed, modified, or neglected by learners through teachers' guidance (Carey and Gelman, 1991). Learning technology should enhance both explicit and implicit levels, give feedback on the learning processes, and support guided individual self-practice (Reychav and Dezhi, 2015). Each learner actively constructs knowledge, which is shared in collaborative learning settings, as in the mobile collaborative learning (MCL) approach. The relation between experts and novices can be sustained by guided learning through intelligent tutors,

---

[16] Mobile learning technology can facilitate learning in formal and informal learning environments, blending them and enhancing interactive experiences and engagement (Bernacki et al., 2020). Formal learning is considered as individual vertical knowledge, linked to teacher-led instruction, classroom presentation of learning materials, and assessment, while informal learning settings are more dynamic and involve unplanned shared actions of horizontal knowledge, which can require self-regulation (Elsafi, 2018). Informal learning is a non-intentional learning process, sustained by social networking, podcast, and e-mail.

tutorials, and educational games. The exploration of multiple social perspectives allows learners to participate in the learning community through continuous socio-constructivist communication activities (Churchill et al., 2016).

Constructivism can be interpreted both by the metaphor of acquisition—if based on learner's individual process of construction—and by the metaphor of participation—if social and cultural practices are considered as the primary step of knowledge processes. Constructivist and socio-cultural approaches have indeed a common plug. The gap between Piagetian constructivism and Vigotskyian socio-culturalism can be reduced through socio-constructivist approaches which re-consider both individual and knowledge as a dynamic whole (Packer and Goicoechea, 2000). Knowledge is indeed synergistically co-constructed and shared within learning communities: shifting from learning processes to knowledge structures, meanings are negotiated to reach collective agreements (Santoianni, 2006).

The socio-cultural critique to constructivism has been based on its intrinsic dualism.[17] Learners' activities have instead to be considered as historically and culturally rooted in contexts, in the intersubjectivity, in the signs and symbols systems (cognitive artefacts), and in the technological tools for learning (peripheral devices). Becoming the member of a learning community doesn't mean anyway to renounce the individual role played within the learning community: knowledge is a transformative activity of mutual construction where both the individual and the context dynamically constitute each other. In this recursive circle, the intrinsic dualism of constructivism between the knower and the known is overcome because it is no longer the individual subjectivity which categorizes a portion of knowledge, but it is the collective intersubjectivity which co-constructs the whole world (or worlds and microworlds)[18] of knowledge.

Tacit Mobile Teaching, Meta-reflection, and Conceptual Change

Knowledge is nowadays considered both explicit and implicit or tacit. While explicit knowledge is formally and verbally expressed, implicit or tacit knowledge is non-formal and non-verbal, having a role in knowledge creation at individual, group, organizational, and inter-organizational levels (Nonaka and Takeuchi, 1995).

The idea of implicit knowledge was raised in the late sixties (Reber, 1967) when unaware processing was focused as an alternative to the cognitive symbolic metaphor of the mind (Sternberg, 1990). In the seventies, implicit learning was still a shadow of the rational mind (Cleeremans, 1997) until, around the eighties, the situated embodied mind included implicit learning as an integrated part of itself (Santoianni, 2011). The value of implicit in the knowledge processes has been underestimated because it has been often interpreted as a ground level of cognition to be dissolved inside explicit through meta-reflection, without considering its

---

[17] In classical constructivism, a dualism still exists between the knower and the known (Dewey, 1992) because the two terms of the epistemic relation are interconnected but independent, since already structured and relatively modifiable (Piaget, 1967, 1972).

[18] World 2 of subjective knowledge and World 3 of objective knowledge (Popper, 1972) are both involved in the transition between learning processes and knowledge structures (Bereiter and Scardamalia, 1996; Chen and Hung, 2002).

adaptive knowledge potential (Santoianni, 2014). Meta-reflective teaching is close to post-cognitivism and to actual educational theories. It is at the border between asymmetric and transmissive classical models and symmetric and collaborative current post-cognitive contemporary models. Meta-reflective teaching checks the quality of learning and can be indeed transversally applied to any kind of teaching and learning, both individual and collective.

Meta-reflective teaching allows learners to reflect on knowledge acquisition in order to acknowledge, organize, and monitor her/his own strategies of information processing and management (Metcalfe and Shimamura, 1994; Reder, 1996; Hacker et al., 1998). It concerns not only the self-control of students' learning, but also the evaluation of the multiple ways through which teachers dynamically develop students' learning. The meta-reflective model relies on self-awareness of learners on how to manage and self-regulate cognitive and emotional processing through acknowledged learners' individual strategies of monitoring and controlling the cognitive flow in order to assess her/his own quality of learning. Monitoring and control processes are, for instance, evaluation of the ease of learning of knowledge to be acquired and the learners' judgment on the chances to recall a learned information or to feel to know it, while the prediction of total recall refers to the total amount of information that could be stored in memory. Teachers review learning strategies in order to stimulate learners' reflection about how to manage their own learning. Tools of metacognitive self-regulated reflection are questionnaires, autobiographies, and narrations.

Relating these concepts to mobile teaching, the possibilities for improving mobile education are dependent on the teachers' challenge to integrate mobile learning inside/outside school. Mobile education implementation and improvement is then related to teachers' 'skill' and 'will' to the possible lack of their professional development on technological infrastructure and/or appropriate pedagogical strategies, and to teachers' beliefs about the usefulness of mobile learning according to classroom management (Christensen and Knezek, 2018).

The meaning of meta-reflection for mobile learning is linked to the concepts of self-efficacy and self-regulation, which are the basic keys for online learning environments. Metacognition processes are composed of knowledge about cognition and monitoring/control of learning. Monitoring is a bottom-up process which allows judgment of learning and can be activated if there is a discrepancy between the ongoing learner's performance and her/his standards. Control of cognitive activities is a top-down process which depends on the outputs of monitoring to intervene (Clerc et al., 2014). Metacognitive strategies and regulation practices are at the core of the relation between network information and learners because they regulate the sense of satisfaction or of disorientation and low self-esteem eventually perceived by students (Pellas, 2014).

Meta-reflection makes explicit the implicit acquisitions of information of the cognitive system since, in the early stages of cognitive development and in the preliminary phases of any cognitive task, implicit learning—the non-voluntary way to learn by environmental experiences—can be activated without personal acknowledgement, in order to cope with emergent problem solving (Santoianni, 2011). Implicit ideas, theories, and concepts developed by individuals about the world around them or others' minds, that is, tacit knowledge, can become aware and verbalized through meta-reflection.

Technology-based learning environments are considered to be the actors of more than knowledge transfer because they support conceptual change (Vosniadou et al., 1995). Emerging theories to explain the surrounding world may indeed be shaped by learners as misconceptions and need to be reviewed by enrichment or by conceptual change (Santoianni, 2003c). Research on conceptual development (Carey, 1985, 1991, 2000; Carey and Spelke, 1994; Nersessian, 1989, 1992; Keil, 1999) studies how theories of mind/world can evolve and transform themselves (Astington et al., 1988; Gopnik and Wellman, 1994; Wellman and Gelman, 1998; Gopnik, 1999).

Innovative Mobile Teaching and the Metaphor of Knowledge Creation

Learning experiences are gradually moving to mobile online learning, even if teachers are not yet prepared to cope with the new technologies and their usefulness for students. How innovative mobile teaching should be then organized?

Within the metaphor of knowledge creation, learning is no more viewed as a delivery of content and shared active participation is considered as a basic issue, alongside the production of knowledge artefacts through well-developed digital literacy. Making meaning means seeking meaning, which is the contribution to the processes of knowledge co-creation within multiple learning pathways. Learners interact through interrelated activities of consuming resources, connecting with learning materials and other users, creating and contributing knowledge content to share because individual and collective are no more divisible (Littlejohn and McGill, 2016).

The teacher is considered only as a co-learner and a facilitator, who guides students through interactive activities to let them be actively engaged with educational contents and other learners. Learning experiences are only designed by teachers so that students can freely enjoy their opportunities to gain content knowledge, practical expertise, and twenty-first-century skills from real-world data. Mobile teaching allows indeed autonomous fruition of materials/resources and encourages real-life experiences. The challenge of coping with learning environments is supported by social interaction and active participation in educational settings of discussion, negotiation, and sharing. At the same time, personal approaches to knowledge are really sustained and multiple measures of mastery are possible (Kraglund-Gauthier, 2019).

Mobile technological devices can be considered as mediators because they connect individuals' thinking and behavior to socio-cultural contexts. Mobile devices promote social participation between mobile learners through co-operative learning—in which learners share work in small groups of different sizes with vary roles/tasks, and are evaluated both on group/individual performances (Slavin, 2010)—and the possibility of learning at the workplace, continuous mobile communication given by social networks, mobile performance support system, e-mail, and web tools 2.1 (Irbi and Strong, 2015).

The two metaphors of acquisition and participation can be both taken into consideration without separating but joining them to co-construct a third metaphor: the metaphor of knowledge creation, which is nowadays the starting point to innovate mobile teaching.

Basic Assumptions of Innovative Mobile Teaching

The actual metaphor of knowledge creation is the emergent result of some shared aspects of the previous ones (Paavola et al., 2004). These aspects can be highlighted in the following issues:

- knowledge is no more seen as a static cultural heritage; it is instead considered dynamic and co-created, thus fostering the metaphor of knowledge creation. This means that knowledge itself can be mobile, continuously changing since negotiated, and shifting from one place to another, real or virtual;

- knowledge creation is a social process because learning is continuously shared within the communities of practices by its members, who are involved in processes of communication, collaboration, internalisation, and externalisation of evolving contents. Mobile technologies support information data sharing and encourage dynamic and synergic relations between learners through real and virtual meetings;

- individuals are the focus of customized educational actions, not in a separate manner but because they are members of learning communities and participate in social activities with their own contribution. Their approach to learning is personalized and enhanced by mobile technologies, which allow management of individually-tailored knowledge;

- dichotomies between mind, brain, and body are invalidated because the process of knowledge creation involves the implementation of many concurrent aspects, each of which plays a role of mediation between individuals and environments. Mobile technologies are designed to overcome the cognitive gap between perceptual and conceptual acquisitions because their affordances meet the whole organism of the learner;

- knowledge is shaped through three different types: declarative or propositional knowledge, which concerns the *know that* of cognition; procedural knowledge, that is embedded in skills, representing the *know how* of cognition; implicit or tacit knowledge, which is knowledge without awareness, exploring informational acquisitions of implicit learning;

- emphasis is given to expansive learning to co-construct grounded materials and practices within networks of activity systems, to knowledge building through conceptual artifacts to co-create content objects and resources within learning communities, and to implicit learning versus explicit knowledge at different levels of individual, group, organizational, and interorganizational interaction.

An innovative theory of learning should join learners' background knowledge, grounding their own behavior with new ideas and practices to integrate 'my inside view' and 'the outside view' of deep learning through interactive mobile processes of collaboration, in which students understand one another's point of view and what they really want to learn in relation to the shared opinions of their own learning community. Integrated mobile education online foresees that teacher's and students' responsibilities and roles may change as empowered and self-directed: management and organization (technology support, task explanation, students' evaluation, reminder of deadlines), personal and professional contextualizing (sharing of personal life events, school happenings, explanation for being offline, suggesting activities on the web, providing examples from work-life, information about professional roles), knowledge building (links to other sites, allowing users' comments with responses, news about recent books, journals, and articles) are all keywords of learning communities (Turbill, 2019).

Active participation and mobile engagement are sustained by collaborative and conversational learning, where interaction occurs through mobile communication between peers and can involve environmental exploration. Related mobile technologies can involve mobile-assisted language learning, interactive voice response, computer-based response mobile system, web 2.0 tools, e-mail, and mobile portals (Sarrab et al., 2013; Dold, 2016). Students' cognitive engagement refers to the learning processes implemented when new knowledge is acquired and analyzed through previous information patterns, to the learning goals achieved by learners' motivation and involvement, and to the active participation and interaction of students. Mobile technologies, as 2D systems like LMS or blogs and, more of them, the 3D multi-user virtual worlds (VWs) narrow learners to real-world situations, while multiple learning performances are foreseen to train learners' skills, and learning objects can simulate authentic educational contents to be actively discovered in real-time interactive simulations within multimedia environments (Pellas, 2014).

References

Astington, J.W., Harris, P.L. and Olson, D.R. (eds.). (1988). *Developing Theories of Mind*. Cambridge University Press, Cambridge.

Bereiter, C. and Scardamalia, M. (1996). Rethinking learning. pp. 485–513. *In*: Olson, D.R. and Torrance, N. (eds.). *The Handbook of Education and Human Development*. Blackwell Publishers, Oxford.

Bereiter, C. (2002). *Education and Mind in the Knowledge Age*. Lawrence Erlbaum, Mahwah.

Bereiter, C. and Scardamalia, M. (2003). Learning to work creatively with knowledge. pp. 55–68. *In*: De Corte, E., Verschaffel, L., Entwistle, N. and van Merriënboer, J. (eds.). *Powerful Learning Environments: Unravelling Basic Components and Dimensions*. Emerald Group Publishing, Bingley.

Bernacki, M.L., Crompton, H. and Greene, J.A. (2020). Towards convergence of mobile and psychological theories of learning. *Contemporary Educational Psychology*, 60: 101828.

Bernacki, M.L., Greene, J.A. and Crompton, H. (2020). Mobile technology, learning, and achievement: advances in understanding and measuring the role of mobile technology in education. *Contemporary Educational Psychology*, 60: 101827.

Block, J.H. and Anderson, L.W. (1977). Mastery learning. pp. 114–117. *In*: Treffinger, D., Davis, J. and Ripple, R. (eds.). *Handbook on Teaching Educational Psychology*. Academic Press, New York.

Brown, J.S., Collins, A. and Duguid, P. (1989). Situated cognition and the culture of learning. *Educational Research*, 18(1): 32–42.

Bruner, J. (1996). Frames for thinking: ways of making meaning. pp. 93–105. *In*: Olson, D.R. and Torrance, N. (eds.). *Modes of Thought: Explorations in Culture and Cognition*. Cambridge University Press, Cambridge.

Cantwell, Smith, B. (1999). Situatedness/Embeddedness. pp. 769–771. *In*: Wilson, R.A. and Keil, F. (eds.). *The MIT Encyclopedia of the Cognitive Sciences*. MIT Press, Cambridge.

Carey, S. (1985). *Conceptual Change in Childhood*. MIT Press, Cambridge.

Carey, S. (1991). Knowledge acquisition: enrichment or conceptual change? pp. 257–291. *In*: Carey, S. and Gelman, R. (eds.). *The Epigenesis of Mind*. Lawrence Erlbaum, Hillsdale.

Carey, S. and Gelman, R. (1991). *The Epigenesis of Mind*. Lawrence Erlbaum, Hillsdale.

Carey, S. and Spelke, E. (1994). Domain-specific knowledge and conceptual change. pp. 169–200. *In*: Hirschfeld, L.A. and Gelman, S.A. (eds.). *Mapping the Mind. Domain Specificity in Cognition and Culture*. Cambridge University Press, Cambridge.

Carey, S. (2000). The origin of concepts. *Journal of Cognition and Development*, 1(1): 37–41.

Carruthers, P. and Smith, P. (eds.). (1995). *Theories of Theory of Mind*. Cambridge University Press, Cambridge.

Chen, D.T. and Hung, D. (2002). Personalized knowledge representations: the missing half of online discussion. *British Journal of Educational Technology*, 33(3): 279–290.

Choi, G.W., Land, S.M. and Zimmerman, H.T. (2018). Investigating children's deep learning of the tree life cycle using mobile technologies. *Computers in Human Behavior*, 87: 470–479.

Christensen, R. and Knezek, G. (2018). Readiness for integrating mobile learning in the classroom: challenges, preferences and possibilities. *Computers in Human Behavior*, 78: 379–388.

Churchill, D., Fox, B. and King, M. (2916). Framework for designing mobile learning environments. pp. 3–26. *In*: Churchill, D., Lu, J., Chiu, T. and Fox, B. (eds.). *Mobile Learning Design*. Springer, Singapore.

Cleeremans, A. (1997). Principles for implicit learning. pp. 196–234. *In*: Berry, D. (eds.). *How Implicit is Implicit Learning?* Oxford University Press, Oxford.

Clerc, J., Miller, P.H. and Cosnefroy, L. (2014). Young children's transfer of strategies: utilization deficiencies, executive function and metacognition. *Developmental Review*, 34: 378–393.

Cochrane, T. (2014). Mobile social media as a catalyst for pedagogical change. pp. 2187–2200. *In*: Viteli, J. and Leikomaa, M. (eds.). *Proceedings of Ed Media: World Conference on Educational Media and Technology*. Association for the Advancement of Computing in Education, Waynesville.

Cole, M. (1996). *Cultural Psychology: A Once and Future Discipline*. Cambridge University Press, Cambridge.

Collins, A. and Greeno, G.J. (2010). Situated view of learning. pp. 335–339. *In*: Peterson, P., Baker, E. and McGaw, B. (eds.). *International Encyclopedia of Education*. Elsevier, Amsterdam.

Davidson, J.E., Deuser, R. and Sternberg, R.J. (1994). The role of metacognition in problem solving. pp. 207–226. *In*: Metcalfe, J. and Shimamura, A. (eds.). *Metacognition: Knowing about Knowing*. MIT Press, Cambridge.

Dewey, J. (1992). *Democrazia e educazione*, La Nuova Italia. Firenze.

Dold, C.J. (2016). Rethinking mobile learning in light of current theories and studies. *Journal of Academic Librarianship*, 42(6): 679–686.

Dunn, R. and Dunn, K. (1978). *Teaching Students through their Individual Learning Styles*. Reston Publishing, Reston.

Egan, K. (1997). *The Educated Mind. How Cognitive Tools Shape Our Understanding*. University of Chicago Press, Chicago.

Elsafi, A. (2018). Formal and informal learning using mobile technology. pp. 177–189. *In*: Yu, S., Ally, M. and Tsinakos, A. (eds.). *Mobile and Ubiquitous Learning. Perspectives on Rethinking and Reforming Education*. Springer, Switzerland.

Engeström, Y. (1987). *Learning by Expanding. Orienta-konsultit*. Helsinki.

Fischer, B.B. and Fischer, L. (1979). Styles in teaching and learning. *Educational Leadership*, 36: 245–254.

Fischer, K.W. (2009). Building a scientific groundwork for learning and teaching. *Mind, Brain, and Education*, 3(1): 165–169.

Gardner, H. (1983). *Formae Mentis, Frames of Mind: The Theory of Multiple Intelligences*. Basic Books, New York.

Gardner, H. (1985). *The Mind's New Science: A History of the Cognitive Revolution*. Basic Books, New York.

Gardner, H. (1995). Reflections on multiple intelligences, myths and messages. *Phi Delta Kappan*, 77(3): 200–203, 206–209.

Gardner, H. (1996). Are there additional intelligences? The case for naturalist, spiritual, and existential intelligences. pp. 111–131. *In*: Kane, J. (ed.). *Education, Information and Transformation*. Prentice-Hall, Engelwood Cliffs.

Gardner, H. (1997). The first seven…and the eight: a conversation with howard gardner. *Educational Leadership*, 55(1): 8–13.

Gardner, H. and BoixMansilla, V. (1997). Of kinds of disciplines and kinds of understanding. *Phi Delta Kappan*, 78(5): 381–386.

Gauvain, M. (2008). Vygotsky's socio-cultural theory. pp. 404–413. *In*: Haith, M.M. and Benson, J.B. (eds.). *Encyclopedia of Infant and Early Childhood Development*. Elsevier, Oxford.

Göksu, D. and Atici, B. (2013). Need for mobile learning: technologies and opportunities. *Procedia Social and Behavioral Sciences*, 103: 685–694.

Gopnik, A. and Wellman, H.M. (1994). The theory theory. pp. 257–293. *In*: Hirschfeld, L.A. and Gelman, S.A. (eds.). *Mapping the Mind. Domain Specificity in Cognition and Culture*. Cambridge University Press, Cambridge.

Gopnik, A. (1999). Theory of mind. pp. 838–841. *In*: Wilson, R.A. and Keil, F. (eds.). *The MIT Encyclopedia of the Cognitive Sciences*. Bradford Book, London.

Grabinger, R.S. and Dunlap, J.C. (1997). Rich environments for active learning: a definition. *Research in Learning and Teaching*, 3(2): 5–34.

Gros, B. (2016). The dialogue between emerging pedagogies and emerging technologies. pp. 3–24. *In*: Gros, B., Kinshuk and Maina, M. (eds.). *The Future of Ubiquitous Learning. Learning Designs for Emerging Pedagogies*. Springer, Berlin.

Haag, J. and Berking, P. (2019). Design considerations for mobile learning. pp. 221–240. *In*: Zhang, Y. and Cristol, D. (eds.). *Handbook of Mobile Teaching and Learning*. Springer, Singapore.

Hacker, D.J., Dunlosky, J. and Graesser, A. (eds.). (1998). *Metacognition in Educational Theory and Practice*. Lawrence Erlbaum, Mahwah.

Hamidi, H. and Chavoshi, A. (2018). Analysis of the essential factors for the adoption of mobile learning in higher education: a case study of students of the university of technology. *Telematics and Informatics*, 35: 1053–1070.

Irby, T. and Strong, R. (2015). A synthesis of mobile learning research implications: agricultural faculty and student acceptance of mobile learning in academia. *NACTA*, 59(1): 10–17.

Jonassen, D. (1999). Designing constructivist learning environments. pp. 215–239. *In*: Reigeluth, C.M. (ed.). *Instructional Design Theories and Models: A New Paradigm of Instructional Theory*. Lawrence Erlbaum, Hillsdale.

Jonassen, D. (2000). Towards design theory of problem solving. *ETR&D*, 48(4): 63–85.

Keil, F.C. (1999). Conceptual change. pp. 179–182. *In*: Wilson, R.A. and Keil, F. (eds.). *The MIT Encyclopedia of the Cognitive Sciences*. Bradford Book, London.

Kirshner, D. and Whitson, J.A. (eds.). (1997). *Situated Cognition: Social, Semiotic, and Psychological Perspectives*. Lawrence Erlbaum, Hillsdale.

Koç, T., Turan, A.H. and Okursoy, A. (2016). Acceptance and usage of a mobile information system in higher education: an empirical study with structural equation modeling. *International Journal Management Education*, 14: 286–300.

Kraglund-Gauthier, W.L. (2019). Learning to teach using digital technologies: pedagogical implications for postsecondary contexts. pp. 589–608. *In*: Zhang, Y. and Cristol, D. (eds.). *Handbook of Mobile Teaching and Learning*. Springer, Singapore.

Kruger, A.C. and Tomasello, M. (1996). Cultural learning and learning culture. pp. 369–387. *In*: Olson, D.R. and Torrance, N. (eds.). *The Handbook of Education and Human Development*. Blackwell Publishers, Oxford.

Lave, J. and Wenger, E. (1991). *Situated Learning: Legitimate Peripheral Participation*. Cambridge University Press, Cambridge.

Lieberman, D.A. (1990). *Learning: Behaviour and Cognition*. Wadsworth, Belmont.

Littlejohn, A. and McGill, L. (2016). Ecologies of open resources and pedagogies of abundance. pp. 115–130. *In*: Gros, B., Kinshuk and Maina, M. (eds.). *The Future of Ubiquitous Learning, Learning Designs for Emerging Pedagogies*. Springer, Berlin.

Ludvigsen, S.R. and Mørch, A.I. (2010). Computer-supported collaborative learning: basic concepts, multiple perspectives, and emerging trends. pp. 290–296. *In*: Peterson, P., Baker, E. and McGaw, B. (eds.). *International Encyclopedia of Education*. Elsevier, Amsterdam.

Mayer, R.E. (2010). Problem solving and reasoning. pp. 273–278. *In*: Peterson, P., Baker, E. and McGaw, B. (eds.). *International Encyclopedia of Education*. Elsevier, Amsterdam.

Metcalfe, J. and Shimamura, A. (eds.). (1994). *Metacognition: Knowing about Knowing*. MIT Press, Cambridge.

Nersessian, N. (1989). Conceptual change in science and in science education. *Synthese*, 80: 163–183.

Nersessian, N. (1992). How do scientists think? Capturing the dynamics of conceptual change in science. pp. 3–45. *In*: Giere, R.N. (ed.). *Cognitive Models of Science*. University of Minnesota Press, Minneapolis.

Nonaka, I. and Takeuchi, H. (1995). *The Knowledge-creating Company: How Japanese Companies Create the Dynamics of Innovation*. Oxford University Press, New York.

Olson, D.R. and Torrance, N. (eds.). (1996). *The Handbook of Education and Human Development*. Blackwell, Oxford.

Paavola, S., Lipponen, L. and Hakkarainen, K. (2004). Models of innovative knowledge communities and three metaphors of learning. *Review of Educational Research*, 74(4): 557–576.

Packer, M.J. and Goicoechea, J. (2000). Socio-cultural and constructivist theories of learning: ontology, not just epistemology. *Educational Psychologist*, 35: 4.

Pandey, K. (2019). Expectations from future technologies in higher education: an introduction. pp. 1061–1065. *In*: Zhang, Y. and Cristol, D. (eds.). *Handbook of Mobile Teaching and Learning*. Springer, Singapore.

Pegrum, M. (2019). *Mobile Lenses on Learning, Languages and Literacies on the Move*. Springer, Singapore.

Pellas, N. (2014). The influence of computer self-efficacy, metacognitive self-regulation and self-esteem on student engagement in online learning programs: evidence from the virtual world of second life. *Computers in Human Behaviour*, 35: 157–170.

Piaget, J. (1967). *Lo sviluppo mentale del bambino, Einaudi*. Torino.

Piaget, J. (1972). *The Principles of Genetic Epistemology*. Basic Books, New York.

Pimmer, C., Mateescu, M. and Grohbiel, U. (2016). Mobile and electronic learning in higher education settings. A systematic review of empirical studies. *Computers in Human Behavior*, 63: 490–501.

Popper, K.R. (1972). *Objective Knowledge: An Evolutionary Approach*. Clarendon Press, Oxford.

Reber, A.S. (1967). Implicit learning of artificial grammars. *Journal of Verbal Learning and Verbal Behavior*, 6: 855–863.

Reder, L.M. (ed.). (1996). *Implicit Learning and Metacognition*. Lawrence Erlbaum, Mahwah.

Retallick, J., Cocklin, B. and Coombe, K. (1999). *Learning Communities in Education: Issues, Strategies and Contexts*. Routledge, Londra.

Reychav, I. and Dezhi, W. (2015). Mobile collaborative learning: the role of individual learning in groups through text and video content delivery in tablets. *Computers in Human Behavior*, 50: 520–534.

Rogoff, B., Turkanis, C.G. and Bartlett, L. (2003). *Learning Together: Children and Adults in a School Community*. Oxford University Press, Oxford.

Santoianni, F. (1997). *Le prospettive educative del modello cibernetico*. pp. 65–77. *In*: Frauenfelder, E. and Santoianni, F. (eds.). *Nuove frontiere della ricerca pedagogica tra bioscienze e cibernetica, Edizioni Scientifiche Italiane*. Napoli.

Santoianni, F. (2003a). *La scienza del comportamento*. pp. 5–23. *In*: Santoianni, F. and Striano, M. (eds.). *Modelli teorici e metodologici dell'apprendimento*. Laterza, Roma-Bari.

Santoianni, F. (2003b). *La scienza della mente*. pp. 24–46. *In*: Santoianni, F. and Striano, M. (eds.). *Modelli teorici e metodologici dell'apprendimento*. Laterza, Roma-Bari.

Santoianni, F. (2003c). *Sviluppo e formazione delle strutture della conoscenza, Edizioni E.T.S.* Pisa.

Santoianni, F. (2006). *Educabilità cognitiva, Apprendere al singolare, insegnare al plurale*. Carocci, Roma.

Santoianni, F. (2010). *Modelli e strumenti di insegnamento, Approcci per migliorare l'esperienza didattica*. Carocci, Roma.

Santoianni, F. (2011). Educational models of knowledge prototypes development. *Mind & Society*, 10: 103–129.

Santoianni, F. (2014). *Modelli di studio, Apprendere con la teoria delle logiche elementari*. Erickson, Trento.

Santoianni, F. (2017). Models in pedagogy and education. pp. 1033–1049. *In*: Magnani, L. and Bertolotti, T. (eds.). *Springer Handbook of Model-Based Science*. Springer, Cham.

Santoianni, F. and Ciasullo, A. (2021). Milestones of bioeducational approach in mind brain and education research, in press. *In*: Rimazei, N. (ed.). *Integrated Education and Learning*. Springer, Switzerland.

Sarrab, M., Al-Shihi, H. and Rehman, O.H. (2013). Exploring major challenges and benefits of M-learning adoption. *British Journal of Applied Science & Technology*, 3(4): 826–839.

Savery, J.R. and Duffy, T.M. (1995). Problem-based learning: an instructional model and its constructivist framework. *Educational Technology*, 35(5): 31–38.

Scardamalia, M. and Bereiter, C. (2006). Knowledge building: theory, pedagogy, and technology. pp. 97–118. *In*: Sawyer, R.K. (ed.). *The Cambridge Handbook of the Learning Science*. Cambridge University Press, Cambridge.

Sharples, M., Arnedillo-Sanchez, I., Milrad, M. and Vavoula, G. (2009). Mobile learning small devices, big issues. pp. 233–249. *In*: Balacheff, N., Ludvigsen, S., de Jong, T., Lazonder, A. and Barnes, S. (eds.). *Technology-enhanced Learning*. Springer, Dordrecht.

Shin, W. and Kang, M. (2015). The use of a mobile learning management system at an online university and its effect on learning satisfaction and achievement. *International Review of Research in Open and Distributed Learning*, 16(3): 110–130.

Sjøberg, S. (2010). Constructivism and learning. pp. 485–490. *In*: Peterson, P., Baker, E. and McGaw, B. (eds.). *International Encyclopedia of Education*. Elsevier, Amsterdam.

Slavin, R.E. (2010). Co-operative learning. pp. 177–183. *In*: Peterson, P., Baker, E. and McGaw, B. (eds.). *International Encyclopedia of Education*. Elsevier, Amsterdam.

Sternberg, R.J. (1985). *Beyond IQ: A Triarchic Theory of Human Intelligence*. Cambridge University Press, Cambridge.

Sternberg, R.J. (1990). *Methaphors of Mind, Conceptions of the Nature of Intelligence*. Cambridge University Press, Cambridge.

Sternberg, R.J. (1997). *Thinking Styles*. Cambridge University Press, Cambridge.

Sternberg, R.J. and Zhang, L.F. (2001). *Perspectives on Thinking, Learning, and Cognitive Styles*. Lawrence Erlbaum, Mahwah.

Sun, S., Joy, M. and Griffiths, N. (2007). The use of learning objects and learning styles in a multi-agent education system. *Journal of Interactive Learning Research*, 18(3): 381–398.

Turbill, J. (2019). Transformation of traditional face-to-face teaching to mobile teaching and learning: pedagogical perspectives. pp. 35–48. *In*: Zhang, Y. and Cristol, D. (eds.). *Handbook of Mobile Teaching and Learning*. Springer, Singapore.

UNESCO. (2012). *Working Paper Series on Mobile Learning: Turning on Mobile Learning in North America*. UNESCO, Paris.

Vosniadou, S., De Corte, E. and Mandl, H. (1995). *Technology-based Learning Environments*. Springer, Heidelberg.

Vygotsky, L.S. (1986). *Thought and Language*. MIT Press, Cambridge.

Walling, D.R. (2014). *Designing Learning for Tablet Classrooms, Innovations in Instruction*. Springer, Switzerland.

Wellman, H.M. (1990). *The Child's Theory of Mind*. MIT Press, Cambridge.

Wellman, H.M. and Gelman, S.A. (1998). Knowledge acquisition in foundational domains. pp. 523–573. *In*: Damon, W. (ed.). *Handbook of Child Psychology*. Wiley, New York.

Yu, C., Lee, S.J. and Ewing, C. (2015). *Mobile Learning: Trends, Issues, and Challenges in Teaching and Learning. IGI Global*, Pennsylvania.

Zhang, L.F. and Sternberg, R.J. (2005). A threefold model of intellectual styles. *Educational Psychology Review*, 17(1): 1–53.

CHAPTER 3
Educational Design of Mobile Learning Environments

Flavia Santoianni

Design is ubiquitous in education.

—Hall et al. (2017)

We have the ambition.
We have the technology.
What is missing is what connects the two.

—Laurillard (2013)

Introduction

Future directions of mobile learning foresee its contextualization and learning personalization, through 'bring your own device'[19] and 'flipped learning'[20] approaches or through user-generated learning contexts, within adaptive learning system, supported by virtual assistants (Pegrum, 2016). Future directions of online learning concern the increase of its implementation in classroom and its institutional support through innovative online degrees, and the use of big data, augmented learning, gamification, learning management systems (LMS), corporate MOOCs, linked and embedded into mobile learning (Martin and Oyarzun, 2018). Learning changes so rapidly through continuously developing trends, hand in hand with technology, that

---

Professor of Education, University of Naples Federico II, Italy.
Email: bes@unina.it

[19] Bring Your Own Device (BYOD) programmes narrow school activities and every-day life experiences through effective use of mobile devices (Mcquiggan et al., 2015).

[20] In flipped learning, classroom lessons—which anyway have to leverage engagement—are flipped and switched with home activities, consisting in learning by doing and active learning, as well as watching instructional videos. In mobile learning approaches, the flipped classroom is extended outside school (Wong, 2016). Flipped learning is a blended approach which requires flexible environments, active learners, personalized application to learning content, and trained teachers (Mcquiggan et al., 2015).

learners may not recognize the world in which they live (White, 2019). All these trends revolve around a key node, represented by mobile learning environments' design, which keeps unchallenged collaborative, experiential, inquiry-based, and problem-based approaches to student learning (Laurillard, 2013b).

Mobile learning is contingent, personalized, situated, context-aware, and authentic. Mobile learning design—a field of study which accelerated its development over the last 15 years in the research area of technology-enhanced learning (TEL) (Bannan et al., 2015)—leverages student-centered learning, in which learners can use mobile resources of devices at their own pace; personalized learning, where learners can choose their preferred tasks and practices; active learning, with which learners can be immediately involved; collaborative learning, to meet each other and share content; informal learning, which occurs in real-life contexts outside formal and institutional settings; and authentic learning, which develops within contexts, using mobile devices.

The general theme of mobile learning environments' design includes consideration of the following aspects: open multimedia resources, which include the learning content, and tools/materials for learning activities; learner-centered activities, focusing on the learning process itself and on learners' reflections on it, more than on its final results; and teachers as facilitators, who regulate scaffolding to students and support of students/community/resources towards students in classroom and online. Teachers become designers of students' learning experiences to integrate both digital and mobile technologies into their education, in order to promote the development of 21st-century skills and digital literacies (Pegrum, 2019). Since learner-centered activities are expected to be authentic,[21] they should refer to real-life experiences, foresee ill-structured problems, and manage tools for both practical performances and theoretical knowledge.[22] Teachers, on the other hand, should better understand technological affordances (Churchill et al., 2016).

Educational affordances of mobile technology—as portability, social collaboration, context sensitivity, synchronous/asynchronous connectivity, and learners' scaffolding (Klopfer and Squire, 2005)—involve how teachers can use devices, and their capturing, analyzing, and representing resources for generating and managing knowledge content during the interaction with adaptive, individual, and collaborative learning environments (Liaw et al., 2010). Since digital learning environments' design depends on many variables, as educational methodologies, available resources situated within formal and informal real-life contexts, learners' development, social collaboration, and technology innovation, they can be considered complex (Hall et al., 2017) and unpredictable in their own evolution (Santoianni, 2006).

---

[21] Authentic activities should involve both personal significance and cultural relevance. Authenticity concerns how students implement activities related to real life, whether they can freely interact with them and feel engaged accordingly to personal goals, and the degree of participatory collaboration (Burden and Kearney, 2017).

[22] The Tell-Ask instruction has to be reformulated in the Tell, Ask, Show, and Do approach, which leverages active learning and motivation, shows learning for demonstration, integrates learners' points of view, and is related to previous knowledge problem-centered strategies, to be applied and experienced in real-world tasks and in authentic contexts (Merrill, 2018).

Even if unpredictable, digital learning environments can be regulated by learning and instructional design technology, whose principles are applicable to the educational design of mobile learning environments, concerning both linguistic and spatial fields.

Learning and Instructional Design Technology

Technology has the potential to meet every learner's needs by engaging students' attention, by motivating them with learning content closer to their interests, to let students learn at their own pace and levels, by personalizing and embedding educational approaches. To implement technological potential, a bridge between digital technologies and educational ambitions is needed, and it can be realized through learning design (Laurillard, 2013a).

Learning and instructional design technology (LIDT)[23] is a field of research born during World War II and fully developed between the 1960s and 1970s (Gustafson, 1991). Even if the theory of instructional design has been considered out of date in itself because of the rapid emergence of its application in the business practices, instructional design models are still theoretically studied.

Learning and instructional design technology is a field of research in which the design process is analyzed according to learners' needs within multimedia technological environments. This area of study represents an intertwining of theory, research, and practice. It is contextually related and an experiential creative work, which requires attention to the individual's preferences (Rieber, 2018). Preferences may vary according to users because students appreciate flexibility, online support for self-directed learning, engagement, and interaction with the online learning community, because students can feel isolated from it or perceive lack of immediacy, while teachers share the need of flexibility and wish to customize their educational offer according to students' differences, even if curricula development may be slowed down by the integration of technological issues.

Instructional Design Processes and Models

Instructional design involves both applied research and educational theories embedded in multiple real-life contexts. Mobile learning settings are indeed situated, and instructional design is asked to be flexible according to their development and to the ongoing interactions. The triangular structuration model of mobile learning comprehends users, their activities, and the contextual interplay of all these aspects, to give technology affordances as interactivity, authenticity, portability, multimodal communication, collaboration, and user-generated content (Bannan et al., 2015). Other affordances allowed by technology are the possibility to generate content, to share comments, and to aggregate communities, to create digital identities and to overcome the gap between online and vis-a-vis setting (Conole, 2013).

---

[23] Nevertheless, different to instructional design, the idea of learning design—a term spread since 2003—is more focused on the learner rather than on the teacher. Core concepts are the notions of activity and environment: epistemic design concerns activities and social design involves social environment (Sloep, 2016).

The process of educational design concerns the concepts of investigation, application, representation, and iteration because at the base of the process lie the individuation of users' needs and related theories, their possible application, and the modeling of design solutions—about learning activities and resources, content materials, curricula, and settings—according to learners' demands through cycles which involve both practice and reflection (Beetham and Sharpe, 2013). Educational design research develops through long-term open-ended cycles of learning technology research, during which different design solutions are integrated, focusing on the processes more than on the products.

Instructional design is a complex process, which involves many different models—classified in product-oriented and process-oriented—emerging from theories, practices, and experiences (Dousay, 2018). Instructional design models can vary according to the chosen delivery format, i.e., synchronous or asynchronous, or the situation in which they are embedded, i.e., inside/outside classroom. They can be referred to the development of different aspects as motivation, technological skills, or multimedia. Modeling, even if based on a common process, means to customize educational resources through operational tools and techniques according to instructional contexts, to the needs of learners and teachers, and to delivery modes.

Learning and instructional design can indeed afford different online delivery methods as asynchronous online learning, where the learning content is just delivered online and learners can approach it on demand, or synchronous online learning, where a part of the online delivered learning content can be enjoyed through real-time online activities, like meetings, thereby empowering users' participation. Other online delivery methods are massive open online courses (MOOCs), with open source content freely accessible to unlimited users, blended/hybrid courses, involving both face-to-face and online asynchronous delivery, but also blended synchronous courses, a mix of face-to-face and online synchronous delivery or multi-modal courses, which combine synchronous and asynchronous online learning (Martin and Oyarzun, 2018).

Mobile learning design may vary according to the levels of interaction between learners and devices. If devices are mobile, but the learners are sitting in classroom at their own desks to read e-textbooks and online dictionaries, to search the web, and to do app-based exercises, interaction is less. If instead, both devices and learners are mobile but learning experiences in digital spaces are not mobile, learning may be not directly affecting contexts, even if aspects as learning environments customization, or peer collaboration, can be enhanced. In a classroom scenario, in which students are moving with their own devices to share digital resources and products through active and collaborative processes, or in an outside-classroom scenario, in which students access, i.e., flipped learning centralized materials and online learning spaces for a task, or distance learners participate in synchronous/asynchronous online courses, learning could be eradicated from its contextual authenticity in real-world situations. In the end, if devices, learners, and learning in digital spaces are all mobile, learners' dynamic activities and experiences effectively develop in continuously changing—digital or real-life—spaces/environments (Pegrum, 2019).

Instructional design is a creative process, based on many instructional techniques—not discrete, but systemic and iterative—through which theory and practice are continuously intertwined. In instructional design, a model gives to the learners the chance to personalize and customize general aspects, while a process represents the chunk of steps needed to reach an outcome. Even if considered a complex and unique process, instructional design can be defined by some recurrent models. The analysis, design, develop, implement and evaluate (ADDIE) model is an educational, basic, underlying process which considers general metadata elements to organize a lesson plan (Gustafson and Branch, 1997; Branch, 2010). The ADDIE model[24] focuses on the concepts of:

- analysis, which includes educational problem identification as a performance gap, the contextual analysis of lessons' expected duration, teacher's prior competences, available infrastructure, and learners' grade level, her/his prerequisites, language to be used, and accessibility;
- design, which comprehends the definition of general and specific educational objectives, through the individuation of subject and topic domain; desired learning outcomes and performances, according to educational curriculum standards; the selection of teaching strategies, approaches, assessment activities, types, and methods;
- develop, which means to generate and validate educational resources, services, and tools, within selected physical delivery settings;
- implement, which continuously refers to real-time delivery of the lesson by organizing the learning environment and engaging learners;
- evaluate, which is sustained by formative and summative evaluation, plus teachers' potential adaptations.

Design-based research (DBR)—intertwined with interdisciplinary design research (IDR), and associated with many different terms[25]—implements principles and methods grounded on theoretical research and applied in practice (Christensen and West, 2018) through the development of dynamic cycles of iterative processes as the ADDIE model alternates practice and theory. Micro levels of research concerning case studies and related instructional strategies or specific technologies coexist with macro levels of research, focusing on online learning global resources or already existing institutional infrastructures. Educational design research is indeed both proximal and distal because it is related to local settings and, at the same time, refer to generalizable guidelines and outcomes. Between distal and proximal design there are intermediary connective resources intertwining processes as interventions, and models as products, which outline their flexibility and adaptability within contexts—design-based research is real-world context-dependent and aims to systematize educational models and approaches (Hall et al., 2017).

---

[24] The plan, implement, evaluate (PIE) model is a variation of the basic ADDIE model which focuses on technology (Newby et al., 1996).

[25] Terms as 'design experiments, design research, design-based research, formative research, development research, developmental research, and design-based implementation research' (Christensen and West, 2018).

Instructional Design Efficacy and Quality

Design-based research develops innovative domain theories and design methodologies (Lee, 2018), focusing on some principles for effective learning environments' design as easy usability of technology, involvement of emotional dimensions, sensorial engagement, and collaborative learning.[26] Mobile technological devices enhance contingent teaching and learning, situated, context-aware, personalized, and authentic learning. The efficacy of mobile technological devices' design is related to their adaptability, scalability, availability, portability, and affordability. Technology affordances of mobile devices are related to the integration of multiple characteristics in only one device, to leverage a wide range of different learning experiences and to encourage mobile devices' usability. Text size and colors of the visual interface, as screen viewing and user-friendly touch screens, play a key role in the processes of customization, while synthetic voices and intuitive features can support inclusion (Major et al., 2017).

Mobile learning design should provide learning environments with appropriate activities, while teachers play the role of facilitators supporting students' learning, giving feedbacks, and assessing the work formatively in classroom and beyond it. Assessment can include choice (true/false, alternate choice, multiple choice), selection and identification (multiple true/false, yes/no with explanations, multiple answer, complex multiple choice), reordering and rearranging (categorising, matching, sequencing, ranking, assembling proof), substitution and correction (within texts, figure drawing, fault correction), completion (short answer, sentence, and flowchart completion, cloze-procedure), construction (open-ended multiple choice, concept map, essay, figural response), presentation and portfolio (project, experiment, performance, discussion) (Timms, 2017).

Quality factors to evaluate the efficacy of the design of mobile learning systems are information quality, system quality, and service quality (Almaiah et al., 2016). Information quality refers to the learning content and to its formats, which can be expressed by text, graphics, multimedia, and sharing resources, designed according to learners' preferences. Information quality implies content usefulness—concerning basic, multimedia, and collaborative content which fits users' needs—and content adequacy—which is attention to accurately select up-to-date, complete, and detailed contents.

System quality is related to systems' functionality in different platforms with which it should be compatible and in which it should allow easy navigation and performant features, as search by text. It is also related to systems' accessibility to learning materials and resources, in order to be quickly connected to the Web for loading pages and downloading/uploading files. The concept of functionality is, moreover, linked to interactivity with teachers and students, comprehending the idea of sharing learning content within the learning community, and to interface design and ease of use. The interface of mobile learning systems should have comfortable

---

[26] Anyhow, some technological devices, as tablets, seem to have been designed more for individual rather than for collaborative use (Major et al., 2017). Tablets guarantee immediate information assistance and personalized teaching and learning (O'Loughlin et al., 2013).

features of page layout designed through attractive colors, graphics, and well-organized menus. To be user friendly, systems should be easy to use and immediately understandable.

Mobile learning systems are effective according to their service quality, which means to meet users' needs through available services provided anywhere and anytime and through personalized learning content and messages,[27] with features allowing users to store their own preferences, to record their own performances, and to be the leader of their own learning process. In the end, service quality involves responsiveness, that is, to give users a prompt service, to assist them, and to provide information on services' availability; safe transactions, trustworthy services, and adequate security features. Other aspects that should characterize the effectiveness of mobile learning systems are the possibility to skip training before use; to share learning resources between students, teachers and students, and teachers only; and to co-create digital learning tools linked to online resources, and to be continuously socially connected through synchronous and asynchronous networks (Mentor, 2017).

Experiential Micro-learning Design

Instructional design can be related to brain functions together with education and cognition in order to design learning experiences within technological environments (Boettcher, 2007). From this point of view, learning experiences develop according to four interacting aspects—the learner, the teacher, the knowledge, and the environment. Each learning experience occurs within a specific environment, which can be simple or complex, depending on available resources and in relation to individual or small and large group activities. Within contexts, learners can be influenced by multiple resources, tools, and other learners; learning tools can be shaped and customized by users, thus co-creating personal learning experiences.

The shared use of resources and tools should be supported by teachers; teaching functions can be flexible and not necessarily identified in only one person, but instead can involve more than one teacher, synergically offering guidance and feedback. To effectively design learning activities, time is needed because learning is no more only listening and answering through an effort of memory. Small chunks of learning should be organized, i.e., shared in classroom and then flipped at home, welcoming learners' generative ideas, which could arise after the interaction with students' contexts of reference and then submitted online to teachers and peers (Looi and Seow, 2015).

Different kinds of instructions can produce various learning outcomes, and this premise can influence learning environments design (Gagné, 1965).[28] Learners are specific individuals, with a personalized knowledge background, which leverages on the synaptic flexibility of each learner, because it enables—the more complex is

---

[27] The process of personalization of learning content is different from the process of individualization because the locus of control is internal as well as external (de Hond and Rood, 2017).

[28] According to the theory of Gagné, teachers first gain students' attention by presenting problems or situations, then describe the goal of the lesson, recall students' prior knowledge, reflect on how knowledge is connected, and present learning materials before finally providing guidance for learning.

the previous knowledge—even more knowledge increase. Knowledge increase is regulated by Vygotsky's zone of proximal development, within which technology can act as a booster. During development, concept formation is organized in knowledge clusters: different from what is generally thought, educational effort should consist in designing a learning environment in which learners can acquire the core concepts and not all the course contents (Santoianni, 2018). Learning is more effective if students spend the more possible time on task, but information should be organized into chunks to be user friendly.

The knowledge gap between increasingly demanding environments and learners' requested performances can lead to forms of cognitive discomfort in students (Santoianni et al., 2013). According to this point of view, teaching models aligned with brain-based education aim at reinforcing core knowledge rather than working on knowledge complexification, by promoting activities that rely on basic conceptual structures. Basic conceptual structures are solicited by the individual need to understand the primary mechanisms of knowledge, both explicit and implicit (Santoianni, 2014). To favor the activation of basic conceptual processing, teaching should simplify learning content.

To design effective mobile learning environments, some key aspects of cognitive processing have to be focused (Mcquiggan et al., 2015). When information is filtered from sensory memory to be temporarily stored in working memory—since working memory is considered as limited in its capacity and duration—the full amount of information cannot be transferred. A mobile learning design should acknowledge the bottleneck between working and sensory memory: when organizing learning content, even if all available information is given to learners, it has to be taken into account that not all information will reach working memory or will be maintained within it, despite the maintenance rehearsal or the links of new knowledge with prior knowledge. Learners can react differently to given information and consequently pay more or less attention to it; sensory information has to be previously selected by designers to avoid that learners do not get distracted by irrelevant sensory information.

The amount of learning content has to be reduced to allow learners to distinguish relevant information, and basic skills should be automatized to practice information and operations, and to reduce the cognitive load. Learning content has to be presented visually and/or interactively, and to be re-presented or easily re-accessed by students if missing from the working memory. To avoid this process, learning content has to be merged in meaningful units and external extensions to working memory have to be developed—like note-taking applications, or working collaboratively for content sharing. Retrieval of information from long-term memory has to be sustained by enhancing associations with previous knowledge and providing learning content at levels that suit learners' skills, in order to encourage them to actively construct meaning.

Learning has to be divided into small chunks, discrete units, named micro-learning or nano-learning, and to be situated in micro-moments (i.e., two or three minutes for each module), tailored on individual preferences. Mobile learning programmes should be then small sized and consequently easy to use, with flexible and personalized contents. Interactive management should be scaffolded—in balance between guidance and game-based learning[29]—to provide individual discovery experiences, and online co-working should be encouraged (Zhang, 2019).

In mobile learning, the main shift from planned instruction of traditional curricula to innovative instructional design is represented by performance support, in order to narrow students to self-directed learning. Performance support can be available in a specific time and place, generally blended with instruction, or it can be present alongside the learning process, continuously blended within classroom activities, which become self-directed learning activities scaffolded by on-demand research. Blended or hybrid learning combines face-to-face and online learning, improving learners' performances, especially if they can share online resources with peers—i.e., audio/video or presentations, followed by questions and feedback— within collaborative learning environments and experiential interactive activities (Palkova, 2019).

Activities, as reading and writing, are different in mobile learning design, according to the small screen sizes of technological devices. Online reading is mainly focused on chunks of short texts, which can be randomly read, through skim-and-scan actions. Online writing is often quick and can suffer from *linguistic whateverism* (Baron, 2008), a kind of disinterest in the correct use of language, which can be intertwined with animated graphics, emoticons, and images (Pegrum, 2014, 2019). Information and communication tools must always be explicitly given as clearly written in a short form to not discourage students to interact with related questions and links. Basic readings, core concepts, key topics, and tailored tasks have to be previously prepared, while assessment has to be fully described as an educational experience in itself (Turbill, 2019).

Designing for Learning and Design Thinking

Designing for learning is a methodological approach which takes into consideration all available learning theories while identifying the common characteristics between them through continuous activities of revision, in order to build new models, to find possible associations with the previous ones, and to provide learners with a learning design toolkit of resources. A toolkit is a decision-making system based on chosen theories and best practices which allows the learner to identify the preferred approaches within a range of educational theories or practices, to select media resources, and to easily access information to personalize learning activities.

Since learning design is continuously evolving, it is difficult to set teaching and learning standards. Nevertheless, analyzing different learning theories may allow building up of a common conceptual framework that focuses on theories' shareable aspects, which can co-exist to co-construct ongoing learning design models (Conole et al., 2004).

Behaviorism is a classic model of teaching; even if overcome by other more recent models, anytime a teaching approach grounds on direct transfer of information

---

[29] Nowadays learners prefer gamified tasks, narratively designed. Specific attention is given to the adaptive challenge related to the user's identity.

from teacher to students—which can be online or in presence—it can be considered behaviorist. The so called 'web page turning mentality' means that learners can search the web only to get feedbacks and assessment; on the other hand, behaviorist teaching promotes repetition and exercises which are also shared by other teaching models as key points of learning. Cognitivist teaching leverages co-construction of knowledge structures, which emerge during continuous interaction between individuals and their environments. Another basic aspect is the idea of personalized learning, which arises on the cognitivist deepening of the multiple diverse facets of the cognitive systems, which can leverage intelligent systems and learning systems.

Systemic computer-supported collaborative teaching improves digital knowledge banks to store, archive, and retrieve shared information, while encouraging adaptation to social feedbacks. Knowledge is nowadays activity-based, relying on the enhancement of communities' development rather than simply on content communication. Networking fosters communities' empowerment and their consequent differentiation according to expertise. Participation is the keyword of dialogical computer-supported collaborative teaching, because learning is socially situated in synchronous and asynchronous environments, which foresee reciprocal interactions between teachers and students, or between users, to co-create knowledge. The idea of participation involves socio-constructivism, which searches toolkits to sustain active learning within engaging environments and to pursue student-centered goals through authentic tasks.[30] Learning becomes more and more experiential, and increasingly supports reflection.

Design thinking is solution-focused even if sometimes solutions are more than one. Socially co-created mental bridges between previously unrelated aspects are needed to solve ill-structured problems. Design thinking is user-focused, to meet learners' needs and to guarantee them effective experiences. Universal design for learning (UDL) provides all the learners the same opportunities to learn, by deep engagement, multiple representation, and multi-faceted activities. Designing means reframing problems and rethinking solutions, and also considering previously-designed knowledge and re-designing it. Design thinking uses cognitive artifacts as prototypes to preview the final result and is configured as a flexible process, which can continuously change (Bower, 2017). Educational design of mobile learning can be implemented through interactive and engaging multimedia activities, with intuitive interfaces and short information units, as apps (Baek and Guo, 2019).

Apps can be *skill-based*, for knowledge building, concerning practical training tasks, which give support for learning, and enhance the teaching of how to apply learning strategies—the design of apps for teaching focuses on effective content, feedbacks, interactions, and adaptability (Cardoso and Abreu, 2019). Apps can be *content-based*, to provide information, involving predetermined instructive tasks, allow experiences through guided discovery; and *function-based*, with an open-ended design to communicate, collaborate, and co-create digital artefacts or shared ideas through brainstorming. Apps foresee a synchronous/asynchronous interaction, which

---

[30] Tasks can be rule-based if they rely on standard procedures; incident-based, if the authenticity of learning contexts stimulates decision-making; strategy-based and role-based, when specific strategies are needed, or roles are played in scenario-based activities (Bower, 2017).

can be uni- or bi-directional, and can have a transactional dimension to purchase goods/services related to users' location. Users' interaction can be limited to specific groups or be open to the entire public (Notari et al., 2016).

Apps can be systematized as *game* apps, competition based or just entertaining; *utility* apps, with basic functions; *administration* apps, to manage and administer education; *tool* apps, which can scaffold learning but have not already prepared learning content; *content* and *reference* apps, which provide specific information in a searchable format; and *social* apps for sharing. Mobile devices let graphics be readily available, but their interfaces should have simple designs and should not be redundant; representations should be contextualized according to their use—in order to facilitate comprehension and communication—and not only integrated for decoration. Learning approach should be intuitive and simplified, avoiding what is unnecessary and segmenting or chunking learning content, which can be directly manipulated by learners and be quickly available to them, followed by continuous feedbacks given by the system (Mcquiggan et al., 2015).

Any design is related to learners' experiences, aspirations, and expectations; it evolves accordingly to technology development and situated contexts. In particular for apps design, feedbacks are basic in order to check and eventually change contents, resources, and community sharing. Learners should be indeed active within the learning process, to contribute to its design and to co-create the added value of mobile learning (Kukulska-Hulme and Traxler, 2013).

Learning and Instructional Design Principles

Instructional design principles are considered to give users the possibility to explore both the features of mobile learning and real-world problems around the learner, to apply gained knowledge; to acknowledge that learning may occur not only anywhere and anytime but also by anyone, because mobile learning can be used individually, or in groups (de Hond and Rood, 2017); to enhance combined forms of learning between mobile devices and other technologies (Herrington et al., 2009).

Mobile design principles aim at student-centered learning. Even if curricula are already structured, learners should be engaged in self-directed learning in formal context through mobile learning. Teaching strategies should be less direct and give instead to learners the possibility to actively represent individual thinking according to their own learning experiences. Learners' ideas—including planning tasks or self-recording learning processes according to these ideas—have to be encouraged and scaffolded through co-construction processes, even if they are still in development and could not be fully correct or accurate. Mobile learning design should consequently encourage learners to use different learning approaches of personalized learning, to explore the surrounding environment for authentic learning, and to socially share their individual learning for community knowledge-building, also through reciprocal teaching with peers, or in family.

Mobile Multimedia Design

Mobile design should leverage on individual interests and needs through learning personalization to encourage creativity and higher order thinking and should provide guidance to learners on the credibility of web resources to strengthen critical thinking. While in a traditional classroom environment, students receive direct instruction and take notes about it—which can be transferred on a mobile device through the support of an app, or stored in a cloud to access it anywhere, or be graphically represented as a map—in flipped, problem-based, and virtual classrooms, lectures can be substituted by slides or student presentations, and students can co-create knowledge collaboratively through digital textbooks, videos, web resources, and multimedia.

Multimedia learning content should be practical, referred to authentic settings, and divided in micro content items and in fragmented time slots; learning activity should be simple and foresee only one action for one micro activity; usability should refer to what information is to be highlighted through interface design—that is menu, button, colors, navigation links: how it can be repeated, and how to give feedbacks to users to enhance their satisfaction (Gu et al., 2011).

Mobile learning design should focus on technological affordances in order to guarantee to learners an always readily engaged, interactive, and collaborative access to digital resources, to facilitate students' learning processes and to prevent students from losing interest in engaging learning. Mobile multimedia learning promotes a kind of assessment which foresees alternative forms of evaluating learners' skills as the mobilized curriculum (MC), which includes paper-based traditional assessment and is also digital, in order to test real-life experiences and to better evaluate learners' differences. A mobilized curriculum doesn't mean only to digitize learning materials, but it can instead represent a deep shift in teaching and learning practices (Looi and Seow, 2015). According to this point of view, instructional design has been stated in 10 principles (Baek and Guo, 2019) which can be applied to multimedia learning:

- learning activities set according to educational goals;
- reliance of learning activities on educational theories;
- need identification of learning styles of both teachers and learners;
- enhanced multimedia interactions between learners and learning content;
- online support to users' technological skills;
- tailoring of learning applications on different mobile operating systems;
- solving of critical aspects of learning applications to increase the system's error tolerance;
- provision of a good quality internet to the online network of mobile devices;
- conducting formative and summative evaluations of the product;
- social interaction of asynchronous learning experiences.

Other principles to design multimedia learning—as grounding elements of mobile learning—are based on Cognitive Load Theory and Multimedia Learning Theory, and can be identified in multimedia, spatial contiguity, coherence, modality, redundancy, temporal contiguity, and individual differences (Mayer, 1999, 2001, 2006). The multimedia principle states that learners prefer to read words together with corresponding pictures rather than words alone; spatial contiguity principle means that putting together words and corresponding pictures facilitates learning because both representations can be stored simultaneously in working memory;

coherence principle focuses on extraneous words and pictures which may represent extra information potentially covering more relevant information; modality principle favors words as narration rather than in visual form, to not overload the visual channel; redundancy principle, for which learning with narration and animation should avoid addition to on-screen text, because a competition of resources may occur; temporal contiguity principle encourages presentation of words and pictures simultaneously and not successively, to enhance the development of more integrated representations; individual differences principle, according to which low-knowledge learners and high spatial-ability learners are empowered by multimedia design (Grimley, 2007).

The interaction with multimedia resources produces multimodal learning experiences, which are more effective if guided by teachers because only the use of technological devices is not education. Expert teachers and pre-service teachers should collaborate in providing support to each other (Boulton, 2017). Moving from teacher-directed towards learner-centered activities is sustained by learners' interactivity,[31] which is a spectrum of various positions involving both teachers and learners, and sometimes developing hybrid teaching and learning results (Rozario et al., 2016). To introduce mobile technology in classroom, teachers and learners should be provided with appropriate instruction, without assuming previous knowledge about technology and regularly checking their feedbacks about technology adoption. Teachers' professional development on mobile technology should be sustained by the community of practices, to share the teaching and learning experiences (O'Loughlin et al., 2013).

Mobile Courses Design Implementation

A basic aspect of instructional design is to motivate and engage learners in applying both theories of learning and usability criteria, in function of content, resources, and related teaching objectives. Learner-centered design is motivational, according to learners' preferences, and adaptively linked to learners' feedbacks.[32] Instructional design may re-structure conceptual representations, also by embedding them within new contextual examples, and adapt to the contingency of specific learning environments, encouraging, if needed, more practice. It asks for concise and systematic information offering, and re-thinking of objectives and assessment, going towards increasing personalization of teaching, tailored on learners' cognitive—and emotional, perceptual, organismic—needs (Haag and Berking, 2019).

Key concepts of courses design are related to courses structure, content presentation, interaction, technology choice, students' assessment, and technical skills, social participation and individual motivation, opportunities of interaction,

---

[31] Interactive learning environments (ILEs) combine educational approaches grounded on learning theories with learning measuring techniques (Timms, 2017).

[32] Adaptive learning systems offer to students personalized learning experiences through courses, activities, and related scaffolding. They aim to facilitate teachers' simultaneous co-working with students, to give them regulatory feedbacks, and to let them have more time to spend with students. Adaptive virtual courses (AVC) are flexible interactive systems which customize instruction through learning basic units, domain topics, educational goals, learning activities and objects—adaptively designed according to learners' characteristics, basic skills, explicit/implicit knowledge levels, learning performances, and contexts (Ovalle et al., 2011).

foreseen time, and ongoing support with timely feedback. To design effective high-quality online courses, courses have to be structured by implementing interactive learning activities, monitored by different control approaches, which can be fully autonomous, with basic guidelines, or highly specified. Interactions can be defined as more or less transactions between learners, learners and teachers, learners and content (Martin and Oyarzun, 2018). Courses' design should develop according to interaction, collaboration, and engagement.

Learning performances can be improved through enjoyable activities because there is a relation between intrinsic motivation, enjoyment, and achievement. Enjoyment can satisfy self-determination and the basic needs—that are autonomy, competence, and relatedness. Enjoyment can easily occur within affinity spaces and informal learning environments where learners can share interests even if they belong to different levels of skills—a significant characteristic, because participants can be of different ages and knowledge backgrounds (Baek and Touati, 2017).

In course implementation, online teaching has to be organized according to learners' engagement and collaboration, and learning facilitation through learners' one-to-one assistance: learning materials of the course should be delivered on time, and teachers should guarantee a continuous feedback to students—a direct feedback, if the communication is synchronous, and also asynchronous feedback, which allows students to reflect more on it. Teachers should encourage the learning community to interact in a supportive way, to give its social presence through discussion forums and group assignments. Moreover, teachers should promote the use of high-quality technology-assisted methods of communication because students may feel lost in cyberspace, especially if they have technological problems (Martin and Oyarzun, 2018).

Principles of a course design and implementation refer to the chosen educational approaches to the ways of interacting and collaborate, to the provided support, to the selected content, and its related activities (Conole, 2013). A learning activity (LA) leverages its context—intended as users involved, skill levels, learning environments, and desired outcomes; adopted teaching strategies and techniques; and contingent resources and tools related to specific tasks (Conole and Fill, 2005).

A lesson plan should be characterized by specific elements to be schematized as title, author, summary, keywords, educational problem, age range, duration, subject and topic domain, educational standards, resources, teachers' and students' prerequisites, accessibility, educational objectives, teaching approach, learning activities, assessment, and adaptations (Sergis et al., 2017). Besides a lesson plan, a courseware should have an effective interface, instructional strategies and task-solving environments with simulation software, hypertexts, examples, and multiple-choice questions. A course structure should be indeed enhanced by the following learning objects (LOs) (El-Bishouty et al., 2015):

- commentaries to introduce a brief overview of the learning content;
- content, that is the course's learning material;
- open-ended questions about the learned content;
- close-ended questions to be used for self-assessment;

- discussion forum, useful for learners' peer tutoring and teacher-students interaction;
- additional resources to deepen learning content through further explanations;
- multimedia formats to clarify the teacher's lesson;
- practical exercises to let students train acquired knowledge;
- concrete examples to show how learned concepts can be interpreted;
- real-world applications to apply how knowledge can be useful in everyday life;
- final summaries to facilitate remembering the learned materials.

Learning Management Systems Design

Web 2.0-based applications as learning management systems (LMS),[33] like moodle or massive open online courses (MOOCs) can support communication, negotiation, and learning between spatial distributed users who can interact with a content-management system and share educational contents and resources through forum messages of type-based applications and e-mails; collaborative learning is enhanced by social navigation and group working (Weller, 2018).

Learning management systems are based on e-learning technology for blended-learning and full online courses. All elements of a mobile learning system are standardized and share the same design, which should be minimalist in shape so as not to distract learners' attention. Their applicability refers to the perceived usefulness related to the levels of required expertise according to content and the possibility to apply it to special learners. As a consequence, mobile learning systems' design should follow some organizing rules.

In a mobile learning course design, factors of influence are learners' already acquired knowledge about the website topics, learners' previous skills and experience with online learning and digital devices, learning resources and foreseen time for learning, type of environment involved—which can be home, workplace, or classroom for blended learning—and technological affordances. In learning design, which can foresee self-paced or instructor-led learning, learners require the teacher's feedback throughout the learning process, to be only 'a click away' from each other through instant messaging and e-mail, social media chats, discussion forums, and video conferencing, which can guarantee an always open line of communication between them and which is reliable enough to meet students' expectations. Apart from e-coaching, peer tutoring can be needed in virtual classroom. Design keys can be identified in the following aspects (Palkova, 2019):

- topic analysis to identify the course content, learning materials, and resources;
- task analysis to define which activities learners have to carry on in order to obtain the desired results, and which expertise is needed;

---

[33] Learning management systems have been criticized because on the one hand, they leverage independent and collaborative learning but on the other, learners can decide what to learn and how to co-construct knowledge domains without the support and feedback of the teacher. Moreover, sometimes educational materials are not based on learners' prerequisites and the learners are not actively involved in gaining needed information resources, and their critical thinking skills and self-regulated learning are not stimulated enough, and learnings are often detached from real-world problems (Pellas, 2014).

- focus on learning purposes to tailor them according to the course objectives as well as to the individual expected levels of performance;
- content presentation to share the narrative storytelling of the learning topics and their subdivisions into chunks of sets of resources, which can be approached autonomously;
- selected samples to let learners practice tasks, procedures, and steps related to interactive tools, as answering questions or making choices.

In a learning management systems design, learning material is naturally interactive and dynamic to keep the users active and motivated, while the system gives instructions and adopts the user's language. Alongside any learning pathway, there are indeed explicit activities and related goals—which can be extrinsic and intrinsic, to motivate and stimulate learners—clearly stated to facilitate the learners' acknowledgement of helpful resources specifically applied to the given tasks, which learners can easily find within the system through free navigation. Navigation options, as links or downloads, are available in narrow relation to the needed resources and can provide feedbacks[34] as right answers or further information, which should be understood and not only memorized (Pensabe-Rodrigueza et al., 2020).

The mobile lesson is one tool for interactively narrowing learners to technology. This method is developed in relation to the desired learning outcomes, according to which mobile lesson's goals are individuated and defined. Mobile learning is then integrated within in-class or virtual activities, thereby co-creating a blended learning environment often linked to the pre-existing content (Vincent-Layton, 2019). The mobile lesson template (Wiggins and McTighe, 2005) is divided into main six steps, which are as follows:

- assign a name to the lesson in order to introduce it to learners;
- prepare an overview to describe the lesson to explain what learners can expect about it;
- foresee the learning outcomes and clearly express both the lesson's objectives and what is expected from students' performance;
- choose the learning materials and select the appropriate resources to meet the learners' effort;
- give detailed step-by-step instructions to let learners follow the lesson, including if any prerequisite or previous skill is needed;
- balance assessment[35] in close relation to what has been asked and is expected as learning outcomes.

Other aspects involve time commitment to complete the lesson's assignment, feedbacks about the activities' processing and related improvements, instructional

---

[34] Systems can be proactive, with continuous guidance; reactive, if they allow learners to make mistakes and then give feedbacks; and requested, when learners can request on demand guidance (Ludvigsen and Mørch, 2010).

[35] Formative assessment should be organized as adaptive to learners' needs and context variability, according to the criteria of verifying the abilities of recall, comprehension, and generalization of already learned notions (Louhab et al., 2018).

examples to be given as models and facilitating tools, and technology challenges. Even if the mobile lesson template design aims to predefine mobile learning experience, some flexible openness has to be left to allow learners to personalize and implement their own learning experience.

Whatever be the device in use, the learning management systems' content should facilitate the target through graphics and interactive tools rather than through texts, which in case should have easy-to-read fonts, sizes, and colors. Learning management systems have to be well-organized and intuitive. High quality learning materials must be functionally implemented in the platform, and be easy to find through navigation. Content is to be organized through a multimedia offer, which allows students to use their preferred learning styles (Palkova, 2019).

Visual Basic Logic Theory for Spatial Design

The mobile learning design educational framework is like a jigsaw comprehending different synergic aspects of teaching and learning, intertwined with many dependent variables related to them (*see* Fig. 1).

This research area has been more often applied to explicit linguistic aspects, to design mobile courses, but it can be implemented into the field of implicit and spatial visualization.

Visualization approaches of learning design use spatial tools and can be based on maps, which show connections between students' conceptual representations and between teacher and learners (Bower, 2017). Spatial approaches can be applied to navigation also because users navigate between windows and resources or apps within the screen, to which they may have a direct focusing access or a sequential access.

A starting point to organize a digital visual interface from a spatial point of view is the literature on spatial representations and knowledge and one which has analyzed concept maps (Novak, 2001; Novak and Gowin, 2001) and mind maps (Buzan and Buzan, 2003), focusing respectively on the sequential and derivative

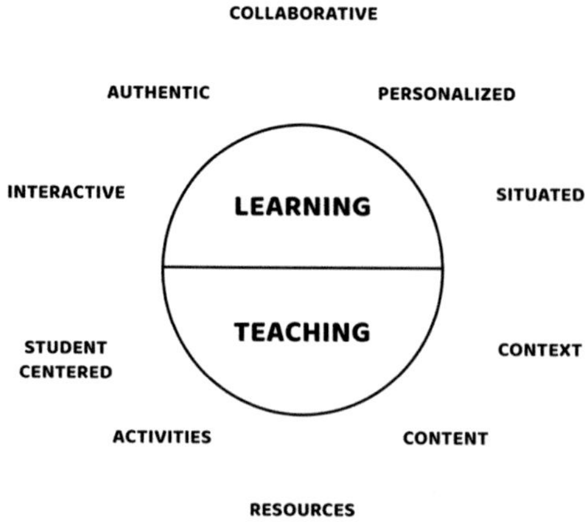

Fig. 1. Mobile teaching and learning.

relationships between spatially represented concepts. Another area of research to be deepened is spatial knowledge, which is influenced by both route and survey navigation (Santoianni and Ciasullo, 2018). Route knowledge consists of sequential itineraries of locations represented through discrete chunks of landmarks' sequential records (Hirtle and Hudson, 1991; Taylor and Tversky, 1992; McNamara et al., 2008), while survey knowledge has been defined as a 'map in the head' (Kuipers, 1982) because it is a global map-like representation (Tversky, 1991), which allows a holistic overall configuration of an environment.

Sequential and derivative logical criteria are not the only possible ones for organizing concepts in spatial representations. According to Basic Logic Theory (Santoianni, 2009, 2011, 2014, 2016a), logical criteria that regulate the organization of concepts can be categorized into six basic logics, which are integration (add), sequencing (chain), individuation (each), comparison (compare), derivation (focus), and correlation (link). Basic logic can be attributed to two macro-classes, sequence and parallelism (Rumelhart and McClelland, 1991), which in turn are divided in three sub-classes—union, separation, and correlation.[36]

Basic logic is derived from the possible relations between linguistics and mathematics' implicit organization of concepts, which can regulate their contextual use and can be considered thinking prototypes since their original characteristics already emerge in a prototype stage, as knowledge prototypes (Lambiotte et al., 1989; O'Donnell et al., 2002). Basic logic shows prototypal relations between implicit and explicit processing working as dynamic patterns which are implicitly activated at the beginning of any cognitive involvement, with the function of explicit knowledge precursors or playing an on-demand role, which consists of simplifying cognitive transactions between the implicit and explicit cognitive collaborations (Santoianni, 2011, 2016a).

If applied to a digital visual interface in mobile learning design, Basic Logic Theory can organize the spatial navigation between linked conceptual nodes of a network/map or between apps in a screen because each basic logic has a specific spatial expression (*see* Fig. 2) and its spatial representation can be a meeting point between implicit and explicit processing.

Personalized spatial learning can be enhanced by customized use of spatial representations, which influence spatial knowledge, according to both individual preferences and knowledge domains (Santoianni, 2016b, 2016c, 2016d).

References

Almaiah, M.A., Jalil, M.M.A. and Man, M. (2016). Empirical investigation to explore factors that achieve high quality of mobile learning system based on students' perspectives. *Engineering Science and Technology, An International Journal*, 19: 1314–1320.

Baek, E.O. and Guo, Q. (2019). Instructional design principles for mobile learning. pp. 717–738. *In*: Zhang, Y. and Cristol, D. (eds.). *Handbook of Mobile Teaching and Learning*. Springer, Singapore.

Baek, Y. and Touati, A. (2017). Exploring how individual traits influence enjoyment in a mobile learning game. *Computers in Human Behavior*, 69: 347–357.

---

[36] Each class comprehends two logic. Union includes integration (add) and sequencing (chain); separation involves individuation (each) and comparison (compare); correlation refers to derivation (focus) and correlation (link).

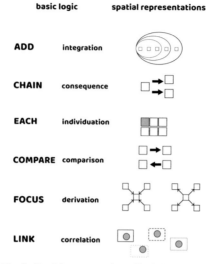

Fig. 2. Spatial representation of basic logic theory.

Bannan, B., Cook, J. and Pachler, N. (2015). Reconceptualizing design research in the age of mobile learning. *Interactive Learning Environments*, 24(5): 938–953.

Baron, N.S. (2008). *Always On: Language in an Online and Mobile World*. Oxford University Press, New York.

Beetham, H. and Sharpe, R. (2013). An introduction to rethinking pedagogy. pp. 1–10. *In*: Beetham, H. and Sharpe, R. (eds.). *Rethinking Pedagogy for the Digital Age: Designing for the 21st Century Learning*. Routledge, New York.

Boettcher, J. (2007). Ten core principles for designing effective learning environments: insights from brain research and pedagogical theory. *Innovate*, 3(3): 1–8.

Boulton, H. (2017). Introducing digital technologies into secondary schools to develop literacy and engage disaffected learners: a case study from the UK. pp. 31–44. *In*: Marcus-Quinn, A. and Hourigan, T. (eds.). *Handbook on Digital Learning for K-12 Schools*. Springer, Switzerland.

Bower, M. (2017). *Design of Technology-enhanced Learning. Integrating Research and Practice*, Emerald Publishing, Bingley, UK.

Branch, R.M. (2010). *Instructional Design: The ADDIE Approach*. Springer, New York.

Burden, K. and Kearney, M. (2017). Conceptualizing authentic mobile learning. pp. 27–42. *In*: Churchill, D., Lu, J., Chiu, T. and Fox, B. (eds.). *Mobile Learning Design*. Springer, Singapore.

Buzan, B. and Buzan, T. (2003). *Mappementali*. NLP Italy, Milano.

Cardoso, T. and Abreu, R. (2019). Mobile learning and education: synthesis of open-access research. pp. 313–332. *In*: Zhang, Y. and Cristol, D. (eds.). *Handbook of Mobile Teaching and Learning*. Springer, Singapore.

Christensen, K. and West, R.E. (2018). The development of design-based research. pp. 542–574. *In*: West, R.E. (ed.). *Foundations of Learning and Instructional Design Technology. The Past, Present, and Future of Learning and Instructional Design Technology*. Ed Tech Books.

Churchill, D., Fox, B. and King, M. (2016). Framework for designing mobile learning environments. pp. 3–26. *In*: Churchill, D., Lu, J., Chiu, T. and Fox, B. (eds.). *Mobile Learning Design*. Springer, Singapore.

Conole, G., Dyke, M., Oliver, M. and Seale, J. (2004). Mapping pedagogy and tools for effective learning design. *Computers & Education*, 43: 17–33.

Conole, G. and Fill, K. (2005). A learning design toolkit to create pedagogically effective learning activities. *Journal of Interactive Media in Education*, 8(1): 1–16.

Conole, G. (2013). Tools and resources to guide practice. pp. 78–101. *In*: Beetham, H. and Sharpe, R. (eds.). *Rethinking Pedagogy for the Digital Age: Designing for the 21st Century Learning*. Routledge, New York.

de Hond, M. and Rood, T. (2017). Flip the school, forget the classroom; how to enable personalised learning with the help of information technology. pp. 317–328. *In*: Marcus-Quinn, A. and Hourigan, T. (eds.). *Handbook on Digital Learning for K-12 Schools*. Springer, Switzerland.

Dousay, T.A. (2018). Instructional design models. pp. 452–479. *In*: West, R.E. (ed.). *Foundations of Learning and Instructional Design Technology. The Past, Present, and Future of Learning and Instructional Design Technology*. Ed Tech Books.

El-Bishouty, M.M., Saito, K., Chang, T., Kinshuk and Graf, S. (2015). Teaching improvement technologies for adaptive and personalized learning environments. pp. 225–242. *In*: Kinshuk and Huang, R. (eds.). *Ubiquitous Learning Environments and Technologies, Lecture Notes in Educational Technology*. Springer, Berlin and Heidelberg.

Gagne, R. (1965). *The Conditions of Learning*. Holt, Rinehart and Winston, New York.

Grimley, M. (2007). Learning from multimedia materials: The relative impact of individual differences. *Educational Psychology*, 27(4): 465–485.

Gu, X., Gu, F. and Laffey, J.M. (2011). Designing a mobile system for lifelong learning on the move. *Journal of Computer Assisted Learning*, 27(3): 204–215.

Gustafson, K.L. (1991). *Survey of Instructional Development Models*. ERIC Clearinghouse on Information Resources, Syracuse.

Gustafson, K.L. and Branch, R.M. (1997). *Survey of Instructional Development Models*. Syracuse University, Syracuse.

Haag, J. and Berking, P. (2019). Design considerations for mobile learning. pp. 221–240. *In*: Zhang, Y. and Cristol, D. (eds.). *Handbook of Mobile Teaching and Learning*. Springer, Singapore.

Hall, T., Thompson Long, B., Flanagan, E., Flynn, P. and Lenaghan, J. (2017). Design-based research as intelligent experimentation: towards systematising the conceptualization, development and evaluation of digital learning in schools. pp. 59–74. *In*: Marcus-Quinn, A. and Hourigan, T. (eds.). *Handbook on Digital Learning for K-12 Schools*. Springer, Switzerland.

Herrington, A., Herrington, J. and Mantei, J. (2009). Design principles for mobile learning. pp. 129–138. *In*: Herrington, J., Herrington, A., Mantei, J., Olney, I. and Ferry, B. (eds.). *New Technologies, New Pedagogies: Mobile Learning in Higher Education*. Wollongong, University of Wollongong.

Hirtle, S.C. and Hudson, J. (1991). Acquisition of spatial knowledge for routes. *Journal of Environmental Psychology*, 11: 335–345.

Klopfer, E. and Squire, K. (2005). Environmental detectives: the development of an augmented reality platform for environmental simulations. *Educational Technology Research and Development*, 56(2): 203–228.

Kuipers, B. (1982). The 'map in the head' metaphor. *Environment and Behavior*, 14: 202–220.

Kukulska-Hulme, A. and Traxler, J. (2013). Design principles for learning with mobile devices. pp. 244–257. *In*: Beetham, H. and Sharpe, R. (eds.). *Rethinking Pedagogy for the Digital Age: Designing for the 21st Century Learning*. Routledge, New York.

Lambiotte, J.G., Dansereau, D.F., Cross, D.R. and Reynolds, S.B. (1989). Multirelational semantic maps. *Educational Psychology Review*, 1: 331–367.

Laurillard, D. (2013a). Foreword to the first edition. pp. XIX–XXI. *In*: Beetham, H. and Sharpe, R. (eds.). *Rethinking Pedagogy for the Digital Age: Designing for the 21st Century Learning*. Routledge, New York.

Laurillard, D. (2013b). Foreword to the second edition. pp. XVI–XVIII. *In*: Beetham, H. and Sharpe, R. (eds.). *Rethinking Pedagogy for the Digital Age: Designing for the 21st Century Learning*. Routledge, New York.

Lee, V. (2018). A short history of the learning sciences. pp. 57–85. *In*: West, R.E. (ed.). *Foundations of Learning and Instructional Design Technology. The Past, Present, and Future of Learning and Instructional Design Technology*. Ed Tech Books.

Liaw, S.S., Hatala, M. and Huang, H.M. (2010). Investigating acceptance toward mobile learning to assist individual knowledge management: based on activity theory approach. *Computers and Education*, 54(2): 446–454.

Looi, C.K. and Seow, P. (2015). Seamless learning from proof-of-concept to implementation and scaling-up: a focus on curriculum design. pp. 419–435. *In*: Wong, L.H., Milrad, M. and Specht, M. (eds.). *Seamless Learning in the Age of Mobile Connectivity*. Springer, Singapore.

Louhab, F.E., Bahnasse, A. and Talea, M. (2018). Towards an adaptive formative assessment in context-aware mobile learning. *Procedia Computer Science*, 135: 441–448.

Ludvigsen, S.R. and Mørch, A.I. (2010). Computer-supported collaborative learning: basic concepts, multiple perspectives, and emerging trends. pp. 290–296. *In*: Peterson, P., Baker, E. and McGaw, B. (eds.). *International Encyclopedia of Education*. Elsevier, Amsterdam.

Major, L., Haßler, B. and Hennessy, S. (2017). Tablet use in schools: impact, affordances and considerations. pp. 115–128. *In*: Marcus-Quinn, A. and Hourigan, T. (eds.). *Handbook on Digital Learning for K-12 Schools*. Springer, Switzerland.

Martin, F. and Oyarzun, B. (2018). Distance learning. pp. 787–816. *In*: West, R.E. (ed.). *Foundations of Learning and Instructional Design Technology. The Past, Present, and Future of Learning and Instructional Design Technology*. Ed Tech Books.

Mayer, R.E. (1999). Research-based principles for the design of instructional messages: the case of multimedia explanations. *Document Design*, 2: 7–20.

Mayer, R.E. (2001). *Multimedia Learning*. Cambridge University Press, New York.

Mayer, R.E. (2006). Ten research-based principles of multimedia learning. pp. 371–390. *In*: O'Neil, H.F. and Perez, R.S. (eds.). *Web-based Learning: Theory, Research, and Practice*. Lawrence Erlbaum, Mahwah.

McNamara, T., Sluzenski, J. and Rump, B. (2008). Human spatial memory and navigation. pp. 157–178. *In*: Byrne, J. (ed.). *Learning and Memory: A Comprehensive Reference*. Science Direct.

Mcquiggan, S., Kosturko, L. and Mcquiggan, J. (2015). *Mobile Learning: A Handbook for Developers, Educators, and Learners*. Wiley, Hoboken.

Mentor, D. (2017). Cultivating digital TLC—teaching and learning communities. pp. 1–10. *In*: *The Proceedings of the MODSIM World, Visualization and Gamification*. Norfolk.

Merrill, M.D. (2018). Using the first principles of instruction to make instruction effective, efficient, and engaging. pp. 428–448. *In*: West, R.E. (ed.). *Foundations of Learning and Instructional Design Technology. The Past, Present, and Future of Learning and Instructional Design Technology*. EdTech Books.

Newby, T.J., Stepich, D., Lehman, J. and Russell, J.D. (1996). *Instructional Technology for Teaching and Learning: Designing, Integrating Computers, and Using Media*. Pearson Education, Upper Saddle River.

Notari, M.P., Hielscher, M. and King, M. (2016). Educational apps ontology. pp. 83–96. *In*: Churchill, D., Lu, J., Chiu, T. and Fox, B. (eds.). *Mobile Learning Design*. Springer, Singapore.

Novak, J.D. (2001). *L'apprendimento significativo. Le mappe concettuali per creare e usare la conoscenza*. Erickson, Trento.

Novak, J.D. and Gowin, B. (2001). *Imparando ad Imparare*. SEI, Torino.

O'Donnell, A.M., Dansereau, D.F. and Hall, R.H. (2002). Knowledge maps as scaffolds for cognitive processing. *Educational Psychology Review*, 1(14): 71–86.

O'Loughlin, A.M., Barton, S.M. and Ngo, L. (2013). Using mobile technology to enhance teaching. pp. 293–306. *In*: Berge, Z. and Muilenburg, L. (eds.). *Handbook of Mobile Learning*. Routledge, New York.

Ovalle, D.A., Arias, F.J. and Moreno, J. (2011). Student-centered multi-agent model for adaptive virtual courses development and learning object selection. pp. 123–130. *In*: The *Proceedings of the IADIS International Conference on Cognition and Exploratory Learning in Digital Age*.

Palkova, Z. (2019). Mobile Web 2.0 tools and applications in online training and tutoring. pp. 609–634. *In*: Zhang, Y. and Cristol, D. (eds.). *Handbook of Mobile Teaching and Learning*. Springer, Singapore.

Pegrum, M. (2014). *Mobile Learning: Languages, Literacies and Cultures*. Palgrave Macmillan, Basingstoke.

Pegrum, M. (2016). Future directions in mobile learning. pp. 413–431. *In*: Churchill, D., Lu, J., Chiu, T. and Fox, B. (eds.). *Mobile Learning Design*. Springer, Singapore.

Pegrum, M. (2019). *Mobile Lenses on Learning, Languages and Literacies on the Move*. Springer, Singapore.

Pellas, N. (2014). The influence of computer self-efficacy, metacognitive self-regulation and self-esteem on student engagement in online learning programs: evidence from the virtual world of second life. *Computers in Human Behavior*, 35: 157–170.

Pensabe-Rodrigueza, A., Lopez-Domingueza, E., Hernandez-Velazqueza, Y., Dominguez-Isidroa, S. and De-la-Callejab, J. (2020). Context-aware mobile learning system: Usability assessment based on a field study. *Telematics and Informatics*, 48(101346): 1–14.

Rieber, L. (2018). The proper way to become an instructional technologist. pp. 16–38. *In*: West, R.E. (eds.). *Foundations of Learning and Instructional Design Technology. The Past, Present, and Future of Learning and Instructional Design Technology*. Ed Tech Books.

Rozario, R., Ortlieb, E. and Rennie, J. (2016). Interactivity and mobile technologies: an activity theory perspective. pp. 63–82. *In*: Churchill, D., Lu, J., Chiu, T. and Fox, B. (eds.). *Mobile Learning Design*. Springer, Singapore.

Rumelhart, D.E. and McClelland, J.L. (1991). PDP, *Microstruttura dei processi cognitivi*. Il Mulino, Bologna.

Santoianni, F. (2006). *Educabilità cognitiva, Apprendere al singolare, insegnare al plurale*. Carocci, Roma.

Santoianni, F. (ed.). (2009). *Costruzione di ambienti per lo sviluppo e l'apprendimento, Il protocollo formativo C.A.S.A. per la scuola primaria, Consorzio Editoriale Fridericiana*. Napoli.

Santoianni, F. (2011). Educational models of knowledge prototypes development. *Mind & Society*, 10: 103–129.

Santoianni, F., Sorrentino, M., Lamberti, E. and Di Jorio, D. (2013). Bioeducational sciences on cognitive discomfort and specific learning disorders. pp. 107–124. *In*: Burgess, E.N. and Thornton, L.A. (eds.). *Cognitive Dysfunctions*. Nova Science Publisher, New York.

Santoianni, F. (2014). *Modelli di studio, Apprendere con la teoria delle logiche elementari*. Erickson, Trento.

Santoianni, F. (2016a). Spaces of thinking. pp. 5–14. *In*: Santoianni, F. (ed.). *The Concept of Time in Early Twentieth-century Philosophy. A Philosophical Thematic Atlas*. Springer, Switzerland.

Santoianni, F. (2016b). Phenomenology and perception of time maps. pp. 35–38. *In*: Santoianni, F. (ed.). *The Concept of Time in Early Twentieth-century Philosophy. A Philosophical Thematic Atlas*. Springer, Switzerland.

Santoianni, F. (2016c). Language and thinking of time maps. pp. 126–128. *In*: Santoianni, F. (ed.). *The Concept of Time in Early Twentieth-century Philosophy: A Philosophical Thematic Atlas*. Springer, Switzerland.

Santoianni, F. (2016d). Science and logic of time maps. pp. 199–202. *In*: Santoianni, F. (ed.). *The Concept of Time in Early Twentieth-century Philosophy. A Philosophical Thematic Atlas*. Springer, Switzerland.

Santoianni, F. (2018). *L'implicito come struttura concettuale. In*: Frauenfelder, E., Santoianni, F. and Ciasullo, A. (eds.). *Implicito bioeducativo, Emozioni e cognizione. RELAd EI Neurociencias y Educación Infantil*, 7(1): 42–51.

Santoianni, F. and Ciasullo, A. (2018). Digital and spatial education intertwining in the evolution of technology resources for educational curriculum reshaping and skills enhancement. *International Journal of Digital Literacy and Digital Competence*, 9(2): 34–49.

Sergis, S., Papageorgiou, E., Zervas, P., Sampson, D.G. and Pelliccione, L. (2017). Evaluation of lesson plan authoring tools based on an educational design representation model for lesson plans. pp. 173–189. *In*: Marcus-Quinn, A. and Hourigan, T. (eds.). *Handbook on Digital Learning for K-12 Schools*. Springer, Switzerland.

Sloep, P.B. (2016). Design for networked learning. pp. 41–58. *In*: Gros, B., Kinshuk and Maina, M. (eds.). *The Future of Ubiquitous Learning. Learning Designs for Emerging Pedagogies*. Springer, Berlin.

Taylor, H.A. and Tversky, B. (1992). Spatial mental models derived from survey and route descriptions. *Journal of Memory and Language*, 31: 261–292.

Timms, M.J. (2017). Assessment of online learning. pp. 217–232. *In*: Marcus-Quinn, A. and Hourigan, T. (eds.). *Handbook on Digital Learning for K-12 Schools*. Springer, Switzerland.

Turbill, J. (2019). Transformation of traditional face-to-face teaching to mobile teaching and learning: pedagogical perspectives. pp. 35–48. *In*: Zhang, Y. and Cristol, D. (eds.). *Handbook of Mobile Teaching and Learning*. Springer, Singapore.

Tversky, B. (1991). Spatial mental models. *The Psychology of Learning and Motivation*, 27: 109–145.

Vincent-Layton, K. (2019). Mobile learning and engagement: designing effective mobile lessons. pp. 241–256. *In*: Zhang, Y. and Cristol, D. (eds.). *Handbook of Mobile Teaching and Learning*. Springer, Singapore.

Weller, M. (2018). Twenty years of ed tech. pp. 88–116. *In*: West, R.E. (ed.). *Foundations of Learning and Instructional Design Technology. The Past, Present, and Future of Learning and Instructional Design Technology*. Ed Tech Books.

White, L.D. (2019). Gatekeepers to millennial careers: adoption of technology in education by teachers. pp. 665–678. *In*: Zhang, Y. and Cristol, D. (eds.). *Handbook of Mobile Teaching and Learning.* Springer, Singapore.

Wiggins, G.P. and McTighe, J. (2005). *Understanding by Design, Association for Supervision and Curriculum Development.* Alexandria.

Wong, G.K.W. (2016). A new wave of innovation using mobile learning analytics for flipped classroom. pp. 189–218. *In*: Churchill, D., Lu, J., Chiu, T. and Fox, B. (eds.). *Mobile Learning Design.* Springer, Singapore.

Zhang, Y.A. (2019). Characteristics of mobile teaching and learning. pp. 13–33. *In*: Zhang, Y. and Cristol, D. (eds.). *Handbook of Mobile Teaching and Learning.* Springer, Singapore.

Part II
Corrado Petrucco

Chapter 4

User and Smartphones Social and Cognitive Interaction

Corrado Petrucco

Introduction

The 1933 Chicago World's Fair was one of the most important ones of the last century, and took as its slogan 'Science discovers, genius invents, industry applies and man adapts himself to, or is molded by, new things'. This is undoubtedly a powerful statement, and one that takes an essentially techno-centric view of progress as something to which man must necessarily adapt: while we continue to create and improve new technologies, we are continually shaped by them (Hyysalo, 2010) in processes of continuous learning bound together with social norms and practices. Our actions have always been mediated by technological artefacts which have become increasingly complex in the course of history: using these tools does not only mean completing a task in order to reach a goal, but also learning to make sense of the world (Bruner, 1996). Cultural-Historical Activity Theory (CHAT) describes these processes effectively: this theory holds that the concept of tool should be interpreted not only as referring to a physical object, but also as a set of collaborative strategies and methods (Engeström, 2001) that enable us to solve concrete problems. CHAT can connect technology and the social context in a single integrated process (Karanasios, 2018) and can be a useful tool for introducing technology in specific settings, such as schools and universities.

Technological innovation inevitably permeates all social structures, including those involved in formal education, though its spread to the latter is generally selective and significantly slower (Firmin and Genesi, 2013): for example, it took considerably longer to introduce personal computers in teaching than it did in work or home settings largely because of factors relating to their cost and the need for training. Psychological factors have been equally important in creating obstacles, in

Associate Professor of Education, University of Padua, Italy.
Email: corrado.petrucco@unipd.it

particular those associated with teachers' refusal to use new technologies (Joo et al., 2018). In this connection, it is interesting to note that one of the best-known models, the technology acceptance model (TAM and TAM2) which attempts to explain the factors influencing teachers' intention to use technologies (Legris et al., 2003), is based not only on the tools' perceived usefulness in the classroom, but also—and above all—on such social factors as subjective norms. In our case, the latter include teachers' perceptions of public opinion or colleagues' attitudes towards adopting educational technologies (Ursavaş et al., 2019). In this sense, there can be little doubt that the technology that has given rise to the debate regarding its use in the classroom is that of the smartphone.

Smartphones as Socially Disruptive Technology

Smartphones as a Technological-cultural Object

Public trust in technological progress has increased in recent years, especially for tools which are most often used (Land-Zandstra et al., 2016): only rarely is our trust in them eroded by the 'digital failures' that can occur. In other words, accepting the fact that technological devices are ever more ubiquitous, and when something critical happens, our reaction, far from being one of techno-resistance, is to increase our investment in technology: if we've lost data, for example, we invest in better, higher-capacity hard disks, or if we struggle to see what's on our monitor, we buy bigger ones, or again, if we have a hard time remembering our password, we try to do without it altogether and use biometrics, such as fingerprints (Skågeby, 2019). In general, we prefer easy-to-use, ergonomic devices that can provide an effective, efficient, and satisfying experience in daily use where the most obvious example is the smartphone. We no longer use it only to communicate, but also—and especially— to surf the Web for information, entertainment, and to learn. In a few short years, the smartphone has gone from being a simple tool to becoming a true cultural object. This is especially true among young people for whom perceptions of the smartphone have gone beyond its pragmatic function to the extent that its use is now a socially stratified cultural practice (Luthar and Kropivnik, 2011). Once only a physical object that meets contingent needs, the smartphone has become essential to establishing and maintaining sociality (Silverstone and Hirsh, 1992) and this continual interaction with the smartphone has problematic side effects, particularly for young people. These range from dependence on the device to attention-deficit disorder triggered by its excessive use.

The Smartphone as a Technological 'Black Hole'

Smartphones are highly flexible tools because, as is the case with the personal computer, new software functions can be added as needed simply by installing an app. To an even greater extent, they are flexible from a hardware standpoint as they have evolved as a major attractor for other technologies that once required a dedicated artifact and which has increased their affordances enormously. Over time, what was a simple means of communication has absorbed the functions of the desktop PC for connecting to the internet, as well as those of the camera, navigator, radio/TV and video recorder. In addition, any standard smartphone also has a variety of sensors that interact with the environment and make it possible to develop apps for specific tasks: accelerometers, magnetic field sensors, gyroscopes, RFID readers and sensors for ambient temperature, heart rate, light, proximity and pressure, to name but a few. They can be used to measure environmental parameters, such as noise levels and temperature, to track and understand map-based mobility patterns, or to share information with online social communities. In recent years, they have also been used in applications for monitoring health and well-being (Cornet and Holden, 2018). In the near future, with the development of 5G and Internet of Things (IoT) transmission protocols, it is very likely that smartphones will be essential for interacting with all the 'intelligent' devices that will be connected to the Web in homes, offices and outdoors (Al-Turjman, 2019). As it happened with the personal computer, the smartphone is now very much a disruptive technology, whose distinctive feature is its portability.

Smartphones and New Affordances: Augmented Reality

A further major disruption in how space/time is conceived in education as well as in everyday life has recently emerged with virtual and augmented reality technologies. Unlike virtual reality, where a headset isolates users from the outside environment, augmented reality (AR) enables us to interact with our surroundings while overlaying planes of information on what we perceive. Simply defined, AR consists of software that can display information and digital artifacts combined with real-world features in a location with precise spatial coordinates. Smartphones, with their inherent mobility, have thus become the ideal tool for AR applications: the view of the real world on the screen can be overlaid with information and/or objects georeferenced in the user's actual location.

Perhaps the most famous application that demonstrated AR's potential and launched the so-called 'augmented reality games' was Pokémon Go. Introduced in 2016, the game introduced millions of children to an alternative world where they could interact with multiple planes of reality superimposed on precise geographic locations. The high levels of engagement with the game was proof of how effective augmented reality could be for mobile game-based learning, and it also yielded interesting feedback about the physical places visited during play, showing that it could positively affect 'place attachment' by stimulating exploration, social relations, and the experience of enjoyment in a place (Oleksy and Wnuk, 2017). Augmented reality applications thus seem to hold considerable promise for classroom use. This is especially true of teaching concepts that are not readily understood from everyday observation, precisely because these applications can combine real-time evidence from the natural world with overlaid digital information (Tobar-Muñoz et al., 2017). As AR teaching with smartphones becomes more common, it is likely that the move towards informal learning approaches will pick up speed, particularly outside the physical school setting, where they will be part of open teaching models, like the flipped classroom.

From Desktop to Laptop to Smartphones: Access to the Web is more Mobile than Ever

Since 2019, the number of Google searches on mobile devices, such as tablets and smartphones, has outstripped the number carried out from desktop computers. In 2013, mobile data traffic accounted for 16 per cent of the worldwide total, while in 2019, it exceeded 53 per cent (Broadb and Search, 2020). In response to this massive uptick, Google has begun to penalize non-mobile-friendly websites in its search rankings, and at the same time to offer assistance for the transition to mobile. The public now sees portability as an essential factor in personal productivity: having the internet always at hand makes it possible to search the Web whenever and wherever needed, and indeed, many people no longer have a landline internet connection. In addition, extreme portability has fueled the development of applications that use georeferencing to enable users to save time and optimize their activities. By now, the smartphone has become a de facto life-logging device, so much so that we cannot even imagine much of what we do every day, whether we're on the move or in a specific place, without one. As Castells predicted, this has shaken loose our traditional space-time references, and we are now projected into the 'space of flows' where space and time become digital and relative so that different planes of cognitive, emotional and physical experience can blend and overlap (Newlin, 2016). Likewise, formal and informal learning processes, once firmly anchored in specific times and places, are fast becoming 'anytime-anywhere learning' (Cross et al., 2019). Little wonder, then, if over 80 per cent of students state that the feature they like most about digital learning technologies, and those using smartphones in particular, is mobility (McGraw-Hill Education, 2016). Whether this attitude benefits their academic performance, however, is still an open question, given the enormous number of variables that may be involved: the overlap between formal and informal settings, for example (Wrigglesworth and Harvor, 2017), the interplay between physical and digital learning spaces (Chai et al., 2016), teachers' choice of pedagogical methods and their attitudes towards this technology. Studies in these areas have yielded interesting, albeit often contrasting, findings.

Smartphones and Cognitive Functions

Smartphones and Memory: Digital Amnesia or Augmented Mind?

'Digital Amnesia' is a recently coined expression which describes 'the experience of forgetting information you trust a digital device to store and remember for you'. Also known as the 'Google effect' (Sparrow et al., 2011), the phenomenon has been addressed in a number of studies, the best known of which is by Kaspersky Lab (2016), whose survey gathered information from 6,000 people, aged between 16 and 65, split equally between male and female across the UK, France, Germany, Italy, Spain and the Netherlands. The results are fascinating: 79 per cent of respondents stated that they use digital devices, and smartphones in particular, to access the information they need more than ever before, 32 per cent admitted that their devices are like an extension of their brain, and only 21 per cent reported that they relied on their memories alone to remember information. Respondents justified their reliance on digital devices by noting that they felt safe in the knowledge that they could easily find information online when and where it's needed. This could to some extent explain what happens when we do not have our cell phones close on hand.

In reality, this process of delegating our memory to outside aids or other people has always existed, even before there were special technologies for the purpose: it is called 'transactive memory', where our own memory is combined with that of someone else who knows the information we need, or even with physical sources, such as books or notes. The internet can be regarded as a sort of transactive memory, both because it enables us to ask other people for information quickly, and because we can consult digital repositories like Wikipedia. However, there is a major difference between the memory we can access with internet-enabled devices and the traditional forms stored in books or provided by other people: in the latter case, response times can be quite long, whereas the Web is much faster and more efficient. As a result, the perceived savings in terms of mental effort are considerable. Thus, around 60 per cent of respondents state that they use external memories so that they can concentrate on more important things. Oddly enough, over 50 per cent of respondents under 35 years of age worry about their reliance on their smartphones to remember things, while only 35 per cent of the over-35s have the same fears. In addition, although all respondents acknowledge the importance of the information stored on their smartphones or other devices, less than 30 per cent regularly backup their precious data.

Effects of Digital Amnesia on Cognitive Processes

Recent research has shed light on how memorization is directly affected by the perception that the information we need is easily available: people tend to forget items they think are readily accessible outside their own minds, e.g., by using online search engines via smartphones or other devices, such as Alexa, Echo or other smart speakers—or is stored on the devices they use every day. Unlike traditional transactive memory, having instantaneous access to the Web means we need to know who to ask or where to go to find the information we need. This creates a sort of 'supernormal stimulus' (Ward, 2013) for transactive memory, making all the other traditional, 'memory partners' less attractive because they are seen as inefficient. Searching for information online becomes a habit, and then dependence, when we rely on it to look for solutions to any and all problems.

The perception of the Web's immediacy and effectiveness appears to have interesting effects on our confidence in our own intelligence and memory, and hence on increasing our cognitive self-esteem (Ward, 2013) to the point where we may think of tools, such as search engines as part of our own daily cognitive activity. Essentially, cognitive self-esteem is based on the positive (or negative) feedback we receive when we must complete a task: thus, succeeding in finding information in an online search increases cognitive self-esteem. But this perception of effectiveness is illusory because there is often a persistent overestimation of information problem-solving ability and performance in recalling information that has been found online on an earlier occasion (Dong and Potenza, 2015).

These aspects are very important because over-confidence and overestimating our ability to find information can be problematic, given that they impede memory retrieval by limiting the functional connectivity and synchronization of associated brain regions (Liu et al., 2018), leading to premature and non-optimal termination of information problem-solving processes (Pieschl, 2019). Summing up, when we have an urgent need for information, we:

1. Turn to the Web to find the information we need, using our smartphone by preference, and
2. If we know that the information is available online, we do not try to memorize it. By contrast, if the information is not readily found online, we put more effort into memorizing it.

The problem here is that when we know that information is available online—because we have already accessed it—we know where it is stored but not its exact content (Heersmink and Sutton, 2020). In this sense, digital amnesia can be defined as the tendency to forget information that we know is stored on a digital device of any kind or, obviously, online. An important side effect is that we may no longer exercise our long-term memory, thus weakening it (Wimber et al., 2015), even though 'forgetting' as a cognitive process is generally regarded as a feature of our brains, rather than a bug, as it has an important function in freeing up working memory resources (Popov et al., 2019). Marsh and Rajaram (2019) have drawn up a list of the internet's properties that have consequences for memory and cognition:

- unlimited scope;
- inaccurate content;
- rapidly changing content;
- many distractions and choices;
- very accessible;
- requires search;
- fast results;
- the ability to author;
- source information is obscured;
- many connections to others.

Some of these properties are particularly interesting because of their influence on meta-cognitive processes: property 4, for instance (many distractions and choices), indicates that how information is presented on Web pages is critical to attention: distractions, such as advertisements and videos surrounding information, have a negative impact on one's understanding. Similarly, having to decide whether or not to share information on social media can lead to cognitive overload, and the processing and comprehension of the information suffers as a result (Jiang et al., 2016).

Effects of Smartphone Screen Size

Even how we read has a major influence: reading on a screen rather than on a printed page appears to have a significant effect on reading-comprehension performance, not only because the screen emits light and thus causes visual fatigue, but also because digital text often has many hypertext elements which, by inviting the reader to follow links, distract attention from the main focus (Mangen et al., 2013). Furthermore, it seems that information displayed on smaller screens induces lower rates of attention and arousal while, conversely, larger screens facilitate greater attention and cognitive access to content (Dunaway and Soroka, 2019). This is an important factor, especially in mobile learning performance: for example, students with the largest screen scored higher and showed greater satisfaction (Park et al., 2018). However, in some cases, using small screen-size smartphones can result in a lower cognitive load, but this depends a lot on other variables, such as the density of the content and the number of interactions required (Alasmari, 2020) (*see* Fig. 1).

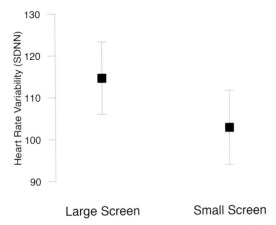

Fig. 1. The arousal level using large and small smartphone screens captured by heart rate variability (HRV) (Dunaway and Soroka, 2019).

Autobiographical Memory

As our memories fade and mutate with time, 'digital memories are unchanging' (Bell and Gemmell, 2009) as we store them in our smartphone: in fact, another interesting aspect concerns our autobiographical memory for people, places, and things. For instance, it has been found that taking smartphone photos of objects in a museum makes them less 'memorable' than if we had only observed them (Henkel, 2014); by now, taking photos and videos with our smartphone is something we take for granted as we try to fix events in our memory that we think we would otherwise be sure to forget (Finley et al., 2018). And posting these photos and videos on social media provides us with a lasting way of sharing our memories, but it also reduces out ability to remember events (Marsh and Rajaram, 2019).

Smartphone Addiction

In Search of a Definition for Smartphone Addiction

'Smartphone addiction' began to attract attention a decade or so ago, and the issue is still widely discussed in the media. By now, it has become a worldwide problem among all cultures, and seems to be most prevalent among young people. Some researchers estimate that between 10–20 per cent of school-age children are affected by it to some extent (Billieux et al., 2015). The concept of 'addiction' to a specific technology is by no means new, and generally refers to excessive use of a particular device, such as a videogame console (Griffiths et al., 2015) or Internet-enabled PC (Weinstein and Lejoyeux, 2010). Though there is no consensus on how 'smartphone addiction' or 'problematic smartphone use' (PSU) (Kuss et al., 2018) should be defined, most researchers in the field tend to regard it as a behavioral addiction which results in psychological dependence and may also be associated with a series of physical symptoms that affect health, including anxiety, depression, nausea, and sleep disorders (Gámez-Guadix, 2014).

The American Society of Addiction Medicine (2011) defines addiction in general terms as a 'chronic disease of brain reward, motivation, memory, and related circuitry', while the APA Diagnostic Classification (American Psychiatric Association, 2013) also lists a series of symptoms that are common to all behavioral addictions (Table 1).

In general, the signs that indicate smartphone dependence include loss of sense of time, difficulty in completing tasks involving work or family commitments, isolation from family and friends, euphoria when online and anxiety and depression when offline, having more than one smartphone or phone numbers, using and/or keeping the smartphone on during the night, and the phantom vibration (or phantom ringing) syndrome caused by continual tension about missing a call or a text. How similar this syndrome is to problems of substance dependence is demonstrated by the fact that researchers have often attempted to predict problematic smartphone use by means of the 'big five personality traits', a model which is also employed to understand vulnerability to substance addiction. The model is based on five dimensions of personality (extraversion, openness to experience, conscientiousness, agreeableness, and emotional instability) (Cocoradă et al., 2018): it appears that these personality traits predict smartphone addiction risk, especially for people with higher neuroticism, lower conscientiousness and lower openness.

The risk of smartphone addiction seems to drop with increasing age and is at its highest for 14–20-year olds (De-Sola Gutiérrez et al., 2016). This suggests that this age group can be particularly vulnerable to the negative effects of cell-phone use due to risks of childhood onset mental disorders characterized by a disturbance in impulse control and cognitive flexibility.

Table 1. Symptoms associated with addiction, based on DSM-5 (American Psychiatric Association, 2013).

Symptoms	
Tolerance building	More and more is needed to fulfill a person's needs
Withdrawal	When an action cannot be performed, anxiety or unpleasant feelings arise
Loss of control	Behavior is not under control
Preoccupation with the addiction	Social activities or work are planned around the addiction
Time planning	Recovering from addiction is controlling life

It should be pointed out that not everyone develops all of these symptoms or develops them at the same time. Smartphones are a special case, given that their extreme portability and the fact they have become indispensable in our daily lives contribute to aggravating dependence and its symptoms. Indeed, smartphone addiction is very similar to Internet addiction but is exacerbated by the device's availability, which was limited when the Web could only be accessed from PCs—necessarily located in a single fixed position—or from laptops, which, though portable, are always relatively unwieldy (Carbonell et al., 2013; Jeong et al., 2020).

We can distinguish between smartphone addiction and Internet addiction *per se* on the basis of how online activities (social networks, for example) or those that can be carried out on the device itself (such as playing games or viewing photos and videos) are accessed: according to this interpretation, the smartphone is a powerful tool that makes it possible to satisfy a need (Panova and Carbonell, 2018). As we have seen, the smartphone has been a major attractor for other technologies, combining all the affordances of video and still cameras, gaming consoles, and Internet-enabled PCs. It has thus been a means of satisfying a range of needs associated with sociality, entertainment, and work. In addition, the smartphone is a highly personal device that can inspire a level of emotional attachment that is not found with tablets and PCs (Brasel and Gips, 2014); similarly, it is seen as having positive connotations that can relieve everyday stress (Cho et al., 2017).

Excessive smartphone use can have negative collateral effects on students' behavior during lessons, for example, or when driving, or in social activities that require us to focus attention on other people (Groarke, 2014). Addictive behaviors can be triggered by internal stimuli, such as the urge to check social media for messages or external stimuli, such as ringtones or notification sounds. What drives these behaviors is the gratification resulting from the release of pleasurable dopamine and endorphins (Montag et al., 2017). But if the smartphone is not at hand for some reason, compulsive users can show signs of restlessness and anxiety that can develop into a full-scale syndrome, known as 'nomophobia', or no mobile phone phobia (Yildirim and Correia, 2015). Nomophobia generally involves four dimensions: not being able to communicate, losing connectedness, not being able to access information, and giving up convenience (Bragazzi et al., 2019).

Causes of Smartphone Addiction

Smartphone use can be categorized as involving either *social* or *process* use:

1. Social use centers chiefly on applications linked to social contacts (social networking, text messaging, voice calls, etc.), whereas
2. Process use includes all other activities that are not directly connected with social relationships (browsing the Web, viewing videos, gaming, etc.).

Both types of use can result in addictive behavior. An interesting approach, known as Compensatory Internet Use Theory or CIUT (Kardefelt-Winther, 2014) assumes that inability to deal successfully with stressful events leads people to compensate through high levels of engagement with Internet-connected devices. Though this mechanism is not in itself unhealthy, it can be for certain sensitive people who already suffer from emotional disorders (Elhai et al., 2019). For example, it appears that people who internalize their emotions tend to use more of the process functionalities of their smartphone out of social anxiety and/or poor social skills (Rozgonjuk and Elhai, 2019). Other studies suggest that Internet and smartphone addiction in younger people is chiefly associated with factors linked to the family and social environments, such as school. Poor family functioning, for example, with conflict and lack of communication is a significant predictor of this kind of addiction in adolescents (Li et al., 2018).

The school environment in particular is not only important to social and relational as well as cognitive development, but is also where the risk of Internet and smartphone addiction is especially high as a result of the negative effects of peer relationships (Jeong et al., 2020): these effects can be critical, as the need to belong has been found to be highly correlated with smartphone addiction (Wang et al., 2017), and low levels of friendship quality can increase the likelihood of excessive smartphone use (Kim et al., 2019). As regards gender differences, some studies suggest that smartphone addiction is more prevalent among women (Chen et al., 2015) because they are more inclined to use their phones as a means of social contact, whereas men's use is concentrated on gaming applications and viewing videos. From the standpoint of economic status and socio-cultural level, it seems that higher levels of parents' education are associated with fewer risks for their children, while low socio-economic standing is often associated with children's problematic smartphone use (Leung, 2008). Thus, many factors, both social and psychological, can lead to smartphone addiction. As all individuals are unique and move in their own unique public and private circles, it is difficult to determine which factors play a leading role in each case.

Measuring Smartphone Addiction

Measuring smartphone addiction is no simple matter: to do so, as completely as possible, attempts have been made to tap both quantitative data, such as usage time or the number of texts, and qualitative data on the various physical, psychological or psychosomatic symptoms people can show.

A number of instruments have been developed over the years, the best-known being the SAS—smartphone addiction scale (Kwon et al., 2013), the SPAI—smartphone addiction inventory (Lin et al., 2014), and the SPAQ—smartphone addiction questionnaire (Al-Barashdi et al., 2014). Each has limitations associated with the sample on which it was tested (e.g., university or high school students). The SAS is probably the most efficient scale as it was developed specifically to gauge smartphone addiction as opposed to Internet addiction. It consists of six subscales with a total of 48 items examining daily-life disturbance, positive anticipation, withdrawal, cyberspace-oriented relationship, overuse, and tolerance. There is also a short version, the SAS-SV, which is more suitable for adolescents (Kwon et al., 2013) and has been adapted in a number of languages and cultural contexts. It has 10 items on a six-point Likert scale from 1 'strongly disagree' to 6 'strongly agree'. Scores range from 10 to 60, with the highest score indicating the highest level of smartphone addiction (Table 2).

Table 2. The items on the SAS—Smartphone Addiction Scale (Kwon et al., 2013).

SAS-SV Scale items, 1 'strongly disagree' to 6 'strongly agree'
1. Missing planned work due to smartphone use.
2. Having hard time in concentrating in class, while doing assignments, or while working due to smartphone use.
3. Feeling pain in the wrists or at the back of the neck while using a smartphone.
4. Cannot tolerate not having a smartphone.
5. Feeling impatient and fretful when I am not holding my smartphone.
6. Having my smartphone in mind even when I am not using it.
7. I will never give up using my smartphone even when my daily life is already affected.
8. Constantly checking my smartphone so as not to miss conversations between other people on Twitter or Facebook.
9. Using my smartphone longer than I had intended.
10. The people around me tell me that I use my smartphone excessively.

The SAS-short version has been widely used because of its ease of administration even among the youngest age group. On the other hand, as it consists of a series of questions that require respondents to think about how they use their smartphones, their responses reflect their perceptions and can thus be highly over- or underestimated. More precise analysis would thus require direct field data, e.g., from a classroom teacher or an observer (Flores et al., 2017). Better yet, the applications that have long existed for monitoring smartphone usage could provide objective data about how, when or where the device is employed and the interactions with it can be compared with self-reported measures. This combined approach could prove highly effective, given that recent studies suggest that people's perceptions about how, and how often, they use their smartphones frequently do not match objectively measured data (Wilcockson et al., 2018). One interesting proposal involves using apps that monitor all interactions that a smartphone user performs by physically touching the screen: an approach that could provide a better understanding of whether app design can contribute to problematic smartphone use, as it would appear to do (Noë et al., 2019). However, collecting personal data from a device, that has become essential in many of our daily activities, raises a number of ethical and privacy issues (Elhai et al., 2018).

The Importance of Self-regulation

What strategies can be used to prevent or cure smartphone addiction? Many approaches have been proposed for treating those who are affected by it and these range from the cognitive behavioral approach to mindfulness behavioral cognitive treatment (Kim, 2013). All are based on the need for individuals under treatment to recognize that they have a problem: often, people who are addicted to something, are unaware of the fact or deny the problem, and thus do not seek help. The next step suggested by these approaches is to implement parental control strategies, especially in the case of children or adolescents, by limiting the time spent on the smartphone while encouraging recreational and outdoor activities and spending time with friends. However, coercive control can backfire, particularly with teenagers (Lee and Ogbolu, 2018), and it is better to take action to build the individual's capacity for self-control. From the psychological standpoint, in fact, it seems that self-control—or rather, self-regulation—is a decisive factor in dealing successfully with emotional impulsiveness and activating metacognitive reflection (Cho et al., 2017), for adolescents, in particular (De Ridder and Lensvelt-Mulders, 2018).

Self-regulation can be defined as the set of self-corrected adjustments originating within the person and taking place as needed to pursue one's current purpose (Carver and Scheier, 2017), or in other words, the entire psychological process whereby a person strives to attain valued outcomes. Low levels of self-regulation are generally correlated with high levels of addiction, and hence also with smartphone addiction (Kim et al., 2019; Van Deursen et al., 2015). Some studies have gone so far as to argue that the real problem is lack of self-regulation itself, given that it positively predicts academic performance (Duckworth et al., 2019), and that the smartphone is merely a temptation that puts self-regulation to the test. Consequently, high levels of in-class smartphone use are only a sign of low self-control and not in a cause of lower academic performance (Bjerre-Nielsen et al., 2020). Improving our capacity for self-regulation calls for a deliberate process of control over our behavior effectively; for example, by delaying immediate gratification. Other more concrete measures include reducing the number of apps on our phone or installing apps that limit access to social media to certain time slots.

Strategies for Preventing Smartphone Addiction

Recent cognitive studies suggest that people with lower working memory capacity are more likely to rely on automatic processes that can activate addictive behavior. This is because working memory aids reflective capacity and limits the execution of automatic processes in response to stimuli (Hofmann, 2008). Measures to make people reflect are undoubtedly useful and may be effective if the individual acknowledges the problem and addresses it. All too often, however, families pay little attention to particularly addictive behavior, such as excessive smartphone use, and the problem then comes to a head in other settings, such as schools or universities. These formal contexts may be the stage for students' problematic smartphone use, but they can also provide an opportunity for programs involving teachers in activities designed to improve self-awareness and self-regulation (Chun, 2018). In some countries, the school and university systems have sought to limit the damage caused by smartphone addiction with 'digital detox programs' (Buctot et al., 2018; Schmuck, 2020), some of which have also used mindfulness-based cognitive-behavioral therapy (Lan et al., 2018). Though experimental results have been encouraging, the problem remains that these approaches can only be applied with a necessarily limited number of students. It should also be borne in mind that smartphone addition is not restricted to students, but can also affect teachers (Ruiz-Palmero et al., 2019).

In many educational settings, programs have been introduced to raise students' awareness of this problem, while others have enforced restrictions that require that smartphones be turned off or left outside the classroom. The latter measures can work for a time, but they do not solve the problem when the student is not at school. Indeed, they can even make it worse by triggering nomophobia—the fear of being without a phone—and the resulting separation anxiety can further interfere with the concentration needed for classroom learning and impair high-order cognitive processes (Hartanto and Yang, 2016). Nevertheless, it is interesting to note that some studies have found that banning smartphones at school improved student performance by over 6 per cent (Beland and Murphy, 2015). Even though these findings refer to performance on traditional standardized tests, which account for only a portion of the assessment methods that can be used, the percentage is significant. More extensive meta-analyses carried out on over 500 undergraduates to determine whether there is a correlation between smartphone use and college grade-point average found a negative relationship, i.e., lower grades with increased smartphone use (Lepp et al., 2015). It should be noted that the earlier studies cited were based on students' self-reported measures of smartphone usage. However, more accurate experimental studies confirm that every hour spent per day using smartphones outside of class reduces students' performance by 3.8 per cent, while for each hour of use during class time this reduction reaches 12 per cent (Felisoni and Godoi, 2018). This means that in four average hours of smartphone use, the drop is around 50 per cent (Rafique et al., 2020) and in the worst case of more than eight hours, a stunning 80 per cent (McCrann et al., 2021).

Other strategies focus on controlling stimuli: studies on thought suppression, or in other words on the ability not to think about external stimuli, and in our case not to think about the gratification afforded by using a smartphone, have found that simply trying to suppress thoughts is ineffective and counterproductive, as it can produce a post-suppression rebound effect leading to even more addictive behavior (Wenzlaff and Wegner, 2000). This seems to be because people are unable to focus sufficiently to distract themselves from a topic they feel is important or interesting. It is thus essential that educators use teaching strategies that succeed in engaging students and prevent boredom, as bored students are likely to reach for their smartphone. Students can also be encouraged to use their smartphones as part of their classwork rather than as a distraction. In this connection, good student-teacher relationships and peer support can also be invaluable in dealing with smartphone addiction in the classroom (Díaz-Aguado et al., 2018; Konan et al., 2018). In any case, the literature does not offer a one-size-fits-all solution, as each school or university setting is unique and calls for its own carefully-modulated approach.

Fighting Fire with Fire: Apps for Smartphone Addiction?

As smartphone addition has been recognized as a problem, many downloadable apps have been developed which claim to help in tackling the syndrome. But proposing technological solutions to problems caused by technology is a paradox that reveals the extent to which such approaches never call smartphone use into question, but merely seek to limit it in various ways. Interestingly, each app proposes different solutions that reflect a particular interpretation of the problem and are based on different assumptions, norms, and discourses present in—and created by—these apps in regard to social, cultural, and political dimensions (Lupton, 2014). One of the most popular of these apps is BreakFree, which measures an addiction score based on interaction time with the smartphone. In addition to making use of gamification—the user can earn rewards by maintaining a low addition score—the app makes it possible to block notifications, disable the Internet connection at scheduled times, and reject incoming phone calls. Another app, Offtime, provides all the functionalities of BreakFree and in addition, memorizes a list of activities that were 'missed' while offline to prevent FOMO, or 'fear of missing out'. A third popular app, Forest, also makes use of gamification, as the user can earn credits during smartphone abstinence sessions that can be used to plant and grow trees in a virtual forest, or even to have reforestation programs in many parts of the world to plant real trees.

It should be pointed out, however, that these and other similar apps limit themselves to providing 'policing' functions and statistics on when and how the smartphone is used, begging the question of how or why usage should be reduced (Dosselaar, 2017). As a result, they thus frame smartphone addiction over-simplistically. In addition, these apps do not offer much context about how the smartphone is used. It is one thing to surf social media during a long bus ride, but quite another to do so during lectures or in bed in the dead of night.

Towards a 'Community of Attention' for Students

The challenge, then, is to pass from teaching based on curricular knowledge to teaching based on social construction of skills that can take place, both in social interactions of everyday life, and through participation in communities of practice, whether in person or virtually. The school can then become a venue for discussing and reflecting on real-world experiences, linking them to the theories and models in the curriculum. This continuing process of exploration and reflection would bridge the gap that Dewey noted between the two settings a century ago. A crucial factor in this effort will be the ability to turn the class into a true 'community of attention', trying to help students to be more attentive to one another (Lang, 2020). In this community of attention, the smartphone can help students focus on the subject matter: no longer a source of distraction, but an important aid to concentration in all of life's settings. For all this to take place, teachers must share a consensus about the opportunities that mobile learning holds, and students must realize that the smartphone can be an important support for life-long-learning, and not just another part of their everyday life and recreation.

References

Al-Barashdi, H.S., Bouazza, A. and Zubaidi, A.A. (2014). Psychometric properties of smartphone addiction questionnaire (SPAQ) among Sultaon Qaboos University Undergraduate students. *Journal of Educational and Psychological Studies*, 8(4): 637–644.

Al-Turjman, F. (2019). 5G-enabled devices and smart-spaces in social-IoT: An overview. *Future Generation Computer Systems*, 92: 732–744.

Alasmari, T. (2020). The effect of screen size on students' cognitive load in mobile learning. *Journal of Education, Teaching and Learning*, 5(2): 280–295.

American Psychiatric Association. (2013). *Diagnostic and Statistical Manual of Mental Disorders (DSM-5®)*. American Psychiatric Pub.

American Society of Addiction Medicine. (2011). Public policy statement: definition of addiction short. *American Society of Addiction Medicine*, 1–8.

Beland, L. and Murphy, R. (2015). *III Communication: Technology, Distraction & Student Performance, Discussion Paper N. 1350*. Centre for Economic Performance Education and Skills Programme. http://cep.lse.ac.uk/pubs/download/dp1350.pdf.

Bell, G. and Gemmell, J. (2009). *Total Recall: How the E-Memory Revolution will Change Everything*. Dutton, New York.

Billieux, J., Schimmenti, A., Khazaal, Y., Maurage, P. and Heeren, A. (2015). Are we overpathologizing everyday life? A tenable blueprint for behavioral addiction research. *Journal of Behavioral Addictions*, 4(3): 119–123.

Bjerre-Nielsen, A., Andersen, A., Minor, K. and Lassen, D.D. (2020). The negative effect of smartphone use on academic performance may be overestimated: evidence from a 2-year panel study. *Psychological Science*, 31(11): 1351–1362.

Bragazzi, N.L., Re, T.S. and Zerbetto, R. (2019). The relationship between nomophobia and maladaptive coping styles in a sample of italian young adults: insights and implications from a cross-sectional study. *JMIR Mental Health*, 64: e13154.

Brasel, S.A. and Gips, J. (2014). Tablets, touchscreens, and touchpads: how varying touch interfaces trigger psychological ownership and endowment. *Journal of Consumer Psychology*, 24(2): 22633.

Broadband Search. (2020). *Mobile vs. Desktop Internet Usage*, Retrieved Sept. 2020. https://www.broadbandsearch.net/blog/mobile-desktop-internet-usage-statistics.

Bruner, J. (1996). *The Culture of Education*. Harvard University Press, Cambridge, MA.

Buctot, D.B., Kim, N. and Park, K.E. (2018). Development and evaluation of smartphone detox program for university students. *International Journal of Contents*, 14(4).

Carbonell, X., Oberst, U. and Beranuy, M. (2013). The cell phone in the twenty-first century: a risk for addiction or a necessary tool? *Principles of Addiction*, 1: 901–909.

Carver, C.S. and Scheier, M.F. (2016). Self-regulation of action and affect. *In*: Baumeister, R.F. and Vohs, K.D. (eds.). *Handbook of Self-regulation: Research, Theory, and Applications*. Guilford Press, New York.

Chai, C.S., Wong, L.H. and King, R.B. (2016). Surveying and modeling students' motivation and learning strategies for mobile-assisted seamless Chinese language learning. *Educational Technology and Society*, 19(3): 170–180.

Chen, L., Chen, T.L. and Chen, N.S. (2015). Students' perspectives of using co-operative learning in a flipped statistics classroom. *Australasian Journal of Educational Technology*, 31(6).

Cho, H.Y., Kim, D.J. and Park, J.W. (2017). Stress and adult smartphone addiction: mediation by self-control, neuroticism, and extraversion. *Stress and Health*, 33(5): 624–630.

Chun, J. (2018). Conceptualizing effective interventions for smartphone addiction among korean female adolescents. *Children and Youth Services Review*, 84: 35–39.

Churchill, D., Fox, B. and King, M. (2016). Framework for designing mobile learning environments. pp. 3–26. *In*: Churchill, D., Lu, J., Chiu, T.K.F. and Fox, B. (eds.). *Mobile Learning Design: Theories and Application*. New York: Springer.

Cocoradă, E., Maican, C.I., Cazan, A.M. and Maican, M.A. (2018). Assessing the smartphone addiction risk and its associations with personality traits among adolescents. *Children and Youth Services Review*, 93: 345–354.

Cornet, V.P. and Holden, R.J. (2018). Systematic review of smartphone-based passive sensing for health and wellbeing. *Journal of Biomedical Informatics*, 77: 120–132.

Cross, S., Sharples, M., Healing, G. and Ellis, J. (2019). Distance learners' use of handheld technologies. *The International Review of Research in Open and Distributed Learning*, 20(2).

DeRidder, D.T., Lensvelt-Mulders, G., Finkenauer, C., Stok, F.M. and Baumeister, R.F. (2018). Taking stock of self-control: a meta-analysis of how trait self-control relates to a wide range of behaviors. *Personality and Social Psychology Review*, 16(1): 76–99.

De-Sola Gutiérrez J., Rodríguez de Fonseca, F. and Rubio, G. (2016). Cell phone addiction: a review. *Frontiers in Psychiatry*, 7: 175.

Díaz-Aguado, M., Martín-Babarro, J. and Falcón, L. (2018). Problematic internet use, maladaptive future time perspective and school context. *Psicothema*, 302: 195–200.

Dong, G. and Potenza, M.N. (2015). Behavioral and brain responses related to internet search and memory. *European Journal of Neuroscience*, 42: 2546–2554.

Dosselaar, C.V. (2017). Smartphone Addiction? There's an App for That! How 'Smartphone Addiction Apps' Frame Smartphone Addiction through Discourse and Affordances, Master Thesis, Utrecht University.

Duckworth, A.L., Taxer, J.L., Eskreis-Winkler, L., Galla, B.M. and Gross, J.J. (2019). Self-control and academic achievement. *Annual Review of Psychology*, 70: 373–399.

Dunaway, J. and Soroka, S. (2019). Smartphone-size screens constrain cognitive access to video news stories. *Information, Communication & Society*, 1–16.

Elhai, J.D., Tiamiyu, M.F., Weeks, J.W., Levine, J.C., Picard, K.J. and Hall, B.J. (2018). Depression and emotion regulation predict objective smartphone use measured over one week. *Personality and Individual Differences*, 133: 21–28.

Elhai, J.D., Levine, J.C. and Hall, B. (2019). The relationship between anxiety symptom severity and problematic smartphone use: a review of the literature and conceptual frameworks. *Journal of Anxiety Disorders*, 62: 45–52.

Engeström, Y. (2001). Expansive learning at work: toward an activity theoretical reconceptualization. *Journal of Education and Work*, 14(1): 133–156.

Felisoni, D.D. and Godoi, A.S. (2018). Cell phone usage and academic performance: an experiment. *Computers & Education*, 117: 175–187.

Finley, J.R., Naaz, F. and Goh, F.W. (2018). Results: behaviors and experiences with internal and external memory. pp. 25–48. *In*: Finley, J.R., Naaz, F. and Goh, F.W. (eds.). *Memory and Technology: How We Use Information in the Brain and the World*. Springer.

Firmin, M.W. and Genesi, D.J. (2013). History and implementation of classroom technology. *Procedia -Social and Behavioral Sciences*, 93: 1603–1617.

Flores, P.Q., Flores, A. and Ramos, A. (2017). The smartphone in the context of the classroom in the primary school and in the higher education. *In*: The *Proceedings of EDULEARN17 Conference*. 3rd–5th July 2017, Barcelona, Spain.

Gámez-Guadix, M. (2014). Depressive symptoms and problematic internet use among adolescents: analysis of the longitudinal relationships from the cognitive-behavioral model. *Cyberpsychology, Behavior, and Social Networking*, 17(11): 714–719.

Griffiths, M.D., Király, O., Pontes, H.M. and Demetrovics, Z. (2015). An overview of problematic gaming. *Mental Health in the Digital Age: Grave Dangers, Great Promise*, 27–45.

Groarke, H. (2014). *The Impact of Smartphones on Social Behavior and Relationships*. Ph.D thesis, Dublin Business School.

Hartanto, A. and Yang, H. (2016). Is the smartphone a smart choice? The effect of smartphone separation on executive functions. *Computers in Human Behavior*, 64: 329–336.

Heersmink, R. and Sutton, J. (2020). Cognition and the web: extended, transactive, or scaffolded? *Erkenntnis*, 85(1): 139–164.

Henkel, L.A. (2014). Point-and-shoot memories: the influence of taking photos on memory for a museum tour. *Psychological Science*, 25(2): 396–402.

Hofmann, W., Gschwendner, T., Friese, M., Wiers, R.W. and Schmitt, M. (2008). Working memory capacity and self-regulatory behavior: toward an individual differences perspective on behavior determination by automatic versus controlled processes. *Journal of Personality and Social Psychology*, 954: 962–977.

Hyysalo, S. (2010). *Health Technology Development and Use*. Routledge, New York.

Jeong, Y., Suh, B. and Gweon, G. (2020). Is smartphone addiction different from internet addiction? Comparison of addiction-risk factors among adolescents. *Behavior Information Technology*, 39(5): 578–593.

Jiang, T., Hou, Y. and Wang, Q. (2016). Does micro-blogging make us 'shallow'? Sharing information online interferes with information comprehension. *Computers in Human Behavior*, 59: 210–214.

Joo, Y.J., Park, S. and Lim, E. (2018). Factors influencing preservice teachers' intention to use technology: TPACK, teacher self-efficacy, and technology acceptance model. *Journal of Educational Technology & Society*, 21(3): 48–59.

Karanasios, S. (2018). Toward a unified view of technology and activity. *Information Technology & People*.

Kardefelt-Winther, D. (2014). A conceptual and methodological critique of internet addiction research: towards a model of compensatory internet use. *Computers in Human Behavior*, 31: 351–54.

Kaspersky Lab. (2016). *From Digital Amnesia to the Augmented Mindat work, the Risks and Rewards of Forgetting in Business*. https://media.kaspersky.com/pdf/Kaspersky-Digital-Amnesia-Evolution-report-17-08-16.pdf.

Kim, H. (2013). Exercise rehabilitation for smartphone addiction. *Journal of Exercise Rehabilitation*, 9(6): 500–505.

Kim, J.H. and Park, H. (2019). Effects of smartphone-based mobile learning in nursing education: a systematic review and meta-analysis. *Asian Nursing Research*, 13(1): 20–29.

Konan, N., Durmuş, E., Ağiroğlu Bakir, A. and Türkoğlu, D. (2018). The relationship between smartphone addiction and perceived social support of university students. *International Online Journal of Educational Sciences*, 10(5).

Kuss, D.J., Kanjo, E., Crook-Rumsey, M., Kibowski, F., Wang, G.Y. and Sumich, A. (2018). Problematic mobile phone use and addiction across generations: the roles of psychopathological symptoms and smartphone use. *Journal of Technology in Behavioral Science*.

Kwon, M., Kim, D.J., Cho, H. and Yang, S. (2013). The smartphone addiction scale: development and validation of a short version for adolescents. *PloS One*, 8(12): e56936.

Land-Zandstra, A.M., Devilee, J.L., Snik, F., Buurmeijer, F. and van den Broek, J.M. (2016). Citizen science on a smartphone: participants' motivations and learning. *Public Understanding of Science*, 25(1): 45–60.

Lang, J.M. (2020). Distracted: Why students can't focus and what you can do about it. *Basic Books*, New York.

Lee, E.J. and Ogbolu, Y. (2018). Does parental control work with smartphone addiction? A cross-sectional study of children in South Korea. *Journal of Addictions Nursing*, 29(2): 128–138.

Legris, P., Ingham, J. and Collerette, P. (2003). Why do people use information technology? A critical review of the technology acceptance model. *Information & Management*, 40: 191–204.

Lepp, A., Barkley, J.E. and Karpinski, A.C. (2015). The relationship between cell phone use and academic performance in a sample of U.S. college students. *SAGE Open*, 5(1): 1–9.

Leung, L. (2008). Linking psychological attributes to addiction and improper use of the mobile phone among adolescents in Hong Kong. *Journal of Children and Media*, 2: 93–113.

Li, J., Li, D., Jia, J., Li, X., Wang, Y. and Li, Y. (2018). Family functioning and internet addiction among adolescent males and females: a moderated mediation analysis. *Children and Youth Services Review*, 91: 289–297.

Lin, Y.H., Chang, L.R., Lee, Y.H., Tseng, H.W., Kuo, T.B.J. and Chen, S.H. (2014). Development and validation of the smartphone addiction inventory (SPAI). *PloS One*, 9(6): e98312.

Liu, X., Lin, X., Zheng, M., Hu, Y., Wang, Y., Wang, L. and Dong, G. (2018). Internet search alters intra- and inter-regional synchronization in the temporal gyrus. *Frontiers in Psychology*, 9: 260.

Lupton, D. (2014). Apps as artefacts: towards a critical perspective on mobile health and medical apps. *Societies*, 4(4): 606–622.

Luthar, B. and Kropivnik, S. (2011). Class, cultural capital, and the mobile phone. *Czech Sociological Review*, 47(6): 1091–1119.

Mangen, A., Walgermo, B.R. and Brønnick, K. (2013). Reading linear texts on paper versus computer screen: effects on reading comprehension. *International Journal of Educational Research*, 58: 61–68.

Marsh, E.J. and Rajaram, S. (2019). The digital expansion of the mind: implications of internet usage for memory and cognition. *Journal of Applied Research in Memory and Cognition*, 8(1): 1–14.

McCrann, S., Loughman, J., Butler, J.S., Paudel, N. and Flitcroft, D.I. (2021). Smartphone use as a possible risk factor for Myopia. *Clinical and Experimental Optometry*, 104(1): 35–41.

McGraw-Hill Education. (2016). *Digital Study Trends Survey*. Retrieved from http://www.infodocket.com/wp-content/uploads/2016/10/2016-Digital-Trends-Survey-Results1.pdf.

Montag, C., Duke, É. and Reuter, M. (2017). A short summary of neuroscientific findings on internet addiction. pp. 209–218. *In: Internet Addiction*. Springer, Cham.

Newlin, A.B. (2016). *Gazing into the Black Mirror: How the Experience of Emplaced Visuality through Smartphones Fundamentally Changes both the Self and the Place*. Doctoral dissertation, The American University of Paris (France).

Noë, B., Turner, L.D., Linden, D.E., Allen, S.M., Winkens, B. and Whitaker, R.M. (2019). Identifying indicators of smartphone addiction through user-app interaction. *Computers in Human Behavior*, 99: 56–65.

Oleksy, T. and Wnuk, A. (2017). Catch them all and increase your place attachment! The role of location-based augmented reality games in changing people-place relations. *Computers in Human Behavior*, 76: 3–8.

Panova, T. and Carbonell, X. (2018). Is smartphone addiction really an addiction? *Journal of Behavioral Addictions*, 7(2): 252–259.

Park, E., Han, J., Kim, K.J., Cho, Y. and del Pobil, A.P. (2018, January). Effects of screen size in mobile learning over time. pp. 1–5. *In: Proceedings of the 12th International Conference on Ubiquitous Information Management and Communication*.

Pieschl, S. (2019). Will using the internet to answer knowledge questions increase users' overestimation of their own ability or performance? *Media Psychology*, 1–27.

Popov, V., Marevic, I., Rummel, J. and Reder, L.M. (2019). Forgetting is a feature, not a bug: intentionally forgetting some things helps us remember others by freeing up working memory resources. *Psychological Science*, 30(9): 1303–1317.

Rafique, N., Al-Asoom, L.I., Alsunni, A.A., Saudagar, F.N., Almulhim, L. and Alkaltham, G. (2020). Effects of mobile use on subjective sleep quality. *Nature and Science of Sleep*, 12: 357.

Rozgonjuk, D. and Elhai, J.D. (2019). Emotion regulation in relation to smartphone use: process smartphone use mediates the association between expressive suppression and problematic smartphone use. *Current Psychology*, 1–10.

Ruiz-Palmero, J., Sánchez-Rivas, E., Gómez-García, M. and Sánchez Vega, E. (2019). Future teachers' smartphone uses and dependence. *Education Sciences*, 9(3): 194.

Schmuck, D. (2020). Does digital detox work? Exploring the role of digital detox applications for problematic smartphone use and well-being of young adults using multigroup analysis. *Cyberpsychology, Behavior, and Social Networking*, 23(8): 526–532.

Silverstone, R. and Hirsch, E. (1992). *Consuming Technologies*. Routledge, London.

Skågeby, J. (2019). Critical incidents in everyday technology use: exploring digital breakdowns. *Personal and Ubiquitous Computing*, 23(1): 133–144.

Sparrow, B., Liu, J. and Wegner, D. (2011). Google effects on memory: cognitive consequences of having information at our fingertips. *Science*, 333(6043): 776–778.

Tobar-Muñoz, H., Baldiris, S. and Fabregat, R. (2017). Augmented reality game-based learning: enriching students' experience during reading comprehension activities. *J. Educ. Comput. Res.*, 55(7): 901–936.

Ursavaş, Ö.F., Yalçın, Y. and Bakir, E. (2019). The effect of subjective norms on preservice and in-service teachers' behavioural intentions to use technology: a multigroup multimodel study. *British Journal of Educational Technology*, 50(5): 2501–2519.

Van Deursen, A.J.A.M., Bolle, C.L., Hegner, S.M. and Kommers, P.A.M. (2015). Modeling habitual and addictive smartphone behavior. *Computers in Human Behavior*, 45: 411–420.

Wang, P., Zhao, M., Wang, X., Xie, X., Wang, Y. and Lei, L. (2017). Peer relationship and adolescent smartphone addiction: the mediating role of self-esteem and the moderating role of the need to belong. *Journal of Behavioral Addictions*, 6(4): 708–717.

Ward, A.F. (2013). *One with the Cloud: Why People Mistake the Internet's Knowledge for their Own*. doctoral dissertation, Harvard Uuniversity. https://dash.harvard.edu/handle/1/11004901.

Weinstein, A. and Lejoyeux, M. (2010). Internet addiction or excessive internet use. *The American Journal of Drug and Alcohol Abuse*, 36(5): 277–283.

Wenzlaff, R.M. and Wegner, D.M. (2000). Thought suppression. *Annual Review of Psychology*, 51(1): 59–91.

Wilcockson, T.D., Ellis, D.A. and Shaw, H. (2018). Determining typical smartphone usage: what data do we need? *Cyberpsychology, Behavior, and Social Networking*, 21(6): 395–398.

Wimber, M., Alink, A., Charest, I., Kriegeskorte, N. and Anderson, M.C. (2015). Retrieval induces adaptive forgetting of competing memories via cortical pattern suppression. *Nat. Neurosci.*, 18(4): 582–9.

Wrigglesworth, J. and Harvor, F. (2017). Making their own landscape: smartphones and student-designed language learning environments. *Computer Assisted Language Learning*, 31(4): 437–458.

Yildirim, C. and Correia, A.P. (2015). Exploring the dimensions of nomophobia: development and validation of a self-reported questionnaire. *Computers in Human Behavior*, 49: 130–137.

CHAPTER 5

Smartphones In and Outside Classroom

Corrado Petrucco

Introduction

We see many young people finding their smartphones as absolutely essential for their social, recreational, and personal lives at all times of the day and continue to use them even in school or university settings. Initially opposed by teachers and banned from classrooms as dangerous distractions, smartphones are slowly gaining ground in teaching experiments, both in the classroom and to a greater extent outside of it. A promising research stream sees the smartphone as a useful boundary object—in other words, an artifact that fulfills a bridging function between formal and informal learning settings (Akkerman and Bakker, 2011; Burden and Kearney, 2017) and can potentially provide seamless learning experiences, both in and outside the classroom environment (Sharples, 2015; Looi et al., 2016). This means that teachers must be able to recognize the learning opportunities offered outside the formal setting and leverage them through activities in which students use smartphones, e.g., during visits to museums, field trips or when they must solve concrete, everyday problems.

Engeström's (2001) Cultural-Historical Activity Theory (CHAT) is an excellent methodological tool for understanding the role that the smartphone can play as a boundary object between formal and informal learning settings. The Activity Theory model sheds light on the actors and elements that make up an 'activity system'. This model can describe the activities carried out within a complex socio-technical system, and in relation to other activity systems that may be involved (*see* Fig. 2). Six components in an activity system contribute to achieving a specific outcome:

- subject – the individual or collective actor engaged in the activity;
- object – the general goal that gives the entire activity system its overall meaning;

Associate Professor of Education, University of Padua, Italy.
Email: corrado.petrucco@unipd.it

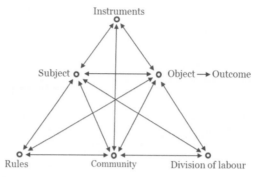

Fig. 2. The interrelations between the components of an activity system according to activity theory (adapted from Engeström, 1987).

- community – the group within which and for which the activity takes place (social, geographical or broader professional community);
- tools – the material and technological artifacts or the language tools used by the actors in the system;
- division of labor – how activities are divided among the subjects and the other actors in the system;
- rules – the implicit and explicit rules, norms and guidelines that regulate the system's activities.

Tensions—referred to as 'contradictions'—can arise between the components of an activity system (Engeström, 2001). If recognized, these contradictions can provide insights into how they can be resolved. In our case, we have a first activity system comprising formal learning environments (schools, universities) and a second activity system made up of informal, external learning environments typical of daily life and of informal and non-formal experiential learning. The example given in Fig. 3 illustrates the process of interaction between the school system and the museum system when students are engaged in a learning activity in which they

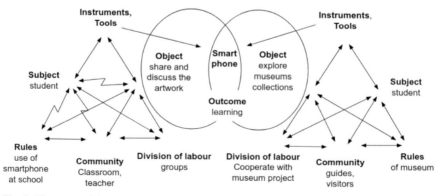

Fig. 3. The two activity systems, school and museum, which share the smartphone as a tool and boundary object.

use their smartphones as a tool for taking pictures of the artwork in a museum and for sharing and discussing art in class. Several contradictions arising from poor use of the smartphone in class are shown in the school activity system.

Smartphones and Learning In and Outside the Classroom

Seamless Learning and Smartphones

One approach to offering a continuous learning experience is encapsulated in the notion of seamless learning (Looi and Wong, 2013; Milrad et al., 2013), which can take place across the many non-formal and informal environments of daily life mediated by smartphone apps. However, it is not easy for teachers to develop programs that can also be pursued outside the classroom, and recontextualize formal learning in an appropriate pedagogical design. This is not only because of the difficulties involved in integration with the constraints of the formal curriculum, but also because seamless learning by its very nature goes beyond the traditional blended learning or flipped classroom approaches, as it is essentially self-directed. Students draw on their own daily experience to make connections with the knowledge and skills gained in formal learning settings. As Garrison suggests (1997, 2011), self-directed learning involves three overlapping dimensions: (1) motivation, (2) self-management, and (3) self-monitoring.

It should be pointed out that the individual student's motivation must be as intrinsic as possible, and thus be accompanied by a good level of engagement, i.e., a willingness to participate actively and co-operate with other students. Self-directed learning is challenging, as its success presupposes that students have sufficient motivation and responsibility (Regan, 2003; Song and Bonk, 2016) without a teacher or other facilitator to provide aid, mediation (Eshach, 2007) or stimulus. An important factor here is the so-called 'transactional distance', or in other words the psychological and physical divide between student and instructor that is inherently part of learning in an out-of-class context (Fuegen, 2012). If the teacher is unable to interact continually and effectively with the student, there may be significant negative repercussions. Very often, students struggle to engage with activities that are not formally linked to the course curriculum (Botero et al., 2019). In addition, learning experiences must be readily transferable from formal to informal settings and vice versa, especially as regards the important cognitive and metacognitive processes that are fundamental to every educational program.

Other issues arise in connection with monitoring students' work with the smartphone. As a tool, it undoubtedly boosts student engagement, but as we shall see, it is also a powerful distraction even for the best-intentioned students. The relationships between self-directed learning, smartphone use, and learning engagement are thus extremely important (Tao et al., 2018), and play a crucial role together with age and the context in which the smartphone is used. A K-12 student will approach mobile seamless learning with intentions and methods that are very different from those of a university undergraduate engaged in the real-world professional practices of engineering, healthcare or the like (Kim and Park, 2019).

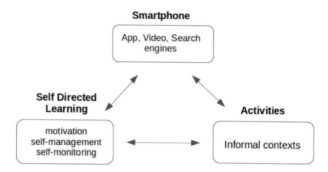

Fig. 4. Interactions between processes and tools in seamless learning activities.

A model summarizing students' typical seamless learning processes is illustrated in Fig. 4. The figure shows the close interrelations between the tool (smartphone), the dimensions of self-directed learning, and the activities carried out in informal contexts.

In this connection, a number of scholars hold that, thanks to mobile devices, the spaces and times of formal learning are no longer particularly important and are increasingly seen by students as irrelevant (Kearney et al., 2019). At the same time, however, these scholars also acknowledge that most formal and informal learning experiences with smartphones in recent years have not been 'disruptive', as mobile pedagogies have essentially replicated traditional classroom-based practices (Kearney et al., 2015).

Towards a Mobile Learning Pedagogy

Using smartphones as a tool in teaching practices calls for making clear choices about which pedagogical models will be adopted. As we mentioned earlier, many teachers struggle to conceive of approaches that differ from traditional practices, and thus replicate methods centering on formal, physical classroom environment. Only a few educators have experimented with student activities in non-formal contexts (excursion sites, museums or the home) or in informal everyday settings (Burden and Kearney, 2017). Precisely for this reason, recent years have seen a call to update teacher education programs to reflect the new pedagogical prospects offered by mobile learning (Herrington et al., 2014), though research has not yet provided feedback about which indicators could be employed to establish best practices for using smartphones in teaching. Of the few studies in this area, most of them focus on individual functions (video, social networks, etc.) or on the impact of mobile learning on student achievement (Crompton and Burke, 2018). Even fewer studies have sought to investigate the type of pedagogy used in mobile learning.

The iPAC scale (Kearney et al., 2019) is an example of a framework that attempts to describe teachers' pedagogical approaches to mobile learning from a socio-cultural dimension, considering the dimensions of personalization, authenticity and collaboration, and their corresponding sub-dimensions of agency and customization; context and task; and conversation and co-creation. Personalization's sub-dimensions of agency and customization enable the student to achieve good levels of autonomy.

Authenticity involves encouraging situated and contextualized learning in order to focus on real-life practices, while collaboration refers to co-creating and sharing digital artefacts and negotiating meanings. Another framework, RASE or resources-activity-support-evaluation (Churchill et al., 2016), takes as its central idea that online resources available via smartphones (multimedia content, apps, learning objects and so forth) are not in themselves sufficient to achieve learning outcomes, but that there must also be activities based on situated, constructivist learning environments, such as problem-based learning. In addition, students must be provided with support through appropriate feedback, and evaluation through continual monitoring. Yet another interesting framework that clarifies the many forms of interaction whereby technologies can be integrated in teaching is TPACK (technological pedagogical and content knowledge), which focuses on the relationship between pedagogical approaches and knowledge about content. The framework includes seven dimensions and components, i.e., seven distinct categories of knowledge essential for the teacher (Table 3).

How the framework can be applied to smartphone use in science teaching is mapped in the example below (*see* Fig. 5), showing the interaction of mobile technology (technology knowledge), inquiry-based learning (pedagogical knowledge) and science concepts (content knowledge) (diagram from Srisawasdi et al., 2018).

Another well-known approach which is often used to integrate technologies in education is the substitution, augmentation, modification, and redefinition (SAMR) framework (Puentedura, 2013). Here, integration is seen as a four-step process where the first step is substitution, which takes place when a technological tool is used to perform a task that could also be accomplished without using technologies. In the next step, augmentation or functional improvements are added to this tool substitution, while modification comes into play when technology allows significant task redesign. In the last step, redefinition, technology is being 'used for learning in a way that could not happen without technology' in new and previously inconceivable tasks

Table 3. TPACK framework of dimensions and components.

Dimensions	
Technology knowledge (TK)	Knowledge about technologies.
Content knowledge (CK)	Knowledge about the subject matter.
Pedagogical knowledge (PK)	Knowledge about teaching methods.
Components	
Pedagogical content knowledge (PCK)	Knowledge that blends both content and pedagogy to develop better teaching practices in the content areas.
Technological content knowledge (TCK)	Knowledge of how technology can present specific content.
Technological pedagogical knowledge (TPK)	Knowledge of how technologies can be used in the teaching process.
Cross Skills	
Technological pedagogical content knowledge (TPCK)	The sum of all the skills needed to integrate technology into teaching in any content area.

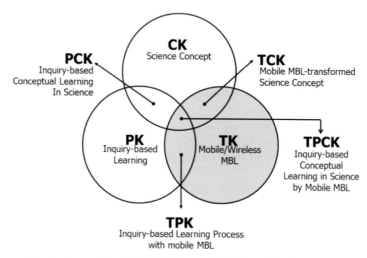

Fig. 5. An example of the TPACK framework (Mishra and Koehler, 2006).

(Crompton and Burke, 2020). SAMR analyses of many mobile learning experiments indicate that few are able to reach the modification and redefinition levels, as most use the smartphone to accomplish tasks at the lower levels of the framework. The approaches to mobile learning rely on mobile technology's affordances but put considerable emphasis on learning design and on situated and authentic learning, as well as on a high degree of teacher involvement in planning and carrying out activities. In this connection, it is important to define how smartphones can be put to use in teaching inside and outside the classroom setting.

Mobile Learning outside the Classroom: A New Learning Theory?

Mobile learning has been something of a buzzword since the early 2000s, but scholarly interest in educational applications increased at the time the first smartphones were introduced around 2010, when lower costs and Internet access made it possible for them to reach a mass market. As we have seen, as smartphones incorporate multiple technologies (camera, video camera, GPS, Internet, TV, etc.), they offer many opportunities for use in mobile learning processes. Though there is no single definition of what mobile learning means, it is clear that the operative word here is 'mobile', as opposed to static: it is learning which takes place 'away from one's normal learning environment', such as the classroom (Sharples, 2002, 2009). Another concept which is often associated with mobile learning is that of contextual learning (Brown et al., 2010) in informal or non-formal contexts. In general, such a context is any place or situation where it is possible to learn and the learner, the relationships, the time, and the place can all be clearly defined (Zimmermann et al., 2007), but which does not have the affordances and the setting typical of formal contexts. Here, it is important to understand whether mobile learning in informal or non-formal contests calls for learning theories that differ from the traditional models

Table 4. Learning theories and mobile learning affordances (adapted from MacCallum and Parsons, 2016).

Learning theories	Mobile affordances
Behaviorism	Portability. Immediate feedback (i.e., quizzes, discussions)
Constructivism	Take pictures, record videos, create new artifacts
Experiential learning	Reflection on experience
Situated cognition	Explore environment, active learning
Communities of practice	Solve problems together, collaborative learning
Connectivism	Share knowledge, networked learning

(MacCallum and Parsons, 2016). Parsons et al. (2016) offer a view of which types of mobile learning activity operationalize which group of learning theories (Table 4).

In view of the smartphone's mobile affordance, situated cognition and situated learning are particularly important theoretical models, given that the latter is defined as learning that takes place in contexts that reflect the way the knowledge will be useful in real life (Liu, 2014). It is clear, however, that any learning theory emphasizes one or more typical elements of the human learning process, and it is difficult to say which theory will best underpin—and to what extent—any given mobile learning activity. In this connection, MacCallum and Parsons give us an example of a radar chart that shows the six learning theories and the degree to which each is involved in a teaching project (*see* Fig. 6). Students were engaged in active learning when discovering aspects about plants and animals living in various habitats in the wood. The theories with the highest degree of involvement are situated cognition and experiential learning.

While the theoretical context is fairly clear, from the practical standpoint the fundamental problem is, as Mayer (2020) suggests: where is the learning in mobile technologies for learning? Or in other words, which instructional features afforded by mobile technology lead to learning? Under what conditions do students learn academic content better with mobile technology than with conventional media? To date, there have been no systematic studies that can give clear answers to these

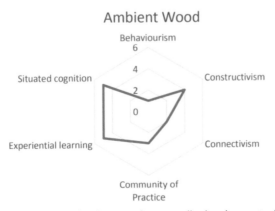

Fig. 6. An example of an educational project to explore a woodland environment using a smartphone.

questions, and above all, help us understand which counts more: the teaching strategy or the tool's affordances. We believe, in any case, that the key may be in the intelligent interaction between the smartphone's specific affordances and the strategies brought into play. Obviously, the mobile device's salient feature is its mobility, which makes it possible to use its functionalities when and where they are needed: exploring contexts, conversations between peers or community members, and collaborative knowledge building, for example, are three important activities made possible by mobile device (Kukulska et al., 2009) which can be applied in formal and informal teaching practices. The mediation of the tool's technological affordances, or in other words, what we can do with it, inevitably affects how these explorations, these conversations, and this knowledge building takes place: for example, if at times it can be difficult to describe a problem in words alone, it comes naturally to use our smartphone to make a short video or take pictures to send to someone who can help us and who is part of our community of learning or of practice. Likewise, we use our smartphone's Internet connection to find solutions to our problem on the Web's search engines.

Not all of these actions can be monitored and mediated by the teacher or covered by rigid educational protocols. To all intents and purposes, the smartphone thus becomes a mediator that enables us to interpret reality and make situated decisions independently. As we will see, however, independence is a great opportunity for personal creativity, but lack of guidance can bring a risk of losing focus and not activating all of the metacognitive mechanisms that formal learning environments can offer. From the pedagogical standpoint then, the teaching strategies we adopt must take these aspects into account in order to integrate formal/non-formal learning with the formal approaches of educational institutions. Precisely for this reason, the teacher's role in managing these processes effectively is even more crucial. But at the same time, it is a role whose foundations are to some extent shaken by these new ways of learning inside and outside the classroom.

Learning opportunities in informal settings are increasingly mediated by technological devices, the smartphone in particular. It is used in informal settings and daily life: in addition to its original purpose of enabling us to make calls, we use it to shoot videos and photos, and for the many other functions offered by its installed apps and sensors, whereby we can manage georeferenced maps, measure distances and movement, and interact with our physical surroundings. And by being connected to the Internet, it adds a whole series of ways of interacting with remote resources, and especially with other people, through access to the Web and social networks. The smartphone thus mediates our actions in a complex system where technology and physical, virtual, and socio-cultural contexts all interact with each other.

We will now present a model which should help to understand the complex interactions that take place in a complex system where the smartphone is the link between the spaces of formal, informal, and non-formal learning inside and outside the classroom, and between teachers, students, and the communities with which students come in contact in their daily experiences (*see* Fig. 7).

A feature that is often noted by researchers attempting to define mobile learning is the importance of social interactions that the tool provides. Obviously, however,

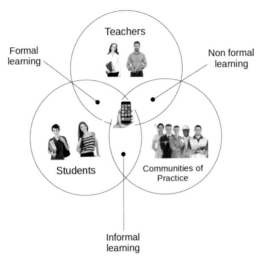

Fig. 7. The smartphone as a learning mediating tool between students, teachers, and communities of practice in formal, informal, and non-formal learning environments.

these interactions vary in type, and not all of them can be seen as potential stimuli for learning. Those that hold out the greatest opportunities are undoubtedly those that take place in communities of practice and communities of learning (Wenger et al., 2002). CoPs in particular are defined by Wenger and colleagues as groups of people who come together spontaneously around problems they have a shared interest in solving, and who share the knowledge that can be used to create artifacts (concrete objects or processes) whereby these problems can be solved. There are no age requirements for belonging to a CoP: while adults can meet in an in-person and/or remote CoP to try to solve a particularly complicated problem, children can also join in a CoP to exchange ideas about the best ways to overcome the difficulties involved in a video game. This can take place either online or in person, or, as is often the case, seamlessly in both environments.

Studies dealing specifically with what students do with their smartphones outside of school (Engel, 2018), however, tend to leave less room for enthusiasm, and report a mostly passive use of the technology (e.g., for watching videos, listening to music or searching the Web). The few constructive and/or creative activities are generally linked to some personal interest, where the students exchange online text and also produce digital artifacts (chiefly videos) to share the solution to a problem they may have encountered in daily life with their community. Such an output is the result of interesting problem-solving processes that teachers could put to good use at school when introducing specific topics in their subject matter. The challenge for teachers is thus to understand how to come to grips with the learning opportunities typical of informal settings and turn them into occasions for thinking about traditional curricula in formal schooling and higher education. In Fig. 7, for example, the teacher can mediate in the educational relationships in formal and non-formal settings in terms of direct action, planning, and monitoring, but cannot readily deal with the learning or the interactions between students and communities of practice outside the classroom.

Smartphone interactions between students can be monitored in formal and non-formal settings, but not in informal settings. One solution could be to stimulate students to share in school the products and processes of their activities in informal environments, in particular those involving their hobbies or personal interests. Here, the teacher's skill will be shown in grasping what portions of these materials can be turned to account in introducing new topics for study or delving further into issues that have already been covered in class.

Mobile Learning and Intrinsic Motivation

As we have seen, an enormous number of variables come into play when a teacher wishes to experiment with mobile learning, and their interactions can at times be extremely complex. There are thus no guidelines that can help us pinpoint the factors that are critical for a mobile learning activity's success. Among these critical success factors, however, there can be little doubt that particular attention should be given to the pedagogical and technological factors (Alrasheedi and Capretz, 2018) which, as we will see in the following paragraphs, must be fine-tuned in every mobile learning activity because of their fundamental role, both individually and in interacting with the settings and socio-cultural systems involved. One of the key pedagogical factors is the student's motivation to learn (Vansteenkiste et al., 2004). This motivation may be intrinsic in cases where students engage in an activity because they find it interesting and enjoyable, or extrinsic when they engage in an activity because they must complete it in order to achieve a specific goal, independently of whatever satisfaction it may give them. In this connection, it is important to understand the role played by technology in stimulating intrinsic motivation in learning processes. Many studies have found that introducing tools, such as tablets, smartphones, and interactive whiteboards in place of their traditional counterparts, meets with considerable student approval (Spector, 2017; Shroff and Keyes, 2017).

Though technology boosts students' intrinsic motivation (Granito and Chernobilsky, 2012), it does not appear to have a positive effect on their performance: extensive meta-analyses comparing teaching strategies with and without technologies found a mean effect size of approximately ES = 0.30 (Tamim et al., 2011), which is statistically significant, but not high. If, however, this figure is broken down by the context in which the technologies are used, we find that the effect size measured for activities planned in informal learning settings is nearly ES = 0.7, especially for K-12 students (Sandberg et al., 2011; Chauhan, 2017). These findings thus appear to reinforce the argument that what makes the difference is having teaching strategies that are alternative to or complement traditional approaches, enabling students to work outside of school, rather than the use of technologies *per se*. Teaching strategies, such as the flipped classroom can help teachers develop smartphone-based activities in informal settings as well. In this connection, a meta-analysis found that the overall size of learning achievement and learning motivation for the flipped classroom is significantly high, ES = 0.66 (Zheng et al., 2020).

Potential Difficulties in Introducing Smartphones in Teaching

There are, however, difficulties that can stand in the way of teachers' decision to use smartphones in learning processes. On the basis of the literature, we can group these potential obstacles into three macro-dimensions (Alwraikat and Mansour, 2017; Papadakis, 2018; Leem and Sung, 2019):

1. Institution problems, i.e., rules or regulations prohibiting or limiting smartphone use in class.

2. Technical problems, for instance, where a Bring Your Own Device (BYOD) model is used, some students' devices may not provide sufficient performance, creating a gap between those who can and cannot afford costly and high-end smartphones. Other problems include limited bandwidth at the school, inadequate apps, and the cost of any paid apps that may be necessary.

3. Problems involving teachers' acceptance of smartphone technology. According to several studies using the TAM (technology acceptance model) as their theoretical framework (Prieto et al., 2014; Papadakis, 2018), many teachers feel considerable anxiety in designing educational actions, given that since these innovative practices can also operate outside the classroom, they are not yet coded and can thus cause significant stress (Sánchez-Prieto et al., 2017).

The history of educational technologies teaches us that every time a new technology is introduced in society (radio, television, computers, the Web, interactive whiteboards, etc.), a group of pioneering teachers has immediately sought to introduce it in their teaching processes with innovative activities. According to the technology acceptance model, these pioneers demonstrate that highly positive attitudes towards technologies, which are strictly dependent on variables, such as perceived ease of use and perceived usefulness, influence behavioral intention to use. PEU is defined as the degree to which an individual believes that using a technology would be free of physical and mental effort, while PU is defined as the degree to which an individual believes that using a specific technology would enhance job performance (Davis, 1989).

Early experiments with technologies generally took place before codified protocols were available. Only later, and often after many years, when the technology had been widely adopted and incorporated in teaching practices, were the teachers supported with training courses that explained how to use the technology, what teaching strategies were most effective, and how to design them, monitor activities, and determine whether specific learning outcomes had been achieved in the subject-matter laid down in the school's or university's formal curriculum. This process meant that many years invariably went by between the time a technology was available for the first experiments and the time it became part of the curriculum. Accordingly, providing teachers with early training in both technical and pedagogical aspects of new technologies helps reduce anxiety and stress, and make teachers more confident. In the past, however, technologies had always been used in circumscribed spaces, such workshops or dedicated classes and were thus thought of as 'anchored' to a specific place where teachers could guide students' interactions with the technology.

This model has been disrupted by the introduction of the smartphone. Its inherently mobile nature, the fact that it has now become virtually part of ourselves, and the multiple affordances of the many technologies it incorporates mean that it cannot be confined to any one space, but is open to all of the informal contexts of daily life.

Teachers' Role in Informal Mobile Learning Processes: Toward a 'Fading' Scaffolding

Faced with the opportunity to introduce mobile learning, today's teachers can choose to:

1. Ignore the smartphone's educational potential and continue using the traditional teaching methods,
2. Include the smartphone only during in-class teaching, or
3. Include the smartphone in both in-class teaching and in activities outside the school, e.g., with the flipped classroom or problem-based learning models.

For teaching activities where the smartphone is used during classroom lessons, the teacher must provide direct mediation, for example, by suggesting how best to use the tool and apps and help in solving any problems encountered when doing so. For activities where the smartphone is also used outside the classroom, the teacher's preparation is necessarily more complex because it generally calls for careful planning of out-of-class work. Such work usually consists of a series of materials to be studied by the pupils, though they may also be asked to document real-life problems (e.g., through videos) that are relevant to the topic to be discussed later in class in accordance with the problem-based learning approach.

During these out-of-class activities, students may inevitably need appropriate guidance (Wanner and Palmer, 2015; Chen et al., 2015) or contextual feedback. This requires considerable commitment on the part of the teacher and is not easily accomplished: out-of-class activities must be painstakingly planned precisely in order to ensure that the student can learn as independently as possible. In this connection, some scholars propose using Vygotsky's notion of the zone of proximal development (Jie et al., 2020) as a framework that makes it possible to distinguish between 'mobile learning' and 'mobile pedagogy'. The ZSP is defined as the distance between the actual developmental level as determined by independent problem solving and the level of potential problem solving under adult guidance, or in collaboration with more capable peers. The outcomes that students are unable to achieve with mobile learning can thus be reached through specific mobile pedagogy strategies involving adaptive support—or scaffolding—provided by the teacher or peers. This support, however, must be temporary, given that from the standpoint of life-long learning, one of the key educational goals must be independence. In mobile learning, then, this scaffolding is a temporary support that is adapted to a student's understanding. Scaffolding is also the process of transferring the responsibility for learning or for a task from the teacher to the student, whose effectiveness depends on whether the teacher's 'fading' of support is timely (Van de Pol et al., 2019; Tuada, 2020).

Smartphones in the Classroom: A Distraction or a Learning Tool?

The use of smartphones in formal settings, such as the classroom, immediately sparked a heated debate: for the most part, criticism centered on what was regarded as the smartphone's power to distract students' attention from the task at hand. In reality, it appears that students engage more with their smartphones primarily when they are bored by the lesson, or less important topics are being covered (Bolkan and Griffin, 2017), though they also do so when the smartphone is used for specific learning activities. In such cases, the student is required to perform two simultaneous and cognitively challenging tasks: listen to what the teacher is saying and at the same time, work with the mobile device. This multitasking is frequent and inevitable, and according to recent studies, can have a negative impact not only on attention, but also on recall, comprehension, and efficiency in performing academic tasks (May and Elder, 2018). Choosing a BYOD program (Johnson et al., 2015), precisely because students use their own smartphones, can also mean that they will lose the concentration needed to complete the assigned tasks because their social apps may at any time call for their feedback—a distraction they can find hard to resist.

The risks for learning processes are high. There is an extensive research literature that demonstrates that students who do not use smartphones in class take notes more effectively, have better recall of lecture content, and show higher overall performance than those who do (Kuznekoff and Titsworth, 2015). Admittedly, threats to students' attention in class did not begin with the smartphone: 'mind wandering' has long been recognized as a problem (Lloyd, 1968; Wammes et al., 2019). However, mind wandering is innate in everyone's mental life and does not require cognitive activity of any kind in interacting and processing feedback. By contrast, using smartphones calls for significant cognitive effort thus aggravates the problem of inattention (Marty-Dugas et al., 2018). In any case, the problem is more pressing than ever, given that several studies have found that on an average, 50–70 per cent of students check their smartphones at least once during a lecture and that few are able to resist the temptation (Atas and Çelik, 2019). This would appear to be true in many cultural contexts around the world (Yang et al., 2019).

We can classify smartphone-related 'interruptions' as either endogenous or exogenous. Endogenous interruptions are those that emerge spontaneously in the users' thoughts in the form of an urge to interact with the mobile device, e.g., to look at social media. If satisfied, these urges lead to immediate gratification. Exogenous interruptions, by contrast, involve the need to use one or more smartphone functions (needing to make an urgent call, for, e.g., to find information in order to finish a task). Though such interruptions are in fact far rarer, if we include cues from the smartphone itself that we cannot control and that require our attention among exogenous interruptions, then the number of potential occasions for distraction rises sharply. Recent studies appear to confirm that even if the smartphone is not actually used, receiving notifications of incoming calls or messages can significantly distract students' attention and prompt mind-wandering, as students may be curious to know who called or wrote, and what the message said (Stothart et al., 2015). It also seems that ringing and notification sounds create a 'halo effect' or cognitive interference effect in the shared physical space that extends well beyond the personal physical sphere and includes the surrounding people as well as the smartphone's owner. Thus, the mere presence of our smartphone—even if it is off—becomes a sort of 'brain drain' (Ward et al., 2017) that depletes our cognitive capacity because a significant portion of our pool of attentional resources is used to suppress the urge to turn on our phone to check if we have received messages.

In formal school and university settings then, the most important problem is that of the processes of attention/distraction that smartphones can disrupt. While the problem can be handled to some extent by the teacher when students are physically present in class, it becomes even worse when lessons are held online. The opportunities for distraction increase exponentially, primarily because students' multitasking increases in online courses (Lepp et al., 2019) and, by dividing their attention among multiple simultaneous activities, saturates and detracts them from their cognitive capacity. Studies in this area confirm that multitasking has a negative impact on students, and in particular, on their comprehension, notetaking, and self-regulation (May and Elder, 2018).

There is little consensus among teachers about the smartphone's potential as a classroom learning tool. While some acknowledge the opportunities it holds for education and seek to include it in their teaching practices, others regard it as risky. Interestingly, not only do many teachers share this unfavorable view, but the majority of parents also show high levels of resistance to the educational risk of smartphones at school (Hadad et al., 2020). By contrast, the teachers who see smartphones as useful argue that if students would rather look at their phones than pay attention to the lesson, the real problem lies in poor teaching practices that fail to engage students. As for teaching activities that are specifically smartphone-based, these teachers believe that if they do not work, it is the fault of the type of app chosen and how it is used. The literature offers many examples in which the smartphone is presented as a tool in a range of traditional and innovative teaching activities, including:

- assessment: student response system & feedback.
- information seeking: finding out more about lecture topics using search engines or links for immediate access to information;
- collaboration: group work and resource sharing;
- new affordances: augmented reality with textbooks and interaction with the classroom environment;
- game-based learning, educational mobile games/serious game/simulations.

For these activities to be effective, they must be carefully scheduled and integrated with the lesson precisely because they are time-consuming and can otherwise offer many opportunities for distraction. In what is perhaps the most successful activity to date, students use their smartphones as student response systems. Using dedicated apps, students can provide immediate feedback with their smartphones when the teacher asks questions during the lesson. The software processes the students' answers in real time and displays the results on the interactive whiteboard or projector. Using SRSs can be extremely useful in maintaining students' attention. Several studies have shown that attention begins to drop after only 20 minutes during a typical class or lecture (Bradbury, 2016). Students using their smartphones as SRSs have been found to be more engaged and attentive, focusing on key points during the lesson (Meguid and Collins, 2017).

What Apps to Use for Mobile Learning?

There are two schools of thought about mobile learning: one holds that for the approach to be as integrated with formal learning programs as possible, students must be provided with purpose-built apps that can be readily combined with the functionalities of the learning management system in use at the school or university. The other maintains that it is simpler and more effective to use currently available educational or general-purpose apps, and then review the results in class or enter them in the LMS. Each approach has its pros and cons. In the first case, developing tailored apps takes time and can be quite costly, to say nothing of the expenses involved in maintaining the app over time. In the second case, there are few or no costs, but problems can arise in converting data formats for use in the LMS. In either case, we believe that in mobile learning—as in any other approach to learning—the most critical factors are not so much as to which technological aids are employed or how educational content is adapted to an LMS, but the educational strategies' ability to stimulate student motivation and engagement.

The question we must ask, in fact, is what types of learner-teacher interactivities does the teacher facilitate when using mobile learning apps? (Rozario et al., 2016). In this sense, choosing apps for use in learning activities can be problematic, both because of the sheer number available, and because of the range of styles (i.e., gamification, etc.) and types of instructional designs that are used (knowledge building, collaboration, assessment, etc.). Some scholars (Notari et al., 2016) have attempted to draw up a classification of apps on the basis of their pedagogical functions (Table 5):

Table 5. Classification of educational apps (Notari et al., 2016).

Typologies	Specifications
Knowledge & skill building apps	Apps with a well-defined setting with a specific instructive design. They most likely address the first two levels from Bloom's taxonomy: remember (identify, recall) and understand (compare, match, classify).
Collaboration apps	Apps to help students produce a collaborative text, media or specific task.
Learning and teaching support apps	Apps for time-tables or homework scheduling, student response systems.
Communication apps	Apps for communicating (WhatsApp, etc.)
Other tools and reference apps	Apps for maps, calculators, etc.

It is important to bear in mind that the choice of the apps to be used also depends on the students' age, as apps that are easily used by older students may pose difficulties for younger age groups. Another interesting factor that seems to be independent of age is gamification. Many learning apps are essentially educational games that stimulate students' intrinsic motivation to learn course content by presenting them with challenges or competitions.

Mobile Learning Usability

While from a pedagogical standpoint the quality of the digital content and methods that teachers wish students to use in the learning process are important (Pocatilu and Boja, 2009), no less important is that all digital tools should comply with the hardware and software quality standards (ISO, 2005) that cover efficiency and effectiveness of use. Researchers have proposed an integrated model (Table 6) for assessing the quality of mobile learning apps which take both hardware and software into consideration (Khan et al., 2018; Alturki and Gay, 2019).

Table 6. Technical quality model (adapted from Khan et al., 2018).

Requirements	
Availability	Available without particular cost or copyright restrictions.
Flexibility, scalability	M-learning applications should be flexible, adaptable to any of the learning environments and scalable.
Quick response, connectivity, efficiency, performance	Must be able to use Wi-Fi and 4G-5G networks and fast feedback.
Reliability	The application should perform its intended functions and operations without experiencing failure.
Functionality, maintainability	The application must include all features that are necessary to provide an improved learning experience and the ability to undergo modifications (including corrections, improvements or adaptation).
Usability, user interface	Ease of use, user satisfaction, attractiveness and learnability.
Security	Personal data protection.

Ergonomic aspects are one of the major issues that make smartphones less efficient than personal computers in interacting with the user. Users frequently complain of neck pain and visual fatigue, largely as a result of smartphones' small screen size (Akurke, 2018). This difference in size also seems to influence how we interact with the apps. The perceived difficulty of navigating on a small screen, for example, translates into a tendency to search less extensively and click on the first link presented by search engines (Chae and Kim, 2004). Consequently, smartphones have grown steadily in size over the years. Smartphones with foldable screens are now in the market as their viewing area is up to twice as big as a standard phone, but costs are still very high.

Another critical problem is the blue light emitted by the smartphone screen. In addition to causing visual fatigue and other eye problems, it also appears to reduce melatonin production in young people, who hold their devices close to their eyes and for longer periods than older users, especially while in bed (Priya and Subramaniyam, 2020). That bedtime smartphone usage is a critical aspect of young people's habits has been demonstrated by researchers who compared the effects of blue light and caffeine (Beaven and Ekstrom, 2013), finding that one hour of exposure to blue light produces the same levels of alerting and psychomotor effects as 240 mg of caffeine. This widespread behavior significantly reduces sleepiness by raising cortisol levels (Heo et al., 2017). As these levels then drop very slowly, they negatively influence sleep quality, with inevitable repercussions on cognitive and physical performance upon waking.

Final Considerations

The new affordances offered by smartphones immediately attracted the attention of teachers and researchers. A single device that can connect to the Web, record videos, and use innumerable apps while georeferencing its activities in space, provides a range of functions with enormous educational potential. The risks are undeniably high, especially when we consider that using these affordances in teaching can encourage or even worsen smartphone addiction, which, as we have seen, can lead to a series of syndromes with repercussions on cognitive, emotional, and behavioral processes. Nor do we know whether attempts in the near future to include mobile devices in teaching practices will be successful because it will not be easy to determine empirically whether any improvements in students' mastery of coursework or their problem-solving abilities in everyday life, and employment settings are in fact due to their learning experience with smartphones.

Smartphones can be introduced in teaching and learning processes in different and intersecting ways, involving their use both in the physical space of the classroom and outside the school, or in other words, in formal learning settings as well as in informal/non-formal settings. While smartphone use in the former can be more readily managed by the teacher as part of organized educational activities, out-of-classroom use in an informal setting involves major challenges and is not easy to integrate with traditional teaching. Open teaching strategies in informal settings, in fact, require that the formal curriculum be redesigned, and in the final analysis call for a rethinking of the teachers' role. The teacher also acts as a guide and mediator in real-world learning processes. To do so, teachers will need training not only in order to acquire technological skills in using specific software and apps, but above all, in order to build the pedagogical and methodological competencies required to design integrated learning environments that combine classroom, informal and non-formal settings. In this connection, it will be important to encourage teachers' participation in professional communities of practice where they can exchange views about mobile learning best practices.

The challenge, then, is to pass from teaching based on curricular knowledge to teaching based on the social construction of skills that can take place, both in the social interactions of everyday life and through participation in communities of practice, whether in person or virtually. The school can then become a venue for discussing and reflecting on real-world experiences, linking them to the theories and models in the curriculum. This continuing process of exploration and reflection would bridge the gap that Dewey noted between the two settings a century ago. A crucial factor in this effort will be the ability to turn the class into a true 'community of attention', trying to help students to be more attentive to one another (Lang, 2020). In this community of attention, the smartphone can help students focus on the subject matter, no longer a source of distraction, but an important aid to concentration in all of life's settings. For all this to take place, teachers must share a consensus about the opportunities that mobile learning holds, and students must realize that the smartphone can be an important support for life-long-learning, and not just another part of their everyday life and recreation.

References

Akkerman, S.F. and Bakker, A. (2011). Boundary crossing and boundary objects. *Review of Educational Research*, 812: 132–169.

Akurke, S.V. (2018). *The Effect of Smartphone Screen Size on Usability and Users' Discomfort*. Thesis of Master of Engineering Science, Lamar University-Beaumont.

Alrasheedi, M. and Capretz, L.F. (2018). *Determination of Critical Success Factors Affecting Mobile Learning: A Meta-analysis Approach*. Preprint arXiv:1801.04288.

Alturki, R. and Gay, V. (2019). Usability attributes for mobile applications: a systematic review. pp. 53–62. *In: Recent Trends and Advances in Wireless and IoT-enabled Networks*. Springer, Cham.

Alwraikat, M.A. and Mansour. (2017). Smartphones as a new paradigm in higher education overcoming obstacles. *International Journal of Interactive Mobile Technologies*, 11(4): 114–135.

Atas, A.H. and Çelik, B. (2019). Smartphone use of university students: patterns, purposes, and situations. *Malaysian Online Journal of Educational Technology*, (72): 59–70.

Beaven, C.M. and Ekstrom J. (2013). A comparison of blue light and caffeine effects on cognitive function and alertness in humans. *PloS One*, 8: 1–7.

Bolkan, S. and Griffin, D.J. (2017). Students' use of cell phones in class for off-task behaviors: the indirect impact of instructors' teaching behaviors through boredom and students' attitudes. *Communication Education* (663): 313–329.

Botero, G., Questier, F. and Zhu, C. (2019). Self-directed language learning in a mobile-assisted, out-of-class context: do students walk the talk? *Computer Assisted Language Learning*, 321(2): 71–97.

Bradbury, N.A. (2016). Attention span during lectures: 8 seconds, 10 minutes, or more? *Advances in Physiology Education*, 404: 509–513.

Brown, E., Börner, D., Sharples, M., Glahn, C., de Jong, T. and Specht, M. (2010). Location-based and contextual mobile learning. *A Stellar Small-scale Study*, the Open University.

Burden, K.J. and Kearney, M. (2017). *Investigating and Critiquing Teacher Educators' Mobile Learning Practices, Interactive Technology and Smart Education*.

Chae, M. and Kim, J. (2004). Do size and structure matter to mobile users? An empirical study of the effects of screen size, information structure, and task complexity on user activities with standard web phones. *Behavior Information Technology*, 23(3): 165–181.

Chauhan, S. (2017). A meta-analysis of the impact of technology on learning effectiveness of elementary students. *Computers & Education*, 105: 14–30.

Chen, L., Chen, T.L. and Chen, N.S. (2015). Students' perspectives of using cooperative learning in a flipped statistics classroom. *Australasian Journal of Educational Technology*, 31(6).

Churchill, D., Fox, B. and King, M. (2016). Framework for designing mobile learning environments. pp. 3–26. *In: Churchill, D., Lu, J., Chiu, T.K.F. and Fox, B. (eds.). Mobile Learning Design: Theories and Applications*. Springer, New York.

Crompton, H. and Burke, D. (2018). The use of mobile learning in higher education: a systematic review. *Computers & Education*, 123: 53–64.

Crompton, H. and Burke, D. (2020). Mobile learning and pedagogical opportunities: a configurative systematic review of preK-12 research using the SAMR framework. *Computers & Education*, 156: 103945.

Davis, F.D. (1989). Perceived usefulness, perceived ease of use, and user acceptance of information technology. *MIS Quarterly*, 13(3): 319–340.

Engel, A., Salvador, C.C., Membrive, A. and Badenas, J.O. (2018). Information and communication technologies and students' out-of-school learning experiences. *Digital Education Review*, 33: 130–149.

Engeström, Y. (2001). Expansive learning at work: towards an activity theoretical reconceptualization. *Journal of Education and Work*, 14(1): 133–156.

Eshach, H. (2007). Bridging in-school and out-of-school learning: formal, non-formal, and informal education. *Journal of Science Education and Technology*, 16(2): 171–190.

Fuegen, S. (2012). The impact of mobile technologies on distance education. *TechTrends*, 56(6): 39–53.

Garrison, D.R. (1997). Self-directed learning: toward a comprehensive model. *Adult Education Quarterly*, 48(1): 18–33.

Garrison, D.R. (2011). *E-Learning in the 21st Century: A Framework for Research and Practice*. Taylor & Francis, New York.

Granito, M. and Chernobilsky, E. (2012). The effect of technology on a student's motivation and knowledge retention. *In: The NERA Conference Proceedings*. Paper 17.

Hadad, S., Meishar-Tal, H. and Blau, I. (2020). The parents' tale: why parents resist the educational use of smartphones at schools? *Computers & Education*, 157: 103984.

Heo, J., Kim, K., Fava, M., Mischoulon, D., Papakostas, G.I., Kim, M., Kim, D.J., Chang, K.J., Oh, Y.L., Yu, B.H. and Jeon, H.J. (2017). Effects of smartphone use with and without blue light at night in healthy adults: a randomized, double-blind, cross-over, placebo-controlled comparison. *J. Psychiatr. Res.*, 8761–70.

Herrington, J., Ostashewski, N., Reid, D. and Flintoff, K. (2014). Mobile technologies in teacher education: preparing pre-service teachers and teacher educators for mobile learning. pp. 137–151. *In: Successful Teacher Education*. Brill Sense.

Jie, Z., Sunze, Y. and Puteh, M. (2020). Research on teacher's role of mobile pedagogy guided by the zone of proximal development. pp. 219–222. *In: Proceedings of the 2020 9th International Conference on Educational and Information Technology*.

Johnson, L., Adams Becker, S., Estrada, V. and Freeman, A. (2015). *NMC Horizon Report: 2015 Higher Education Edition*. The New Media Consortium, Austin.

Kardefelt-Winther, D., Heeren, A., Schimmenti, A., van Rooij, A., Maurage, P., Carras, M. et al. (2017). How can we conceptualize behavioral addiction without pathologizing common behaviors? *Addiction*, 112(10): 1709–1715.

Kearney, M., Burden, K. and Rai, T. (2015). Investigating teachers' adoption of signature mobile pedagogies. *Computers & Education*, 80: 48–57.

Kearney, M., Burden, K. and Schuck, S. (2019). Disrupting education using smart mobile pedagogies. pp. 139–157. *In: Didactics of Smart Pedagogy*. Springer, Cham.

Kearney, M., Burke, P.F. and Schuck, S. (2019). The iPAC scale: A survey to measure distinctive mobile pedagogies. *TechTrends*, 63(6): 751–764.

Khan, A.I., Al-Khanjari, Z. and Sarrab, M. (2018). Integrated design model for mobile learning pedagogy and application. *Journal of Applied Research and Technology*, 16(2): 146–159.

Kim, J.H. and Park, H. (2019). Effects of smartphone-based mobile learning in nursing education: a systematic review and meta-analysis. *Asian Nursing Research*, 13(1): 20–29.

Kukulska-Hulme, A., Sharples, M., Milrad, M., Arnedillo-Sánchez, I. and Vavoula, G. (2009). Innovation in mobile learning: a european perspective. *International Journal of Mobile and Blended Learning*, 1: 13–35.

Kuznekoff, J. and Titsworth, S. (2015). Mobile phones in the classroom: examining the effects of texting, twitter, and message content on student learning. *Communication Education*, 64(3): 344–365.

Lang, J.M. (2020). *Distracted: Why Students can't Focus and What you Can Do about It*. Basic Books, New York.

Leem, J. and Sung, E. (2019). Teachers' beliefs and technology acceptance concerning smart mobile devices for SMART education in South Korea. *British Journal of Educational Technology*, 50(2): 601–613.

Lepp, A., Barkley, J.E., Karpinski, A.C. and Singh, S. (2019). College students' multitasking behavior in online versus face-to-face courses. *Hypothesis*, 3: H3.

Liu, M., Scordino, R., Geurtz, R., Navarrete, C., Ko, Y. and Lim, M. (2014). A look at research on mobile learning in k-12 education from 2007 to the present. *Journal of Research on Technology in Education*, 46: 325–372.

Lloyd, D.H. (1968). A concept of improvement of learning response in the taught lesson. *Vis. Educ.*, October, 23–25.

Looi, C.K. and Wong, L.H. (2013). Designing for seamless learning. pp. 146–157. *In:* Luckin, R., Goodyear, P., Grabowski, B., Underwood, J. and Winters, N. (eds.). *Handbook of Design in Educational Technology*. Routledge.

Looi, C.K., Lim, K.F., Pang, J., Koh, A.L.H., Seow, P., Sun, D. and Soloway, E. (2016). Bridging formal and informal learning with the use of mobile technology. pp. 79–96. *In*: *Future Learning in Primary Schools*. Springer, Singapore.

MacCallum, K. and Parsons, D. (2016). A theory-ology of mobile learning: operationalizing learning theories with mobile activities. pp. 173–182. *In*: Dyson, L.E., Wan, N. and Fergusson, J. (eds.). *Mobile Learning Futures. Sustaining Quality Research and Practice in Mobile Learning*. 15th World Conference on Mobile and Contextual Learning, mLearn.

Marty-Dugas, J., Brandon, C.W.R., Oakman, J.M. and Smilek, D. (2018). The relation between smartphone use and everyday inattention. *Psychology of Consciousness: Theory, Research, and Practice*, 5(1): 46.

May, K.E. and Elder, A.D. (2018). Efficient, helpful, or distracting? A literature review of media multitasking in relation to academic performance. *International Journal of Educational Technology in Higher Education*, 15: 1–14.

Mayer, R.E. (2020). Where is the learning in mobile technologies for learning? *Contemporary Educational Psychology*, 60: 101824.

Meguid, E.A. and Collins, M. (2017). Students' perceptions of lecturing approaches: traditional versus interactive teaching. *Advances in Medical Education and Practice*, 8: 229.

Milrad, M., Wong, L.H., Sharples, M., Hwang, G.J., Looi, C.K. and Ogata, H. (2013). Seamless learning: an international perspective on next-generation technology-enhanced learning. pp. 95–108. *In*: Berge, Z.L. and Muilenburg, L.Y. (eds.). *Handbook of Mobile Learning*. Routledge, Abingdon.

Notari, M.P., Hielscher, M. and King, M. (2016). Educational apps ontology. pp. 83–96. *In*: *Mobile Learning Design*. Springer, Singapore.

Papadakis, S. (2018). Evaluating pre-service teachers' acceptance of mobile devices with regards to their age and gender: a case study in Greece. *International Journal of Mobile Learning and Organisation*, 12(4): 336–352.

Parsons, D., Wishart, J. and Thomas, H. (2016). Exploring mobile affordances in the digital classroom. pp. 43–50. *In*: Arnedillo-Sánchez, I. and Isaias, P. (eds.). The *Proceedings of the 12th International Conference on Mobile Learning Mobile Learning*. IADIS.

Pocatilu, P. and Boja, C. (2009). Quality characteristics and metrics related to M-learning process. *Amfiteatru Economic*, 11(26): 346–354.

Prieto, J.C.S., Miguéláñez, S.O. and García-Peñalvo, F.J. (2014). Mobile learning adoption from informal into formal: an extended TAM model to measure mobile acceptance among teachers. pp. 595–602. *In*: *The Proceedings of the Second International Conference on Technological Ecosystems for Enhancing Multiculturality*.

Priya, D.B. and Subramaniyam, M. (2020). A systematic review on visual fatigue induced by tiny screens (Smartphones). *IOP Conference Series: Materials Science and Engineering*, 912: 062009.

Puentedura, R.R. (2013). *SAMR: Moving from Enhancement to Transformation* [Web log post]. Retrieved from http://www.hippasus.com/rrpweblog/archives/000095.html.

Rafique, N., Al-Asoom, L.I., Alsunni, A.A., Saudagar, F.N., Almulhim, L. and Alkaltham, G. (2020). Effects of mobile use on subjective sleep quality. *Nature and Science of Sleep*, 12: 357.

Regan, J.A. (2003). Motivating students towards self-directed learning. *Nurse Education Today*, 23(8): 593–599.

Rozario, R., Ortlieb, E. and Rennie, J. (2016). Interactivity and mobile technologies: an activity theory perspective. pp. 63–82. *In*: Churchill, D., Lu, J., Chiu, T. and Fox, B. (eds.). *Mobile Learning Design*. Springer, Singapore.

Sánchez-Prieto, J.C., Olmos-Miguéláñez, S. and García-Peñalvo, F.J. (2017). M-learning and pre-service teachers: an assessment of the behavioral intention using an expanded TAM model. *Computers in Human Behavior*, 72: 644–654.

Sandberg, J., Maris, M. and De Geus, K. (2011). Mobile english learning: an evidence-based with fifth graders. *Computers & Education*, 57(1): 1334–1347.

Sharples, M. (2002). Disruptive devices: mobile technology for conversational learning. *International Journal of Continuing Engineering Education and Lifelong Learning*, 12: 504–520.

Sharples, M., Arnedillo-Sánchez, I., Milrad, M. and Vavoula, G. (2009). Mobile learning. pp. 233–249. *In*: *Technology-enhanced Learning*. Springer, Dordrecht.

Sharples, M. (2015). Seamless learning despite context. pp. 41–55. *In*: *Seamless Learning in the Age of Mobile Connectivity*. Springer, Singapore.

Shroff, R.H. and Keyes, C.J. (2017). A proposed framework to understand the intrinsic motivation factors on university students' behavioral intention to use a mobile application for learning. *Journal of Information Technology Education: Research*, 16(1): 143–168.

Song, D. and Bonk, C.J. (2016). Motivational factors in self-directed informal learning from online learning resources. *Cogent Education*, 3(1): 1205838.

Spector, J.M. and Park, S.W. (2017). *Motivation, Learning, and Technology: Embodied Educational Motivation*. Routledge.

Stothart, C., Mitchum, A. and Yehnert, C. (2015). The attentional cost of receiving a cell phone notification. *Journal of Experimental Psychology: Human Perception and Performance*, 41(4): 893.

Tamim, R.M., Bernard, R.M., Borokhovski, E., Abrami, P.C. and Schmid, R.F. (2011). What forty years of research says about the impact of technology on learning: a second-order meta-analysis and validation study. *Review of Educational Research*, 81(1): 4–28.

Tao, Z., Yang, X., Lai, I.K.W. and Chau, K.Y. (2018). A research on the effect of smartphone use, student engagement and self-directed learning on individual impact: china empirical study. pp. 221–225. *In: International Symposium on Educational Technology*. IEEE.

Tuada, R.N., Kuswanto, H., Saputra, A.T. and Aji, S.H. (2020). Physics mobile learning with scaffolding approach in simple harmonic motion to improve student learning independence. *Journal of Physics: Conference Series*, 1440(1): 012043.

van de Pol, J., Mercer, N. and Volman, M. (2019). Scaffolding student understanding in small-group work: students' uptake of teacher support in subsequent small-group interaction. *Journal of the Learning Sciences*, 28(2): 206–239.

Vansteenkiste, M., Simons, J., Lens, W., Sheldon, K.M. and Deci, E.L. (2004). Motivating learning, performance, and persistence: the synergistic effects of intrinsic goal contents and autonomy-supportive contexts. *Journal of Personality and Social Psychology*, 87(2): 246–260.

Wammes, J.D., Ralph, B.C.W., Mills, C., Bosch, N., Duncan, T.L. and Smilek, D. (2019). Disengagement during lectures: media multitasking and mind wandering in university classrooms. *Computers & Education*, 132: 76–89.

Wanner, T. and Palmer, E. (2015). Personalizing learning: exploring student and teacher perceptions about flexible learning and assessment in a flipped university course. *Computers & Education*, 88: 354–369.

Ward, A.F., Duke, K., Gneezy, A. and Bos, M.W. (2017). Brain drain: the mere presence of one's own smartphone reduces available cognitive capacity. *Journal of the Association for Consumer Research*, 2(2): 140–154.

Wenger, E., McDermott, R. and Snyder, W. (2002). *Cultivating Communities of Practice: A Guide to Managing Knowledge*. Harvard Business School Press, Boston.

Yang, Z., Asbury, K. and Griffiths, M.D. (2019). Do Chinese and British university students use smartphones differently? A cross-cultural mixed methods study. *International Journal of Mental Health and Addiction*, 17(3): 644–657.

Zheng, L., Bhagat, K.K., Zhen, Y. and Zhang, X. (2020). The effectiveness of the flipped classroom on students' learning achievement and learning motivation. *Journal of Educational Technology Society*, 23(1): 1–15.

Zimmermann, A., Lorenz, A. and Oppermann, R. (2007). An operational definition of context. *In*: Kokinov, B., Richardson, D.C., Roth-Berghofer, T.R. and Vieu, L. (eds.). *Modeling and Using Context, Lecture Notes in Computer Science*. vol. 4635, Springer, Berlin, Heidelberg.

Part III
Alessandro Ciasullo

CHAPTER 6

Sound as a Mediator between the Body and the Virtual Technological Educational Approaches

Alessandro Ciasullo

|||||||||||||||||||||||||||||||||||||||||||||||||||||||||||||||||||||||||||||||||||||||||||||||||||||||||||||||||||||||||||||||||||||||||||||||||||||||||||||||||||||||||||||

When [...] man invented words and music he altered the soundscape and the soundscape altered man. The epigenetic evolution interacting progressively between humanity and his rugged landscape has been profound.

—Fuller, 1966

Introduction

'*In the beginning was the word, and the word was with God, and the word was God*' (Jn 1:1): thus opens the Prologue of the Evangelist John who, in this powerful statement, through some syllogisms defines the Christians' God, in the preamble, as a God who is the word. Therefore, the word represents the expression of a thought that emerges, starting from a sound expressed in its form. To simply rephrase John's statement, it would be said that our civilization would not have evolved as it has over the years without the sound being capable of expressing the word.

What appears to be a religious and confessional quotation, although powerful and full of meaning, actually allows introduction of an essential concept in this work: how much the rational expression related to man is one of the leading evolutionary conditions that allowed humans to climb the chain of animal development, putting us in a position of superiority over other species.

Sound and its physical components are one of the central elements in the relationship between the subject and environment. The physical characteristics of

Researcher of Education, University of Naples, Federico II.
Email: alessandro.ciasullo@unina.it

sound are aspects of the individual's wise nature and therefore stand as a real vector of information that identifies man's communicative and relational nature.

The communicative characteristic of sound is given by its physical nature that is expressed by the systematic and regular vibration of frequencies in several Hertz (a measurement of sound vibrations explained in a note). This system can reach, by moving air particles, the hearing organs which in turn translate these vibrations into sounds that are subsequently translated into stimuli and in the end, coded into messages that may or may not have meaning (Fitch, 2016). Hence, the physical process of sound happens through vibrations and then, in the course of its evolution, takes on different communicative characters translated into symbolic and semiotic codes, which give them meanings (Blasi et al., 2016).

This evolutionary power offered by sound allowed man to determine himself as a communicative being, first according to a non-coded sound communication code and then by elaborating, in the course of evolution, a process of translation of sound into verbal code. On account of this, the transition from nature to culture invested processes of transformation of sound that form a spatial clue, an element of emphasis of a state of mind; the primary approach of communication of moods has become a primary element of intra-species communication (Ciasullo, 2020b). One moved from the primary function of sound as a natural element of most animal species to sound as an expression of languages encoded in alphabets and, thus, cultural terms of particular peoples and cultures. All of this is often defined within spaces and groups of the same ethnicity, not always located within national areas and thus limited by regulatory conventions (just think of linguistic minorities and bilingual regions).

Therefore, the sonority of the 'word' assumes a connotation of mediation with the world, essential enough to allow it a cultural expansion such as to express control over the world. It follows that sonority is an advantageous element as well as an essential component of animal species, representing at various evolutionary levels their communicative, expressive, and descriptive potential (Vicky et al., 2020).

In this chapter, it is tried to indicate how, even today, the sound, although often considered complementary to the sense of sight, continues to express a significant conciliating possibility between the subject and the virtual/real environment, which cannot by any means be replaced by the predominance of the visual. In certain circumstances or conditions, the immediacy of sound codes or verbal language can act as catalysts for the visible message. Nonetheless, this can also function as an implicit code for mood perception, as indicator of the spatial dimensions of a given place, and also replace the visual indications when they, due to system unavailability, should produce them and, last but not least, with those individuals with visual disabilities (Brown et al., 2011).

It can then be stated that sound is a descriptor of the universe. Hence, evolutionary characteristics are closely linked to the expression of human intelligence. Its strength can surely be expressed about language, paraverbal and good communication but not in verbal or musical enjoyment, which are indications of the subject's location in space-specific sounds of a given place. The sound is, therefore, in itself a perceptual/sensory and communicative element which is crucial and signifiable in the phylogenetic and cultural development of animal species and, in particular, of man who represents the highest link in the evolutionary chain.

Human species' importance is certainly not given by physical elements, since in nature there are other animal species which are structurally able-bodied than man, but by the development of intelligent characteristics, the result has been the brain/mental evolution of the human species (Bribiescas, 2009). With the advantageous evolution of human intelligence, the relationship between the subject and the environment has undergone elaborative forms of transition. This process, 'sophisticated' and refined extensively with/by the response between man and the surrounding environment, certainly not with an abolition or cancellation of the sensory stimuli, allows them to interact with the context. This means that in order to realize an interactive process with the world, the sensory and sensitive relationship between the subject and the environment always remains crucial.

In this sensitive connection dimension between the subject-environment, there is also an implicit aspect that encourages a relevant role for man, who has to learn and interact with the world (Santoianni, 2014). In this dimension—under trace—of the human mental functioning, there is a series of information related to genetic memory that allows an individual to manage a considerable amount of data mind-processed and in a completely 'unconscious', but continuously active way. In this genetic memory, the primary sensory information allows individuals the innate adaptive possibility that enables them to be a part of the world and interact with it. This memory also includes the sensory clues of hearing and even the sound meanings that are transferred through the auditory canal. Therefore, it is believed that sound is the perceptual object that gives sense to hearing and as a sensory peculiarity, it is part of an implicit and profound dimension of the human mind (The Functional Phylogenies Group, 2012; Blasi et al., 2016; Fitch, 2016; Schafer, 1993).

On account of these considerations, it is possible to understand that any mediation between the real/virtual environment cannot simply view the sound dimension in all its evolutionary force. If the possible recognition of mediation between real and virtual environments is eliminated, it would mean a systematic reduction of crucial sensory information that does not include one of the leading deep dimensions of man, but precisely the discreet relationship determined by hearing and sounds.

Sound Environments, Soundscapes, and Mobile Learning

Listening, as well as listening training becomes a powerful means to realize more significant social interaction elements in time, like the present one. A person is thought to be more and more immersed in the sound jungle made of good chaos and real sound pollution. An active listening resulting from the ability to discriminate sound characteristics can harmonize social interactions, contribute to a proper awareness, and educate to recognize sound scenarios.

The study by Carles et al. (1999) is very interesting to make us understand the power that a sound environment can have from a cognitive point of view and the relationship between visual and acoustic stimuli on environmental perception as has been experimented. The course was built by presenting natural and semi-natural environments and urban green spaces, measuring the level of pleasure experienced by about 75 people involved in the project. Some sounds, natural, artificial, and alarm sounds, were presented in an association. What emerged is that, in general, the alarm sound can free itself from the sound range of background sounds and therefore it stands out of any context, be it isolating and prevailing. This is attributed to the sound frequency and the prevalence spectrum it has over natural sounds.

The research was very significant as it was discovered that all levels of pleasure that emerged from listening to certain sounds in association with images shown happened mainly when there was coherence between background and displayed images. Individuals found more satisfying a close correlation between images and sound—this would support the thesis that sound is an essential element of the environment. The need to maintain a level of coherence between images and sound landscape emerges even more strongly (Carles et al., 1999). These levels of correlation between visual environment, sound environment, and pleasure illustrate an operational perspective of further strengthening training through mobile devices that consistently include precise indications of the soundscape (Schafer, 1993, 1969).

The link between individual, sound and environment can be interpreted in two ways: the first would see sound as a different form of mediation between the individual and the environment; the second would see sound as an element of the environmental relationship with a substantial difference. While the senses, such as sight, smell, taste, hearing and touch have a connotation of passive perception, i.e., they are senses that allow us to perceive things external to the subject and make them our own without an active projection on the surrounding environment (except for the eyes in exchange of glances, but this would already belong to a symbolic system, secondary to the visual function), the only projective actions of a subject on the environment are the physical action and the emission of both verbal and simple sound outputs.

This is why action and verbal sounds could be understood as the two essential components that an individual has to exercise for their presence in the environment.

Adequate educational processes can stimulate the transition to new educational perspectives and new possibilities of transforming the subjects. In this case, the biological, physiological, and evolutionary dimension of individuals would not be bent to the logic of technology without considering technology's embodied nature. Technology could be a way to emphasize and stimulate the evolution of educational processes towards further horizons. The temporal and spatial dimension, for now, entrusted to real physical places, could expand to decentralized areas typical of virtual environments (VLE) or learning developed on technologies. To do this, the technological universe should include and, above all, consider analyzing the forms, specificities, evolutions, and also all the intrinsic and extrinsic perceptual aspects.

Considering the biological nature—in technology—means giving technologies a development closer and closer to the real needs of human subjectivity made of social relations, bodily functioning, and co-existence with the external environment. Not considering this possible alliance means continuing to deny the evidence of the technological world increasingly present in society's overall development in its productive, social, organizational, and management mechanisms. The school must find the alliance aimed at endowing the individual, the person, and the subjectivity with tools that will allow them to better manage in a completely personalized, and therefore inclusive, transformation, and training process.

The prospect of a school without classrooms, made up of de-spatialized and de-temporalized distance education and training systems, seems distant, but the idea that there cannot be a complementary growth strategy entrusted to an efficient and respectful technology for education and learning intensely conflicts with the evolution of reality beyond school. The real problem, therefore, is to act so that the technologies including those that form learning are actually able to stimulate the implicit and explicit dimensions of the students, of those who are formed, that is, they are able to form the subject fully or to support adequate learning characteristics so as to be constituted as a further and not exclusive learning mode.

As in all transformations, the transition takes place through an adaptive way, i.e., through the pressure that already but indirectly technologies exert on society and therefore wait until the time is autonomously ripe for the entry of technologies into the educational systems (especially in the scholastic reality) or prepare the field for a respectful transition, scientifically prepared, adequately tested, and substantially validated in its functions, guided by science and tested in its possibilities of real development.

One is thought to be leaning towards a path that is the second described because it is believed that a lousy transition that takes place for contingent reasons must be replaced by a reasoned change, supported by hypotheses and experiments that demonstrate the effectiveness of tools, and especially the organization of the contents. The awareness should replace reasoning from the perspective of a long-distance war between traditional and innovative didactics that the time is ripe to reach a new, extraordinary, epochal transition of the educational paradigm.

Soundscapes

The problem of the definition of music as a set of sounds can be found with an extraordinary description in a pioneering text of analysis written by R. Murray Schafer in his book, *The New Soundscape* (1969) which describes how the transformation brought by the American composer John Cage inside the musical fruition is that of having brought back the sounds 'outside' the street, the environmental reality, the everyday reality inside the concert halls.

This novelty reported by Schafer in his text opened the way to a series of essential meanings concerning the concept of music—first of all, that music and its highest expression became the entire sound universe, that new music and new musicians were all those who could produce sounds, that a further corollary opened up for music education which at that point became sound education. The idea that sound was the essential basis of any interaction between the elements of nature and man and also a specific characteristic of the features that crowded the universe opened the way to the previously reported concept of soundscapes (Schafer, 1993, 1969), intended as an overall sound scenario that opened up to acoustic, historical, anthropological, and unfailingly educational interpretations. In his 1993 text, *The Soundscape: Our Sonic Environment and the Tuning of the World,* the author argues with historical and anthropological analyses that the growth of society's overall development corresponded to a growing modification of the rugged landscape. The suffering induced by acoustic information's over-abundance was inversely proportional to our ability to perceive the differences, nuances, and subtleties of sound. In that text, Schafer (1993) argued that those interested in educating the richness of sound landscapes were precisely that of educating to listen, analyze, and make sensory distinctions and discriminations.

Schafer hypothesizes the idea that sounds have a matrix aimed at the psychic well-being of the subjects. To this end, he imagines a series of sounds which are able to accompany man, feed him, and recreate sound environments healthier. It does so by assuming education of what is called 'sound catwalks' to train people to sound quality. The idea of sounds being able to accompany, create healthy contexts, and guide and lead the subjects in training is the basic idea that pushes us to imagine sounds as natural and artificial mediators for the use of mobile technology, aimed at learning. But equally important is the line of studies that deals with the so-called ecology of the rugged landscape proposed (Truax and Barrett, 2011) as a new synthesis that leverages on two critical fields of study: landscape ecology and acoustic ecology. Schafer's vision, which is defined by anthropologist Stefan Helmreich of MIT (Helmreich, 2010) as 'pastoral conception', is, however, contrasted with an alternative vision that I would call incidental (Ingold, 2007).

Helmreich suggests that the rugged landscape would be nothing more than a mix of the aesthetics of contemplation with the acoustic observation of a reality that is both subjective and objective and therefore cannot be considered in its whole dimension precisely because it is the result of a mediated version of external and internal sounds. In this regard, he introduces technological sounds, such as those produced by telephony, phonology, and sound architectures that cannot be considered less relevant to understanding the world. In his article, he mentions how much diving through the submarine '*Alvin*' and how the sound phenomenon produced by that experience of diving in the deep sea was able to determine the wonder of transduction. In that subjective perception given by transduction, Helmreich states that sound truth mediated by the submarine walls could not be considered secondary to direct listening to the soundscape, which, especially in an underwater environment, would be impossible to achieve. Therefore, from this vision emerges a nature of environmental sound as an element that can be neither totally natural nor subjective, since in the mediation between subject and environment, and in the possibility of listening to the unheard through some instruments, evidence of a sound character is realized.

The idea of sound landscape as a system of sound relations and not as a split sum of individual isolated sounds is present in an article by Solomos (Breitsameter, 2018). In his article, Solomos suggests the existence of a strong link between the vibrating object (able to emit sound with regular vibrations), the environment in which the vibration spreads, and the subject is listening. He suggests that all current research in music, sound art, and other disciplines with a useful object implicitly seek to go beyond the concept of sound in favor of compounds that combine sound with other elements. Beyond, therefore, the notions of 'sound spaces' and 'sound environments', Solomos introduces the concepts of 'sound landscapes', 'sound environments', and 'environments' and 'sound atmospheres' (Solomos, 2018).

Mobile Learning

The vision of sound relations understood as the sound relationship between objects, allows one to look at mobile training systems' design to consider the singular matrix of sound and the relationship that it makes explicitly in the complicated relationship with other sounds.

From complexity as a condition of existence itself to complexity as a paradigm of organization, the elements can involve subjective reality in its learning plasticity. The design of learning environments is intended as a scientifically balanced formula to involve the entire matrixes that affect the subject in training. It is a question of looking at the technological component as an element of other programming and stimulating solicitation and not only as a tool to mediate some already systematized educational formulas. The work of redesigning the learning systems related to M-Learning must be thought of from the foundations. It must have a series of operational indications supported by well-organized and well-thought-out scientific positions.

The idea that among the primary characteristics, the relationship between action and sound has led to the hypothesis that sound design can be a real technological science with its technical aspects and precise operating conditions (Hug, 2020). Therefore, it is a question to of giving to the entire organization of learning processes related to the technological characteristics of M-learning, the feasibility determined by a careful, well calculated, designed, and studied interaction between sensory development, organization of learning, verification system, and flexibility of use time.

The idea of combining organizational processes with the on-demand dimension is a further opportunity to complete the training and self-training process. It is a matter of bringing within the educational design of static, dynamic, mobile learning environments, all those educational experiences of experiential matrix and above all, what has developed a strong bond with individual subjects' reality in the expression of their singularity also territorial. In this regard, numerous studies on the role of sound take up the 'local' dimension of training through sound and show how the real interest related to aspects of training mediated by music has a size strongly linked to the natural and social environment that produces it (Mark Reybrouck, 2020; George-Walker et al., 2010; Janata, 2009; Kukulska-Hulme, 2009; Lave and Wenger, 2006; Paynter, 2002; Ritchie, 2020; Schiavio et al., 2019; Solomos, 2018).

The sound has not only the function of involving the formation of the subject in a broad background. The issue cannot be thought of as an entire biological base, and the environmental sound does not have the exclusive task of providing it with more management elements than the absence of this condition within technological systems. In reality, the sound has such characteristics that can be expressed precisely because they are associated with the action of the subject to stimulate the embodied knowledge and motor characteristics to develop learning characteristics not directly associated with sound itself. This is the case of the experiment carried out by Register (2001). It demonstrated that in pre-school children, the guided use of listening to certain sound sequences and some specific music could develop in such pupils a greater predisposition to reading/writing (Register, 2001).

Therefore, the sound and the music can be constituted as elements on which to build a dense web of aspects and design to form the interactive basis of multiple training processes. In a detailed scientific analysis, entitled *Spatiotemporal Music Cognition* (Toiviainen and Keller, 2010), it is stated that the elaboration of sound and music in the human dimension includes a wide range of perceptual/active/cultural actions that involve cognition, emotion, learning, interaction, and inculturation. However, the problem analyzed by the two scholars (Toiviainen and Keller, 2010) is that research on human music processing has traditionally focused exclusively on the auditory domain, thus dissociating perception and action and largely ignoring multimodal interactions. This limitation can be a push to imagine that sound and music can not only be the background elements of the development of processes related to proper musical learning, to determined movement in dance, to musical training in a broader sense, but they can be realized as comprehensive training in relationships and sound worlds.

Sound has a vehicle of meanings as a communicative expression of contexts, as a promoter of movement, as an essential element of musical production, as a therapy for specific dysfunctions, as an aspect on which verbal communication is built, as an element on which emotional elements are constituted and conveyed in the paraverbal.

Sound, understood as a situated and distributed description of knowledge, cannot but have dignified and significant representation in realizing a system of training and education. Sound environments and their declination into soundscapes, music processing as an art of culturally regulated combination of sounds, the indications outlined to indicate some choices through the coding of some artificial sounds, the increasingly widespread tendency to use the so-called ASMR (Autonomous Response of the Sensory Meridian) (real sound scenes capable of reproducing familiar environments), remarkably serene contexts tell us that sound is beginning with a progressive force to make its way into all contexts of life as it is thought to begin to glimpse, or rather to feel, all the benefits it can have in the setting of all human activities (Chae et al., 2020; Poerio et al., 2010; Smith et al., 2020).

The role of sound becomes as central as the rediscovery of sound understood as a formative activity that must be taught diffused, carefully elaborated, and provided with those spaces of attention necessary for its development. Sound can therefore become a complementary and accessory element and also a mental place for those who are not able to have adequate visual stimuli about sensory disabilities; that is, it can be both a means to convey appropriate processes of development complementary to other learning and an autonomous element whose use can lead to more effective listening skills, better understanding of physical phenomena underlying the vibrations, and the essential elements of a culture of action focused on the correlation sound/movement.

For these countless reasons, the relationship between M-learning and sound, technology and virtual environmental sound, and an indication of movement in space, sound, and music are crucial elements for developing, mediating, and designing new frontiers of learning mediated and located in digital specialized training tools. It is not a matter of using digital technologies in the world but of bringing back into digital technologies part of the world with its adaptive phylogenetic dynamics and all those sensory elements to recreate a strong synergy between real and virtual, between real and digital, between natural and organized and organic supplement offered by technologies. The educational challenge is in full realization, the goals are ambitious, but the reality is already a part of the digital, and the digital is already a part of reality.

Technology and Subjectivity

At this point, however, it should be defined what M-learning is and how the sound dimension can be contemplated within the development of an educational research strand in substantial expansion and healthy development but still is far from being finely defined and structured primarily in light of the countless resistances that new technologies have always triggered in schools and training processes, especially those of an institutional nature.

The fact that the perspectives of M-learning are all to be prospected and defined in their epistemological nature, even before the operational one, prods us to think about the tools in a more detached way and to be guided by a theoretical reconnaissance that sinks however in the observation of a daily practice due to the massive use of smartphones, tablets, notebooks, and all other mass communication tools currently used. The problem addressed by Kearney et al. (2012) is the pedagogical possibility offered by mobile learning in a perspective that considers the temporal dimension condensed into a single temporal space. What emerge are the three main options provided by M-learning: authenticity, collaboration, and personalization. The feature of authenticity highlights the opportunity for contextualized, participative, situated learning; the collaboration-function captures aspects related to conversation and relationships; the customization feature has substantial implications for the possibility of opening fully customizable and autonomous paths in what is a path of personal organizational choices. In this case, the condition that plays a role of evolution or anchorage to reality is always the educational perspective's spatial and temporal dimension of the educational view (Kearney et al., 2012).

A wide variety of operational possibilities related to the use of M-learning appear from this vision; however, what is not clarified is 'how' and 'what' contributes to pushing this educational perspective towards operational hypotheses that can be used in training not just by merely relying on the use of technological tools and objects. The problem is still evolving and does not open to easy hypotheses that can be solved since the debate is substantially developed on the media and their ability to make communication fast, flexible, interactive, delocalized but with the possibility of continuing to remain within a specific and defined learning community or open, depending on the educational need one has and feels the urgency for it (Sharples et al., 2015).

Clearly, there is intention to define the function that M-learning could assume through the observation that some species-specific evolutionary elements are constituted as an internal substratum to the electronic development of a given product. Therefore, it is not our intention to investigate, as extensive literature already does, what use is made of mobile technology aimed at learning, but to verify which processes, neurophysiological characteristics could implement the use of these tools to develop appropriate design strategies to support sustainable development and strategy of commitment, presence, and flexibility (George-Walker et al., 2010) suitable to these systems.

This means that technology, together with culture and society, cannot be considered as elements external to the subject and, therefore, cannot be considered factors external to the training processes. The real effort lies in combining the biological, phylogenetic component of the issue, the one that tends to expand educational strategies through the use of some technological tools, such as M-learning. In this case, M-learning becomes a process through the use of a product which can bring back in its holistic logic: bios, logos, implicit adaptive responses, training support to transformation. It represents a strong possibility to transform the reality of the subjects and become a tool for educational change if thought in its technological nature as well as in the biological, evolutionary, social, cultural and playful is in a process that is no longer centered in a specific geographical place, but makes use of decentralized systems of participation, that enable people to communicate in different areas and at different times.

What would appear at first as the dissolution of the spatial and temporal dimension of the educational process, in reality, becomes a way to encourage the emergence of learning communities united by processes of interaction no longer strictly localized, but concretely linked to the learning interest of individual subjects, their specific training needs, their needs of the relationship concerning a given theme object of learning and especially in the overall time of life (McQuiggan et al., 2015; Ally, 2009; Innocenti et al., 2019). The subjects are released from a dimension of centralization of the learning process made of specific places and times equally organized, to bring them to an extent made of centers of interest, flexible organization, capacity for action, and reorganization of their learning and tools to manage them. The knowledge is no longer a dynamic process given in static contexts, but becomes an active process given through dynamic means.

In this way, the digital transformation of the world and its increasingly faster and more organic networks are closely correlated to companies' development with the need to combine that development with a capacity for management, organization, and use by training systems. Assuming closed systems in which cultural transmission is contemplated as an element of perpetuating tradition without allowing technological development to enter and transform the learning processes, in a context increasingly pushed towards digital transformation, means to realize static cultural methods within static structural elements.

One intention is to realize also in favor of tradition and its need to be a substratum on which to build the future learning processes capable of providing that tradition with elements of connection and active organization to achieve in an immaterial way distance learning communities, support them in their need for transformation, and provide them with tools of more significant contact with the social and cultural reality that lies beyond individual learning.

The Evolution of Linguistic Processing, Sound, and Verbal Adaptive Evolution

Our approach, therefore, has to do with the characteristics of the bio-educational sciences[37] that verify which relationships there are and there can be between the subject and the environment and which consequences of epigenetic development are determined in the training courses. The thorny topic to be addressed is therefore how the primary elements of human evolution can play a fundamental role in the development of education, training, and education systems that can envelop the subject in its biological and cultural characteristics and in turn, how they can be included in the development of electronic and digital technologies for training.

In a 2001 article, Mark Reybrouck argued that even the processes of musical understanding could not be built from the idea that music was a general ontological system in which one takes part through a path of individual and cultural growth, but that it had to be understood as an adaptive elaboration process that starts from sounds and arrives at their complexity (Reybrouck, 2001). Suppose an 'adapted' organism has found the way to manage and live the world in which it lives, understanding the processes of co-existence. This means that the adaptive process is not only identified with the relationship with the 'real world', but it is self-structuring. In fact, it implies elaborating its strategies of subsistence, management, and organization in the process of response and balance concerning the external world, yet with a process of reworking, structuring and organization within the subject itself. This results in knowledge, adaptability, and organizational strategy as a suitable tool to organize responses to an external system internally.

Therefore, these rational processes are said to be a conceptually strong basis for explaining the implicit principles that lead to the explicit adaptive development of the subjects in formation and, more generally, of the human race since sound itself represents the species' bio-cultural product. In the ability to balance one's inner dimension in response to the external environment, evolutionary processes capable of refining sound, communication, action on the world are realized, as it is precisely from the evolution of increasingly refined responses that a better ability to manage the elements external to the subject is recognized. This observation, already highlighted by Von Glasersfeld in the text, *Homage to Jean Piaget* (Glasersfeld, 1997), leads us back to a constructivist vision of the evolutionary dynamics of the subjects, in a path of relationship between subject and environment, though with a predominance in the organization of responses by the individual over the world.

The intellectual, evolutionary, and formative properties of the subject would result in an organizational and constructive capacity of pattern responses to the external environment (Piaget, 1977). On consideration of these hypotheses, it becomes evident that the organizational capacity of the single subject concerning the world—on which the subject acts and that at the same time imposes its concrete presence around the subject—determines that the whole of the adaptive characteristics

---

[37] Bio-educational sciences study the links between individuals' learning and the environment, investigating the biological dimension of subjects in the processes that support each individual's cognitive skills' epigenetic development.

intraspecies become an informative, genetic patrimony, able to contemplate a series of answers preorganized, already working and so useful to the management of the external world complexity. The possibility of generating pre-organized responses, and building new ones, states the extraordinary ability of man to be in continuous evolution, in perennial transformation, as promoter of management and harmonization elements between the subject and the environment.

In this perspective, the discourse carried out by Piaget (1977) is far from what one believes, as it destines to the subjective ability of the individual the power to 'solve' the world around him according to intrinsic adaptive skills in order to arrive at a species memory (Cavalli-Sforza and Feldman, 2003) able to solve, through a genetic structure, subjective behaviors. It is not a question of considering genetic memory as the absolute determinant of the form that the subject tends to assume in his personal development, but of observing how an intraspecies intelligence has allowed in the course of phylogenetic development to represent the resolution and adaptation to the world, transferring those evolutionary achievements to his genetic heritage. After all, the elements that make it possible for us to verify a strong structuring of man's adaptive responses to the world, allow us to build man culturally but with the advantage that some neuronal dynamics are a conquest result of the adaptivity of the species.

The evolution of linguistic processing, for example, can no longer be simply associated with the verbal ability to emit sounds and articulate the movement of the mouth but must be connected and thought of starting from general properties of sense-motor integration. Linguistic evolution, which could represent a moment of absolute priority in the adaptive process of the species, takes place when all the underlying processes or linguistic development find ways and moments of co-existence, such as the integration of acoustic, visual, articular, and temporal evolution of these relations with the recovery of referential memory (Greenberg, 1998).

This would confirm once again that the development of higher faculties on the one hand is the result of an adaptive response to the surrounding world that determines a process of epigenetic character; while on the other, it is the organized and organic relationship between split functions. Man evolves further when he can unify and give unity to interaction with the outside world. Genetic constraints and phylogeny in mammals would play the role of channeling homologous processes in an extensive series of species, whose brain dimensions are different. This hypothesis elaborated by Manger (2005) tries to tell us that the process of brain organization follows similar primary organizational logics in some mammalian species and consequently represents an implicit organizational formula, which is able to elaborate more and more sophisticated adaptive strategies, integrating systems, dividing them, correlating brain areas, structuring complex responses (Manger, 2005).

These hypotheses should make us understand two substantial elements: the first is that our genetic make-up contains in itself evolutionary factors without which one cannot do and often determine their form on the environment due to the adaptive capacity of individuals refining them; second, the environment which has transformed its natural elements can and indeed continue to change from a social, cultural, and technological point of view. From these two shreds of evidence, it is assumed that any educational process cannot but take into account a genetic, implicit, pre-organized nature of the responses that the subject has borrowed from his species and that these responses are firmly present in all learning dynamics, to the point of being strongly 'contaminating' the teaching-learning processes. The other observation tells us that formal, non-formal, and informal educational processes cannot occur by excluding from educational and training practice everything that happens in the world and around the formed subjects. Educational processes, where the external component (social, cultural, and technological) can be extraneous to the topic being created, cannot be hypothesized.

The reasons are quite evident: when social and cultural evolution passes through men, it undergoes a shaping/modeling process in a dichotomous way within which the individual contributes to modify it and is limited, technology and technological artifacts are catalysts of transformation as they perform the function of reducing complexity and/or management of the same, to encourage an adaptive response and support the need for a playful dimension by the subject.

Sound, Corporeity, Verbal and Paraverbal Communication

In the scientific field, concerning the implicit dimension of sound and musical perception, one speaks of cognitive modules related to this activity with precise temporal functions and located in exact human brain areas. These considerations are based on the observation that sound often stimulates many primary elements, including emotional development and emotional expression and musical perception, which can potentially influence emotions, affect the autonomic nervous system, the hormone and immune system, and activate (pre)motor representations (Koelsch and Siebel, 2005).

In one of their researches, Koelsch and Siebel (2005) state that sound perception, and in particular that which is in its most evolved form, which is music, can involve complex and numerous brain functions. They concern the acoustic analysis that regards the sensory dimension, auditory memory, spatial analysis of the auditory scene, and only later do they work on elaborating musical syntax and semantics. This transition from sensitive to cultural data helps us understand the evolutionary nature of sound (Koelsch and Siebel, 2005).

Therefore, there is a close correlation between biological data and sound as evidenced/proven by elements of the brain's physiological structure that allow us to say that sound and its evolutions are elements that have transformed the physical nature of the mind itself. The subsequent growth of sound as a complementary element to language, a characteristic proper to nature, an indicator of danger, an aspect of communication between subjects has assumed in the phylogenetic transition a more and more symbolic and semantic connotation to the point of evolving into a verbal language to express the linguistic evolutions of a given social group, and allow it to communicate, and in other variants grow into language becoming musical language (The Functional Phylogenies Group, 2012). The musical lexicon collects secondary elements more and more evolved to become a real autonomous language diversified in different genres, specific cultural evolutions, instrumental and vocal sound representations.

However, the characteristics of sound are not limited to the simple field of perception/sound production but significantly call for the body of the subjects. Research in the field of embodied musical cognition, conducted at IPEM, a research laboratory in systematic musicology at the University of Ghent in Belgium, is increasingly addressing the relationship between corporeity, knowledge, and sound/music (Leman and Maes, 2015). It is referred to coding/decoding based on musical expressiveness, synchronization, and dragging, and the effects that actions have on musical perception. The empirical results discussed show that the embodied dimension of knowledge is part of an interconnected network of sensory, motor, affective, and cognitive systems involved in music perception. These researches show that embodied cognition is a theory based on a dynamic network of relations and sound and music (Leman and Maes, 2015).

Therefore, the corporeity—and more generally man in his holistic complexity—cannot amend, in the learning and training development of subjects, the emergence of multiple relationships between perception, emotions, verbal development, bodily actions, active approach in the relationship between subject and environment. In short, the issue is overall in its physical, sensory, communicative expressions, and this strongly involves sound in its embodied, organismic, subjective, and plural dimension (Greenberg, 1998; Burunat et al., 2015; Maes et al., 2014a; Jensenius, 2007; Pennycook, 1985; Rosenboom, 2020; Ciasullo, 2015, 2020a).

The active process of listening in our societies is considered more linked to the act of communication and therefore, to the content of the communicative relationship than to the selection, distinction, and deep recognition of the sound meaning. It should be accordingly re-established that the link between sound as a connecting element between the individual and the environment is already in the natural composition of sound. Expressivity, emotion, fluency are elements proper to language but are always accompanied by paraverbal communication; that is, the set of tonal sounds that give speech different meanings. The same words could be used with another sound-linguistic registerable to transmit to the listener the linguistic object of what one is communicating and an 'implicit sound' which is able to transfer emotions, expressivity, and character data of the source.

In the text, *Paraverbal Communication with Children: Not through Words Alone* (Heimlich and Mark, 2012), the authors deal with the difficulties of communication that afflict some children of school age and especially those with problematic children; the text refers to the great possibility offered by the use of rhythm, the pleasure of communication promoted by sound and how this can activate, integrate, and promote the development of emotions and ideas. The sound of the paraverbal drags and the rhythmic presence furthers/promotes learning elements in the subjects involved to the point of stimulating them in the physical action. It is the dimension that Maes (Maes et al., 2014a) defines as the coupling between action and perception in the formation of musical meaning. In the study of Maes et al. (2014a), there is an attempt to demonstrate that embodied experience of music is an essential factor in people's use and understanding of musical metaphors.

Sound and Its Evolution in Music

More often than it is thought, when people listen to a song, a melody, a sound, or musical composition, they match that same listening to the actions just mentioned to describe the melodic and rhythmic contours or the gestures that produce the sound from which it originated (Godøy and Leman, 2010). That's why, through the use of a model of description of the associations between movement and cognitive metaphors, they represent Effort/Shape Theory by Laban Maes et al. (2014). It shows that there is a very close correlation between the music listened to, physical movement (which the listener uses to describe it) and meaning it represents for the subject itself. All this happens in all topics considered by the study, whether they are musically literate or without previous specific musical training. This confirms that the sound dimension, linguistic meanings, and corporeity can co-exist at an implicit level in the individual subject but it also supports that the sound expresses its own form, spatiality, vision, and action. Therefore, it exists both in the sensitive sphere as well as in the physical body. It's not a bold statement to say that sound is 'implicit action'.

Rethinking music with a more dynamic approach can certainly lead to the study of the musical phenomenon that is capable of influencing research, well-being, healing, sport, musical engineering, and brain studies (Leman and Maes, 2015). Musical perception has long been seen as an element supporting the body dimension to the point of considering musical and bodily expression as in a dual system of mutual collaboration. This allowed Leman and Maes (2015) studies to find broad elements of concordance to explain, through musicality, bodily expressions. This is possible because embodiment presupposes the existence of mirroring processes that facilitate the encoding of expressive gestures into sounds and the decoding of sounds into expressive gestures.

Therefore, it can be affirmed that the action determined by musical listening is fully rooted in the sense-motor component and is intrinsic to human physiological dynamics. Consequently, sound and music result in direct expressions of human subjectivity. The translation of sound into gesture can also have, according to Maes et al. (2014b) studies, a level of inverse translation, i.e., from a fluid and coordinated movement, it is possible to define which type of sound or sound it is understood from. Therefore, musical expression and body expression act in an implicit communication channel, though not always bound to emotional aspects, to determine a series of two-way relationships to form a substantial alliance between action and sound. At this point, it follows that the hypothesis of music semiotics is no longer only bound to aspects of a conventional character and by codified and shared rules, but would be in some way a direct expression of a series of implicit correspondence mechanisms at the basis of perception/action/movement.

This close correlation between sound and action implies the possibility for sound indications to be mediators for real action or virtual action as through their direct exchange with the body of the subject, they determine the gesture, the movement, and the mobility towards dynamics of a physiological character. This leads us to consider sound as an element of indirect relationship with the subject's biological base and, therefore, as an element linked to the human being's phylogenetic processes. The sound would no longer be only an element of communication, but a direct, expressive potential based on the subject's relationship with the world.

The idea of realizing learning processes mediated by technology cannot be admitted without an adequate possibility to discover education and sound training. Even more, imagine a technological development made of tools for educational and training mediation without the contribution of sound and music (Gordon, 2007; Whitman, 2005). Imagining a curricular development of skills and sound perception enhancement only through short and occasional music lessons it not enough. Not making a proper use of the right training processes that enhance the relationship between sound and action seems inappropriate. Therefore, there is a need to rediscover sound, understand the aspects related to gestures, recover the implicit in the relationship between sound meaning, and the active correspondence of gestures and movements. This means rediscovering and decoding the gestures linked to sound and, if possible, to decline a theory of the sound gesture and proceed to reconstruct a series of formative actions aimed at reprogramming the relationship between sound and consequent real/virtual actions, moving to rewriting and re-education to sound.

For the time being, sound is mainly a communicative mediator linked to verbal language and, in its musical dynamics, an element of socio-cultural representation. Sound is thought to be a relationship of signification, mainly related to word and music and one which does not enhance sound in its singular and peculiar aspects. Instead, one should have a series of curricular elements that sanction the priority value of sound as a primary element of communication and direct mediation with its autonomy, status of meaning, autonomous structure, and epistemology. It is a matter of recovering its nature and design, verifying its motor correspondences, clarifying the aspects of equality, codifying its peculiarities, and proceeding to a systematic definition of the sound element. These possibilities could quickly lead to an educational rewriting of sound and of a sound/action grammar by no more lengthy proceedings according to perspectives without an adequate scientific support and not based on substantial aspects of the development's sound/action relationship.

Centrality of Sound

Giving centrality to sound does not certainly mean not considering the role of verbal communication and how much it is substantially made up of sound and stands on it. The peculiarities of sound allow the whole phonological and phonatory system to perform the great evolutionary function of verbal communication that has allowed man to evolve in his cultural, social, and existential dimension (Berent et al., 2016; Peyrin et al., 2012). However, what is evident in this scale of elements is that sound represents the basis on which verbal language constitutes its plot, its substance, its strength through a series of secondary elements defined as 'paraverbal', that include kinesics, proxemics, and emotional tonality that tends to give similar words different meanings, intentions, moods, depending on the type of tone attributed to the emission of the same name.

The gestural expression, the facial expressions that accompany or not the vocal sound, the mirrored interfacial sync between the interlocutors, the so-called eye contact, the use of space and the control of spatiality as an implicit category, the touch and its tactile and motor dimensions, the aspects of sound modification of the voice and silence, play a crucial role in human interaction and are evolutionary powerful, and all the elements are catalyzed regarding some main features: sound, action, sight (Pennycook, 1985). All dimensions related to paralanguage described by Pennycook in a study of 1985 tend to integrate an extent of co-existence necessary between an explicit communicative dimension and a correlation dimension related to paralanguage. These premises therefore, also involve a substantial possibility of elaboration of the relationships between perception and interpersonal communication on the visual channel, sound, and therefore, even concerning the implicit connections between action and sound (Leman and Maes, 2015; Maes et al., 2014a; Maes et al., 2014b).

There can be no communicative act that does not consider the sonority that until now would seem to be evolutionary, always associated with the dimension of the verbal without considering an autonomous and complementary function. Understanding the independent nature of sound and educating to sound before music means elaborating strategies of explicit decoding of the role that sound takes on in human communication, in the position it could take on for the development of adaptive and enriched learning environments (Santoianni, 2010) that could take advantage of an 'explicit theory of sound'. This characteristic of real contexts enriched with all the components of sensory stimulation, including sound, in our text also takes on the function of opening up to the integration and processing of sound models to integrate technological tools, including mobile technology with training purposes.

The Evolution of Sound between Learning Groups and Technologies

Lave and Wenger (2006) state that language acquisition, of communicative evolution, results from iteration and co-operation. The process of language learning, but above all, the development of the strategies connected to it, including sound, would be elements produced by interactions and not by an internal representation of the subjects. After all, if communication, sound, and action are considered aspects of exchange and management with the outside world, it is inevitable that the linguistic construction and all other forms of cultural construction would only be responses to relations, including human social links. What is defined as *legitimate peripheral participation would be nothing* more than the minimum contribution given to social interactions to co-construct meanings; however, even in this marginal partiality, there would be elements of operational transformation of practices, meanings, vocabulary, and actions of a learning community. According to these theses, 'learning' would be more complicated than 'understanding' because the first process is always mediated by the group within which these dynamics of transformation are defined (Lave and Wenger, 2006).

These concepts lead us to identify in the educational relationship, in the phylogenetic organization, in the cultural sophistication of the increasingly detailed responses to the world's growing complexity, the background within which sound and action find reasons for convergence, union, and evolution. Therefore, the dynamic process of transformation of meanings is no longer a simple individual and subjective evolutionary characteristic but becomes more and more a vehicle of internal/external transformation in a perpetual correlation. What changes is the use and mediation that certain 'genetic constructs' determine in certain 'cultural constructs'.

Educational Mediations

Education can prepare elements that facilitate and take into consideration. In reasonable priority, the subjects' evolutionary component and the parts proper of the subjective corporeity can happen only when correlating the organismic subjectivity of the topic with the environmental dimension (virtual, natural, social) in a relationship of continuous osmosis. It is not a matter of compressing the subject towards the ecological component, external to his subjectivity, because it would cancel the implicit, character, individual instances. Still, it means to allow the individual to express in the relationship with the environment his constitutive diversity equipping him with management and organization tools of the external reality itself.

In this vision of mediation between the subject and reality, a relevant role should be played by the technologies that by now fill, minute after minute, our daily life, but that seems once again to find excellent resistance in the school communities of many countries of the world. The problem already widely addressed and well documented by Papert (1993), is namely that the technological advancement within the school always takes place concerning teachers' resourcefulness to mediate between the external demands of society and those of change within the school reaffirmed. According to Seymour Papert, already in 1993 a substantial and anachronistic resistance to technological progress was demonstrated, as schools continue to isolate the computer in separate classrooms or at most use the interactive whiteboards and/or resident technology in an entirely improper way. The need to use the technological tools is revealed; they are always aimed to include modalities and programs proper of the traditional didactics. In Papert's vision, there was already the proposal of the computer understood like a part of all the learning as for the past it had been the books, the pens, and the pencils. The idea that guided his theories was linked to the fact that everyone should become autonomous authors of the learning style regarding their learning dynamics (Papert, 1993). These proposals are still current today, and their request is once again unfinished and lacking in organizational feedback, such as to imagine a real revolution in learning and schooling.

There is intention to mediate between the needs of a traditional vision of the school in presence and the opportunities to build appropriate processes representative of a school made of innovation, technological mediation, respect for subjective corporeity, in line with the needs of transformation and evolution that concern the human dimension as such.

Clearly, what is really concerning in this transition of educational possibilities offered by technologies, is the substantial distant position that is determined by the needs of 'conservation', poor attitude to change, reduced capacity for qualification, and retraining of teaching staff, and the inadequate possibility of investment in quality technological transitions. Suppose this appears as a risk of a system extended to the whole dimension of formal education. In that case, it is quite substantial compared to the construction of curricular approaches aimed at enhancing sound education before music.

However, the curricular approach to music hides even more than the transition to new technologies, a delay in preparing the teaching class, especially in elementary school. The time dedicated to listening, to the identification of the spatial depth of sounds, to recognize the characteristics of pitch, intensity, and timbre are increasingly hidden within school curricula dedicated to music composed mainly of ditties, low-level choral practices, and short hours for the structuring of musical curricula, focused first in the enhancement of sound elements and then in the understanding of musical genres (Paynter, 2002). All this appears as a reverse process in which you start from reading a great classic of literature without knowing either the grammar or the word's meaning. The actual paradigm should be turned upside down, not only for those who in a specialized way believe that they need to deepen the musical discourse for preparation as musicians or music lovers, but also for rewriting adequate training to sound.

It's fascinating how Green's (2008) idea is to bring back into the classroom informal learning practices so as to recognize and promote a range of musical skills and knowledge that have long been neglected in music education. In other words, the aim is to reshape music education, and I would add 'sound', broadening the range of recognition of musicality, intrinsic to each subject. Rewriting such a curricular representation would open the possibility of rediscovering intrinsic sonority within all issues capable of linking action and perception. Such training would allow us to have sound as a unifying mediator between corporeity, technology, and the design of both the real and virtual learning spaces; a sound as an active grammar of sound and body understanding of the subjects (Green, 2008).

Sound Education between Musical Instruments and Real Scenarios

It is very interesting, in this regard, that the experimentation carried out by Ritchie (2020) who, through the use of real string instruments (violins, violas, cellos), actively investigates the sound production by the subjects involved through a series of phases in which they are brought into a narrative dimension of the approach to the use of instruments, to verify how direct learning of something new can offer opportunities for personal growth. People are thus involved in experimenting with the negotiation of complex physical and mental processes, integrating them quickly and efficiently into practice with non-verbal cues, and above all, they experience a new way of training in obvious actions and community learning practices (Ritchie, 2020).

The intent to participate in the processes of realization of concrete experiential formative acts should not be misunderstood as an educational experience, but rather as a process of deciphering and unifying actions, formative mediations, and technical outcome of such activities. This working hypothesis, focused on enhancing and rediscovering the primary qualities of sound as a crucial element of human evolution, would have the merit of bringing back training to improve the very essence of sound significantly. The rewriting of a sound alphabet would also allow the reconstruction of a relationship of greater awareness with the external environment.

Experiments in acoustic education were reported by Teruyo Oba (2006) where an automated auditory recognition system, called Kikimimi-Zukin, allowed children involved in a JST 2003/2004 project to recognize the sound environment and thus realize what a useful scenario they were living in. I think the idea of a sound landscape (Schafer, 1993) could represent the most exciting study hypothesis because it creates a scenario of interaction between the subject and the surrounding environment, starting from observing sounds and giving to each one characteristic, source, function. Oba's idea was to determine in children an initial cognitive effort to identify the sounds and their nature of belonging, all mediated by their school community to allow a sharing and mediation on the interpretation of the surrounding sounds.

The Kikimimi-Zukin instrument was first used as a step-by-step guide through the auditory process of scanning, focusing, characterization, and association. This opportunity to identify the soundscape (Schafer, 1993) has strongly oriented children towards a greater awareness of the local music scene. Above all, it has allowed them to trace a more spatial and three-dimensional understanding of the sound phenomenon (Oba, 2006). The sound dimension can, therefore, be summarized in some substantial elements:

1. Sound is a primary acoustic process associated with movement and gestures.
2. Sound clues constitute a very interesting and robust base of space clues.
3. The sound landscape helps to identify several well-located environmental peculiarities and characteristics of a given place.
4. Sound perception, as well as other characteristics of human education, can be encouraged and trained through appropriate training processes.
5. There can be no organization and technological mediation to realize adequate learning processes that do not consider sound, soundscape, and music as elements that compress other perceptual features.

From these considerations, it follows that the organization of a system of interaction between the subject, learning, and technology cannot do without elements of sound representation to stimulate the issue to motor actions, to signal the presence in a place through spatial indications also within VLEs (Virtual Learning Environments), to allow him to identify the characteristics of a specific location, situation, context (Santoianni and Ciasullo, 2020; Ciasullo, 2020a; Santoianni and Ciasullo, 2019).

All these elements, on the one hand, belong to an implicit dimension of the phylogenetic endowment of the human species, and on the other, as important vehicles for the development of the subject, they must adequately involve the training processes in the structuring of a sound education curriculum, before any musical training.

References

Ally, M. (2009). *Mobile Learning: Transforming the Delivery of Education and Training.* Athabasca University Press, Athabasca.

Berent, I., Zhao, X., Balaban, E. and Galaburda, A. (2016). Phonology and phonetics dissociate in dyslexia: evidence from adult English speakers. *Language, Cognition and Neuroscience.* Routledge, 31(9): 1178–92.

Blasi, D.E., Wichmann, S., Hammarström, H., Stadler, P.F. and Christiansen, M.H. (2016). Sound—meaning association biases evidenced across thousands of languages. *The Proceedings of the National Academy of Sciences*, National Academy of Sciences, 113(39): 10818–23.

Breitsameter, S. (2018). Soundscape. pp. 89–95. *In*: Morat, D. and Ziemer, H. (eds.). *Handbuch Sound: Geschichte – Begriffe – Ansätze*. J.B. Metzler, Stuttgart.

Bribiescas, R.G. (2009). *Men: Evolutionary and Life History*. Harvard University Press.

Brown, D., Macpherson, T. and Ward, J. (2011). Seeing with sound? Exploring different characteristics of a visual-to-auditory sensory substitution device. *Perception*, SAGE Publications Ltd, STM, 40(9): 1120–35.

Burunat, I., Brattico, E., Puoliväli, T., Ristaniemi, T., Sams, M. and Toiviainen, P. (2015). Action in perception: prominent visuo-motor functional symmetry in musicians during music listening. *PloS One*, Public Library of Science, 10(9).

Carles, J. L., López Barrio, I. and de Lucio, J.V. (1999). Sound influence on landscape values. *Landscape and Urban Planning*, 43(4): 191–200.

Cavalli-Sforza, L. and Feldman, M.W. (2003). The application of molecular genetic approaches to the study of human evolution. *Nature Genetics*, Nature Publishing Group, 33(3): 266–75.

Chae, H., Baek, M., Jang, H. and Sung, S. (2020). Storyscaping in fashion brand using commitment and nostalgia based on ASMR marketing. *Journal of Business Research*, March.

Ciasullo, A. (2015). Armonie Bioeducative, *Scale e Arpeggi Pedagogici, FrancoAngeli*. Milano.

Ciasullo, A. (2020). *Opportunità degli ambienti virtuali tra sonorità, simbolizzazione e spazialità. In*: Panciroli, C. (ed.). *Animazione Digitale per La Didattica, FrancoAngeli*. Milano.

Degli Innocenti, E., Geronazzo, M., Vescovi, D., Nordahl, R., Serafin, S., Ludovico, L.A. and Avanzini, F. (2019). Mobile virtual reality for musical genre learning in primary education. *Computers & Education*, 139(October): 102–17.

Fitch, W.T. (2016). Sound and meaning in the World's languages. *Nature*, 539(7627): 39–40.

George-Walker, L.D., Hafeez-Baig, A. Gururajan, R. and Danaher, P.A. (2010). Experiences and perceptions of learner engagement in blended learning environments: the case of an Australian University. *Cases on Online and Blended Learning Technologies in Higher Education: Concepts and Practices*. IGI Global, Pennsylvania.

Glasersfeld, E.V. (1997). Homage to Jean Piaget (1896–1982). *The Irish Journal of Psychology*, 18(3): 293–306.

Godøy, R.I. and Leman, M. (2010). *Musical Gestures: Sound, Movement, and Meaning*. Routledge.

Gordon, E. (2007). *Learning Sequences in Music: A Contemporary Music Learning Theory*. GIA Publications, Chicago.

Green, L. (2008). *Music, Informal Learning and the School: A New Classroom Pedagogy*. Ashgate Publishing, Farham.

Greenberg, S. (1998). A syllable-centric framework for the evolution of spoken language. *Behavioral and Brain Sciences*, 21(4): 518Cf–518.

Group, T.F.P. (2012). Phylogenetic inference for function-valued traits: speech sound evolution. *Trends in Ecology & Evolution*, 27(3): 160–66.

Heimlich, E.P. and Mark, A.J. (2012). Paraverbal Communication with Children: Not through Words Alone. *Springer Science & Business Media*, Berlin, Heidelberg.

Hug, D. (2020). How do you sound design? An exploratory investigation of sound design process visualizations. pp. 114–121. *In: The Proceedings of the 15th International Conference on Audio Mostly*. New York, Association for Computing Machinery.

Ingold, T. (2007). Movement, knowledge and description. pp. 194–211. *In*: Ulijaszek, S. and Parkin, D. (eds.). *Holistic Anthropology: Emergence and Convergence*. Berghahn Books, New York.

Janata, P. (2009). Music and the self. pp. 131–141. *In*: Vaas, R. and Brandes, V. (eds.). *Music That Works: Contributions of Biology, Neurophysiology, Psychology, Sociology, Medicine and Musicology*. Springer, Vienna.

Jensenius, A.R. (2007). *Action-Sound: Developing Methods and Tools to Study Music-related Body Movement*. Ph.D. thesis, Department of Musicology, University of Oslo.

Kearney, M., Schuck, S., Burden, K. and Aubusson, P. (2012). Viewing mobile learning from a pedagogical perspective. *Alt-J. of Research in Learning Technology*, 20(1).

Koelsch, S. and Siebel, W.A. (2005). Towards a neural basis of music perception. *Trends in Cognitive Sciences*, 9(12): 578–84.

Kukulska-Hulme, A. (2009). Will mobile learning change language learning? *ReCALL*, 21(2): 157–65.

Lave, J. and Wenger, E. (2006). *L'apprendimento situato. Dall'osservazione alla partecipazione attiva nei contesti sociali, Edizioni Erickson*. Trento.

Leman, M. and Maes, P.J. (2015). The role of embodiment in the perception of music. *Empirical Musicology Review*, 9(3-4): 236–46.

Maes, P.J., Van Dyck, E., Lesaffre, M., Leman, M. and Kroonenberg, P.M. (2014a). The coupling of action and perception in musical meaning formation. *Music Perception*, 32(1): 67–84.

Maes, P.-J., Leman, M., Palmer, C. and Wanderley, M. (2014b). Action-based effects on music perception. *Frontiers in Psychology*, 4.

Manger, P.R. (2005). Establishing order at the systems level in mammalian brain evolution. *Brain Research Bulletin, Evolution and Development of Nervous Systems*, 66(4): 282–89.

McQuiggan, S., McQuiggan, J., Sabourin, J., Kosturko, L. and Shores, L. (2015). *Mobile Learning: A Handbook for Developers, Educators, and Learners*. Wiley-Blackwell, Hoboken.

Oba, T. (2006). The sound environmental education aided by automated bioacoustic identification in view of soundscape recognition. *The Journal of the Acoustical Society of America*, 120(5): 3239–3239.

Papert, S. (1993). *The Children's Machine: Rethinking School in the Age of the Computer*. BasicBooks, New York.

Paynter, J. (2002). Music in the school curriculum: why bother? *British Journal of Music Education*, 19(3): 215–226.

Pennycook, A. (1985). Actions speak louder than words: paralanguage, communication, and education. *TESOL Quarterly*, 19(2): 259–82.

Peyrin, C., Lallier, M., Démonet, J.F., Pernet, C., Baciu, M., Le Bas, J.F. and Valdois, S. (2012). Neural dissociation of phonological and visual attention span disorders in developmental dyslexia: FMRI evidence from two case reports. *Brain and Language*, 120(3): 381–94.

Piaget, J. (1977). *The Development of Thought: Equilibration of Cognitive Structures*. Viking, Oxford.

Poerio, G.L., Blakey, E., Hostler, T.J. and Veltri, T. (2018). More than a feeling: autonomous sensory meridian response (ASMR) is characterized by reliable changes in affect and physiology, public library of science. *PloS One*, 13(6): e0196645.

Register, D. (2001). The effects of an early intervention music curriculum on prereading/writing. *Journal of Music Therapy*, Oxford Academic, 38(3): 239–48.

Reybrouck, M. (2001). Biological roots of musical epistemology: functional cycles, umwelt, and enactive listening. *Semiotica*, 134(1/4): 599–633.

Reybrouck, M. (2020). Experience as cognition: Musical sense-making and the 'in-time/outside-of-time' dichotomy. *Interdisciplinary Studies in Musicology*.

Ritchie, L. (2020). Images of learning through music: the sounds of cognition. *International Journal of Management and Applied Research*, 7(3): 257–66.

Rosenboom, D. (2020). Active imaginative reading and listening. *Leonardo Music Journal*, September, The MIT Press.

Santoianni, F. (2010). *Modelli e strumenti di insegnamento, Approcci per migliorare l'esperienza didattica*. Carocci, Roma.

Santoianni, F. (2014). *Modelli di studio, Apprendere con la teoria delle logiche elementari*, Centro studi Erickson, Trento.

Santoianni, F. and Ciasullo, A. (2018). Digital and spatial education intertwining in the evolution of technology resources for educational curriculum reshaping and skills enhancement. *International Journal of Digital Literacy and Digital Competence*, 9(2): 34–49.

Santoianni, F. and Ciasullo, A. (2020). Teacher technology education for spatial learning in digital immersive virtual environments. *In: Examining the Roles of Teachers and Students in Mastering New Technologies*. IGI Global.

Schafer, R.M. (1969). *The New Soundscape*. BMI Canada Limited, Don Mills.

Schafer, R.M. (1993). *The Soundscape: Our Sonic Environment and the Tuning of the World*. Simon and Schuster, New York.

Schiavio, A., Gesbert, V., Reybrouck, M., Hauw, D. and Parncutt, R. (2019). Optimizing performative skills in social interaction: insights from embodied cognition, music education, and sport psychology. *Frontiers in Psychology*, 10(July).

Sharples, M., Taylor, J. and Vavoula, G. (2005). Towards a theory of mobile learning. pp. 1–9. *In: The Proceedings of M-Learn*, 1(1).

Smith, S.D., Fredborg, B.K. and Kornelsen, J. (2020). Functional connectivity associated with five different categories of autonomous sensory meridian response (ASMR) triggers. *Consciousness and Cognition*, 85 (October).

Solomos, M. (2018). *From Sound to Sound Space, Sound Environment, Soundscape, Sound Milieu or Ambience ... Paragraph*, 41(1): 95–109.

Toiviainen, P. and Keller, P.E. (2010). Special issue: spatiotemporal music cognition. *Music Perception*, 28(1): 1–1.

Truax, B. and Barrett, G.W. (2011). Soundscape in a context of acoustic and landscape ecology. *Landscape Ecology*, 26(9): 1201–1207.

Vicky, A.M., Dorey, N.R. and Ward, S.J. (2020). Front matter. pp. I–XXVII. *In*: *Zoo Animal Learning and Training*. John Wiley, Hoboken.

Whitman, B.A. (2005). *Learning the Meaning of Music*. Thesis, Massachusetts Institute of Technology.

CHAPTER 7

Interactions between Technologies, Body, and Sound Production

Alessandro Ciasullo

From the most elementary sensorimotor actions to the most sophisticated intellectual operations, which are interiorized actions carried out mentally, knowledge is constantly linked with actions or operations, that is, with transformations.

—Piaget (1976)

Introduction

According to numerous studies, the link between body and sound, body, movement, and music is powerful. The reasons take us back both to a physiological dimension as well as a socio-cultural explanation. The relationships between music and body can be thought to be the evolutionary aspects of movement and sound. These multiple correspondences between music and body also exist because both activities evolve within specific cultural contexts: one needs specific instruments, devices and/or equipment in the contingency of some environments with particular characteristics and both are determined by some social factors, which can both improve the cognitive and artistic skills of those involved (Schiavio et al., 2019).

There is then a close correlation, which we could define as cause/effect of motor action and sound in its evolved relationship which is music. This link has led cognitive neuroscience and humanities to interpret the relationship between effort and movement, action and musicality and to be analyzed, based on representations and perspectives of what is called 'embodied knowledge'. From Schiavio et al. (2019) study, we know that by observing the social interactions that intervene in

Researcher of Education, University of Naples, Federico II.
Email: alessandro.ciasullo@unina.it

motor action and sound, both contribute significantly to understand and implement appropriate learning processes in the so-called social groups or learning communities. In addition, the skills are often acquired together, in groups, whether they are activities for musicians or athletes. What emerges is how the performance in either body activity or music is strictly connected to the development of distributed forms of bodily memory (Schiavio et al., 2019).

These scientific findings lead to understand that the collective dimension of mediation between corporeity, movement, sound, and digital technologies can only be developed from particular, codified, specific forms of collaboration. This vision changes the order of learning processes that would see the acquisition of skills as a subjective, autonomous, and solitary process of improvement. This means that the expressive possibility mediated by the use of technology can be oriented to develop, improve, and create adequate processes of self-efficacy and autonomy of the subjects; in fact, in order to achieve it, it must be initiated from collective cognitive construction developments.

Therefore, cognitive construction is an element of collective mediation, while the acquisition of instrumental skills and the development of one's knowledge and skills can be subjective. Moreover, especially in music or high-level body actions, the perturbative element determined by unforeseen situations during a performance and the answers given to solve the emergency cannot be considered as an expression of a regulatory and resolving instinct and therefore as an adaptive response. In reality, learning also regulates the adaptive response in certain conditions of apparent momentary disadvantage; even the most complex ones, since they appear as the phylogenetic evolution of a genetically stratified collective memory, also mediated by social relationships that determine in the individual the acquisition of social skills in the immediate and in the memory of species' long-term genetic skills.

This peculiarity linked to bodily adaptability mediated by social skills and learning is undoubtedly an evolutionary characteristic. It is precisely through co-operation that man can solve problems in the community that cannot be achieved alone. The collaborating aspect, however, as in all the characteristics of transition between nature and culture, cannot be limited to the management of phenomena that concern the simple power of everyday life, of the simple biological dimension of subjects, but must evolve, in this path of cultural transition, including the digital tools, to become support, prosthesis, instrument of new processes of subjective liberation. Before representing these tools as an element of further subjective realization and therefore as a push to autonomy, they should be considered as instruments in favor of distance and/or in community learning processes. This would be so to proceed from the plural to the singular, from the learning community to the possibility of formative autonomy, from collectivity to singularity. Differently, the risk would be that digital tools for training may become anomic places of isolation and reduce the individual's sociality. Just think of the apparent sociality resulting from the use of social networks that lead to an allegedly extended network of relationships, but which nevertheless contribute to substantial isolation of the subjects.

Is Mobile Learning Ready to Host Sounds?

In studies conducted some time ago on the state of the art of *M-learning,* the main categories related to this technological possibility were:

- *Technology-driven mobile learning* describes mainly a series of innovative technologies to demonstrate the characteristics of flexibility and didactics of some tools;
- *Miniature but portable e-learning,* portable mobile technology, wireless, are handy to move on the smartphone or some virtual learning environments (VLE) technologies to better enjoy the contents of e-learning, overcoming the desktop computer's stativity;
- *Connected classroom learning,* mobile technologies used for collaboration within a classroom are perhaps connected with other classrooms in the interactive whiteboard presence;
- *Informal, personalized, situated mobile learning,* the same mobile technology enriched with localization and video recording, is aimed at developing educational experiences that would be difficult or impossible to achieve in other ways;
- *Mobile training/performance support* technologies are used to increase workers' productivity and efficiency as they need to have an immediate just-in-time resolution of some problems;
- *Remote/rural/development mobile learning* technologies are used to address infrastructure development to implement appropriate training and evolutionary processes where traditional e-learning has failed (Traxler, 2005; Traxler, 2007, 3; Grimshaw-Aagaard, 2019; Ally, 2009, 12–13).

In the ordinary meaning attributed to M-learning, a vision emerges, still entirely focused on observing technology, technological-centric and which is not even that entrusted to technical analysis about the sensory, technological, supportive, and immersive components to be used for its greater effectiveness, but the role that technology can have and play in today's society. According to Traxler (2007), we cannot attribute centrality to the technological universe if we talk about M-learning. We would risk making what is an opportunity a simple exchange of logical opinions on the role of technologies in educational processes. Traxler states in his writings, that taking up the proceedings of the MLEARN[38] and WMTE conference, for example, the most interesting words found are 'personal', 'spontaneous', 'opportunistic', 'informal', 'pervasive', 'situated', 'private', 'context-aware', 'bite-sized' and 'portable'. These words appear, according to the author, in contrast to the conventional e-learning words, like 'tethered', 'structured', 'media-rich', 'broadband', 'interactive', 'intelligent' and 'usable' (Traxler, 2007).

At this point, the real definition that M-learning will have to undergo is related to the usability and the ability to project the subject in a dimension of particular

---

[38] MLEARN is World Conference on Mobile, Blended and Seamless Learning; WMTE is an International Workshop Mobile and Wireless Technologies in Education.

immersion and not so much in that of resolving as mobile technology for e-learning. The integration between mobile learning and e-learning can be solved to the extent that mobile technologies, in addition to being able to put the contents of e-learning in rapid fruition, are also able to realize a series of customized approaches that can provide training opportunities built and self-built on the subject. Therefore, the theme of M-learning is not solved either by looking at the technological nature of the learning experience that users will make or by looking only at the contents which that technology will have to convey. In this way, the centrality of these processes involves all the substantial aspects of mediation and content. However, if we operate without recognizing the significant value of the stimuli, and as far as we are concerned, we would impoverish the possibility offered by technology so young but already powerfully pervasive in everyday reality that we are given to live (*see* Fig. 1).

The translation interface of human actions into actions read by the computer involves a long series of mediations, ranging from interpretation of the motor input, the understanding of activities through gesture translation interfaces, the processing of the information obtained, the graphic output through the screen and/or the required technological response. The passage between natural interfaces and interpretation of human actions combined with the speed of calculation for processing that motor input sanctions the technological evolution and its scientific-evolutionary advances. These processes, which are still evident today through a series of hardware objects, are increasingly moving towards miniaturization and integration into unique tools, such as smartphones, tablets, note-books. This process of miniaturization, of overall reduction and integration of input-calculation-output systems, is increasingly allowing us to get quick answers to work, playful, educational and informative needs, allowing us to build an electronic bubble material (technological objects and physical memories) and immaterial (software, Internet networks for telecommunications, cloud computing, cloud storage), to follow us, accompany us, enhance our intangible presence through the remote translation of graphical, textual, video outputs.

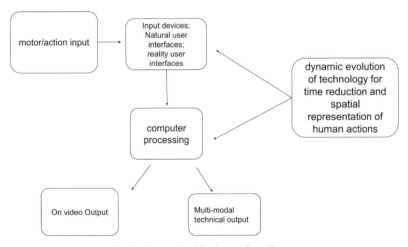

Fig. 1. Human/machine interaction scheme.

We are in the era of real connection where wi-fi networks, Bluetooth standard connections, and 5G signal technologies allow us to manage everything necessary to our relationship—man/machine, technology/reality. Within what we call *input devices* and *natural input interfaces* (NUI), a significant role is played by the most used tangible interface of the last decade—the tangible screen of our smartphones and tablets (Norman, 2010; Wigdor and Wixon, 2011). The tactile pressure on our screens allows a direct mediation with the video content, the proposed image and this direct mediation between image and touch, this overall reduction of space between thought/action/touch/video response, and the general reduction in hardware input interfaces, such as the mouse and keyboards, has led to an evolved process of intermediation between man/machine. So the transportability, ease of use, the possibility of direct intermediation with content, video, and sound output—all concentrated in a single object—have allowed us a digital revolution that, in a decade, is changing the rules of the game of information technology, telecommunications, technological evolution, and especially all the possibilities that this revolution represents in terms of use of the content, entertainment, work, and distance learning.

Transition through Technology

This revolution, therefore, passes through its primary technological tool. This object has become a digital alter-ego for each of us, which, even if we wanted to, in today's society would make it difficult for us to do without, that is, the smartphone. We went from tools that give us modest video and audio results, not wholly satisfactory but useful for telephone communication. Display of multimedia content to real technological concentrates to collect biophysical parameters of the owner through the recognition of fingerprints (first with special sensors, now directly from tangible screens), sensors to recognize facial parameters and its movements in 3D, to LIDAR (light detection and ranging or laser imaging detection and ranging) sensors that through laser projection can determine distances and build textures of the surrounding environment and then virtualize it, define it in measurements, observe it, and encode it (Aldhaban, 2012; Bort-Roig et al., 2014; Xu et al., 2015).

This interactive evidence is defined under the macro category of so-called 'intelligent interfaces' (Heylen et al., 2006). The so articulated set of elements integrated within a single handy device bring back with further evidence the theme of hyper-complexity that drives the systems that govern them and also represent how necessary it is not to reduce, trivialize, close in small technical analysis the great theme of the development, use, and implementation of mobile systems for training. The idea of creating adequate tools for representation and mediation between man/machine often derives from formative reflections, even when they refer to issues that have to do with the development of hardware/software integration systems. As a matter of fact, in the studies of Heylen et al. (2006), the questions from which one starts to generate action/response representation systems are mainly questions associated with the degree of satisfaction students have in the lessons, the level of tutoring they manage to receive, and the preparation and training of their teachers. Their research group, called *Human Media Interaction*, aims to verify students involvement and participation in educational relationships at a distance and their degree of participation, and what strategies they put in place concerning the mistakes they make.

What kind of discomfort or frustration do they feel? Are they bored, or do they feel stimulated to do better? Do they understand the instructions? Are things too easy or too difficult for them? (Heylen et al., 2006). This crucial demand to determine what is the actual human behavior in recognizing the vocal sound of the language, the tone, the degree of interaction that exists between people involved in various exchanges, determines the need to create systems capable of widely interpreting these characteristics in order to ensure that everything is as reactive as possible to the requirements produced by the response of the machines. The idea of integrating an M-learning system to access virtual lessons remotely, or to meet with their responses the spirit of the demand, also being aware of environmental answers adapted to the formulated demands, can increase the level of 'presence', of 'immersion', of fluid fruition of the virtual process (Grimshaw-Aagaard, 2019; Slater, 2003). The intent is certainly not to replace reality but to provide it with further integrations, supports, augmented reality systems, technological tools that increase presence and immersion.

The topic of the relationship between augmentation and immersion prompts a series of reflections, among which the first is to understand how augmented reality (AR) is a system of reality enrichment as it presents itself to our eyes (and I would add to 'our ears') to which, through the use of interfaces and integration tools, we can assign digital features, captions, references, quality to enrich reality as it presents itself to our eyes through the interaction between location systems (geo-tags), camera/video camera of our smartphone, tablet, etc. (Luangrungruang and Kokaew, 2018). On the other hand, the immersion tends to bring within an environment, characteristics to involve in the best possible way the subject which/who benefits from it. In such cases, the quantity and quality of stimuli can influence, within a given environment, the level of the capacity of involvement and presence. However, our work of immersion cannot possibly be achieved within a real environment in which we integrate additional stimuli to recreate some characteristics that are useful in that context.

The main issue concerning M-learning remains rooted in a series of questions to which several theoretical, phenomenological, and interpretative answers are sought. The items are the same as those Maria Uther proposes in her text, *Mobile Learning: 'Why mobile? Mobile devices help or hinder learning?' 'How can mobile learning be evaluated?'* (Uther, 2020, 2)? It is essential to know that a global theoretical framework on the role of M-learning has not been completely defined, and it is precisely for such reasons that systematization of the phenomenological components, together with the theoretical reflection that passes through the analysis of data and reference studies on this topic, appears to be a necessary condition for defining its contours.

One of the main problems identified by Wang Ng and Nicholas (2013) lies in the fact that much of the research associated with the implementation of M-learning within formal education systems, such as schools, are often funded through technical sponsors and therefore do not have the necessary time length to see what level of impact they might have on learning and the required systematicity to give pedagogical elements greater clarity on the sustainability of such systems within schools. Economic sustainability use remains a crucial element for a complementary and additional redefinition of the role of M-learning in learning (Wan Ng and Nicholas, 2013). In a study conducted by Elphick (2018) on iPad to complement classroom activities, the most critical emerging aspect is reflection and awareness about digital skills. The study found that digital skills have increased significantly since the use of these tools. This testifies to three main things:

- the first is that technology is used as a support and implementation of traditional didactics, stimulating the processes of literacy, and growth of digital competence (required among other things by the EU as one of the eight key competences of citizenship);[39]
- the second that digital competence and the use of technology, like all other 'traditional' skills, undergo a more mature and organic development and are promoted by the school;
- the third is that learning about 'digital skills' is particularly significant for young people who have not yet developed adequate skills in using digital tools (Elphick, 2018).

This last scientific observation allows us to confirm that M-learning, integrated with traditional didactics also in formal school contexts, can encourage—as a secondary effect—a more effective digital literacy process, supporting inclusion processes in favor of those subjects in a socio-cultural, linguistic, nonetheless cultural disadvantage. Language learning mediated by mobile devices appears equally significant in this context as it assigns a positive response in terms of cognitive growth (Yurdagül and Öz, 2018). Especially in Yurdagül and Öz's study, it would seem that all things were equal in groups using mobile devices for language learning, and that there was a considerable difference in their learning quality compared to students in the control groups. However, what also emerged from these studies was the impossibility of using standardized tests for evaluation of the actual outcomes of the impact that M-learning can have on learning. This aspect cannot be analyzed through the use of experimental research as their effect could not recognize the development of language and digital skills (Yurdagül and Öz, 2018; Kukulska-Hulme, 2009).

Spatiality and Sounds

Critical studies reveal how spatio-temporal representations are relevant clues in sound perception, to give sounds and perceive from sounds the significant references of localization in space. Casati and Varzi (1999) address, in one of their works, *Parts and Places: The Structures of Spatial Representation*, the complex interpretation and definition of spatial perception through a holistic vision that contemplates mereology, namely the relationship between the individual parts and the whole

---

[39] On 22 May 2018, the European Council accepting the proposal made on 17 January 2018, by the European Commission, launched the *Recommendation on Key Competences for Lifelong Learning* and the *European Framework of Reference* Annex, replacing the Recommendation of the European Parliament and the Council of 18 December 2006, and its Annex on the same subject.

(through the description of Leśniewski's systems) (Marsonet, 1981) and the topology that according to Casati and Varzi would be the study of spatial continuity and compactness. These hypotheses lead to a unified picture of spatial representation as a theory of representation of spatial entities (Casati et al., 1999).

It is no coincidence that in audio systems—aimed at high fidelity (hi-fi) and now at the so-called Hi-end—stereophony is one of the central elements because it is determined by a dense network of audio recordings in different points of a place where a performance is recorded and which can be achieved by using a wide range of speakers suitable for reproduction that allow re-creation of a filmed environment to 'immerse' the subject in the listening. Stereophony and all its evolutions are only ways to widen the range of sound frequencies but also, and above all, a way to determine spatial effects in music reproduction, to represent as faithfully as possible the placement of a listener in a given context—a concert hall, a room, a place—clues that allow the listener to perceive a localized and spatialized representation of sounds and their origin.

Therefore, the spatial clues determined by sound constitute an essential basis for elaborating sound paths and musical representations capable of attributing greater fidelity and better quality to sound representations. It is important to emphasize that natural sounds are characterized by waveforms (vibrations in Hertz) that can segment the sound and give meaning to spatialized communication (Herdener et al., 2013). These processes, however, for a long time have not been well known, mainly show how they are represented physiologically in the auditory cortex of the subject. Now, thanks to the studies of Herdener et al. (2013) and the use of functional magnetic resonance imaging (fMRI)—optimized for the course of the auditory system—it has been shown that there is a topographically ordered spatial representation of temporal rates of sound perception in the human auditory cortex. This topographically ordered representation of temporal sound signals in the human primary auditory cortex is complementary to maps of spectral signals from functional magnetic resonance imaging frequencies (Herdener et al., 2013).

It is important to understand the nature of sounds and how they are structured in the perceptual action because they allow us to realize that the perceptual-spatial structuring, useful for the structuring of sound paths within digital systems and M-learning, relates to specific physical peculiarities of the sounds emitted and how the auditory system can perceive and organize them in their substance. Essential studies on sound dynamics and its characteristics have led to development techniques, called DirAC (directional audio coding). They tend to verify the relationship between psychophysical assumptions and energy analysis of the sound field. These studies intend to generate a strong and organized scientific response to achieve spatial synthesis within virtual worlds, to attribute through a series of reverberations those spatial clues that are necessary to create sound replies within digital virtual environments (VE and VLE) (Laitinen et al., 2012).

It is important to adequately define the effects of the spatial representation of sounds as this can help us better understand the sound characteristics attributed to any context, even from a strictly technical point of view, and respond to specific physical characteristics and neuronal responses. This constitutes a substantial basis for the realization of adequately involving and, in some cases, immersive fruition systems. However, proceeding in the sound organization of multimedia contents used through mobile technologies one has to consider the technical and scientific characteristics of sound production and perception (Cohn and Hazarika, 2001).

Sounds and Environments

We can also experience fantastic, unreal realities. We can benefit from actions which in real contexts would not be feasible. This does not take anything away from the central concept of presence because the subjects' physiological nature can live and adapt to the stimuli of that context. Ours is a predominantly adaptive nature but also, and above all, a relational nature. We could also simulate flight, crossing walls and doors in a virtual world; what would never change is our ability to respond actively and mentally to that situation. The human response does not change and therefore, its presence does not change either.

This imposes that virtual reality does not have objective evaluation systems, but it can be evaluated for comparison of subjective and intersubjective experiences. The immersive systems have the merit of allowing significant functional responses in the subjects who experience it. Therefore, it can have its experiential possibility by simulating different worlds, unusual contexts, and experimenting, in any case, the experience of presence (Slater, 2003).

The environment emerges as an active perception of what I can listen to and isolate from the context and is, therefore, a dynamic characteristic produced by this existing as a sound stimulus and what I can bring out from listening. However, my presence is anchored to what exists, even concerning the sound dimension. Even if we proceed to define the Exosound (exosonus) as external sound and the Endosound (endosonus) as internal sound, the environment as an organic system, both real and virtual, although conceivable as exo-environment and endo-environment, would always emerge from an environmental aggregate formed by external sensations and internal knowledge.

Body, Musical Activity, and M-Learning

Why can sound and music be useful tools to design and implement M-learning development, linked to body understanding and response? The answer is that because the virtual can be closer to reality if it can complete a series of stimuli that can stimulate the entire sensory field, especially from the visual to the auditory. M-learning allows us to combine mobility, the smart use of technology, and the sensory range involved in implicit and explicit learning processes.

In the musical activity, corporeity can manifest itself through dance—a codified corporeal performance of rhythm and music. There is a close correlation between physical movement and the process of musical enjoyment. It is known that the rhythmic structure of a given musical composition tends to arouse in those who listen to conscious or unconscious physical movements of accompaniment. The action, in fact, through a rhythmic de-structuring made of consecutive gestures, tends to emphasize the rhythmic division of the beats. This consequent gesture from the rhythm helps us to better understand the musical structure, rhythmic basis, and segmentation of its internal structures, allowing us to subdivide it and then assimilate it (Toiviainen and Keller, 2010).

In DeNora's essays entitled, *Music-in-Action: Selected Essays in Sonic Ecology,* there is a transition towards the observation of an evolved corporeity, not expressed in the analysis proposed in the first essays (DeNora, 2017). In fact, an initial perspective showed music and sound as cultural data realized in the light of a mature compositional ability and an adequate understanding of acoustic and realizing phenomena; on the other hand, in recent essays, the decisive role in the processes of cultural relations and musical production of corporeity is an element of representation, both of the historical-social world and an implicit, biological, instinctive dimension of the musical output. Sound, music, and corporeity react and act as an organic and holistic representation of man in his cultural and biological dimensions. Its embodied knowledge is already the sum of intelligence involving historical-cultural measurement, corporeity, and higher cognitive processes.

The educational theme returns to the relationship between brain and mind, nature and culture, unique environment, *bios,* and *logos*. This perennial dichotomy is apparently antinomic. For this reason, it is challenging to solve the question according to the classic scheme of the division into clear and well-defined categories. Therefore, the social class should include the phenomena both in their biological dimension and in their biological evolution of rational reasoning. To explain this last interpretative equation, we could say that the body, its physicality has its reasonable characteristics, even when we give to the instinct a completely irrational value; at the same time, the interpretations of the theoretical order, of speculative-reflexive character, can be considered a way of doing, conditioned by processes of neurophysiological character and therefore oriented by the biological nature of man. We believe so in a certain way, even for the role played by our physiological functions and body nature.

Corporeity and Music: The Motor Theory of Perception

Many studies still report a close correlation between the experience of sound and gestures (Godøy and Leman, 2010). The authors, Godøy and Leman, scientifically urge us to understand the link between music and movement, starting from the observation that today's society is thoroughly bombarded daily by musical performances, dance, music videos, TV commercials, computer games, or consoles, animations, films. The gesture-movement relationship is first and foremost a concrete expression of how society is evolving to such an extent that they argue on how the new frontier of development of this sound-movement relationship is increasingly shifting towards new digital interfaces to realize meanings through mediation between music and movement. However, neuroscientific, psychological, and educational research is increasingly moving towards the analysis of bodily movement and thus of embodied knowledge. This, in addition to changing the perspective with which we observe the phenomena of learning and neuronal dynamics, adds a paradigm shift concerning musical understanding that is passing from forms of observation and abstract analysis, for example, related to notational systems, to the holistic experience of sound/movement linked to the body (Godøy and Leman, 2010).

The described relationship of strong interdependence between gestures and music production is traced back to a different relationship between the mental image of sound and sound (Godoy and Jorgensen, 2012). The idea behind this evidence was to verify the mental capacity to imagine sounds in the absence of sound itself, to imagine sounds we had already experienced, or to invent new sounds through our 'inner ear'. This led to a new vision, split between perception and imagination, between listening to music and imagining music. However, it is quite clear that there is an inseparable relationship between what we listen and how we prepare ourselves to hear or anticipate what will come after listening. Memory and cognitive processes, therefore, perform a substantial work of anticipation in musical scenarios, and this is a primary process that builds an implicit dimension over which we have no full control but whose forms we can imagine to the point that a sound, not connected to the sound images of who produced it, what instrument determined it, involves for us a vision of that 'impure' sound (Godoy and Jorgensen, 2012).

A sound to have its substance should, therefore, be associated with mental images. There seems to be a unique link between motor actions and the musical perception; in fact, the musicality and the inner expression of that musicality seem to have broad elements of similarity to motor acts. All these actions show parameters of similarity between subject and subject; it is not difficult to see and discern an individual predisposition to body movement during a performance, or the listening of any composition, or piece of music and see what rhythmic response or motor description is realized in him. The Motor Theory of Perception (Godoy and Jorgensen, 2012; Godøy and Leman, 2010; Godøy et al., 2016) discovered that the relationships between action and musicality are well rooted in human cognitive faculties and that people tend to perceive and have a musical representation mentally simulating body movement as if they were involved in the production of sound. These bodily dynamics related to sound were demonstrated through contact tracing, recording actions in relation to the type of music and sound that users were made to listen to (Godøy et al., 2016).

These numerous considerations lead us to understand how much the close link between these 'physiological-cognitive operations' can be used to create mobile learning tools capable of involving the subject in deeper processes than the simple informative use that would just be weak and transitory and not able to affect the learning processes with a decision.

Sensory World and Salient World, Immersivity and Presence

In virtuality, the sound is an 'emerging perception', but it is essential to establish a clear difference between audio and sound. Audio allows solicitation of sound waves to generate sound through its virtual potential. Therefore, the sound is an effect of the audio, not a precedent; neith can it be assimilated in nature. A further distinction when we describe the virtual worlds must be made between virtual 'world' and virtual 'environment'.

Grimshaw-Aagard (2019) defined environment as something built perceptively rather than sensorial and real. The environment becomes a subset of sensory things within the world, be it natural or virtual. In any world, there are a numerous series of sensory objects not always fully perceivable by the subject; however, they are still present. Therefore, the sensory world cannot be acquired; yet, inside the same world, we can highlight 'salient' sound realities which are thought to enrich virtual reality's sensory experience. This subdivides the virtual world into the sensory things/aspects which can either be general and therefore 'sensory world' or 'salient world' associated with to the choice of assigning more significant meaning to particular/definite sonorities. With more studies carried out about virtual environment, more emphasis is placed on the constructions of dynamic perception of the 'salient world', in order to achieve a greater individualization of it (Grimshaw-Aagaard, 2019).

However, further scientific attention must be paid to the concept of immersion as opposed to presence. Slater (2003) stated in his famous essay that the term 'immersion' served to indicate what technology provided in terms of sensory expressions and tracings capable of maintaining a good fidelity concerning their equivalent sensory modes in the real world. The more it could wrap the subject in sensory terms, the more immersive they could be considered. This peculiarity is, in fact, objective because it speaks precisely of the amount of involvement it is given, based on the abundance of sensory stimuli. However, it refers to issues other than how humans perceive the same stimuli. I would like to make a distinction here similar to that of the science of colors. This means that the degree of immersion can somehow be objectively calculated without this, telling us enough about how much the subject involved feels as part of that world (Slater, 2003).

The scholar, through a series of analogies, makes us further understand what 'presence' represents. The presence is the ability of a virtual sound context to make us perceive a quality, such as making us 'present' virtually in that environment. Slater cites the example of listening to an orchestra in quadrophony and the sensation it produces, to the point of immersing us almost as if we were in that context. That sound immersion, however, has nothing to do with attention, appreciation, the quality of the musical style or taste of the listener, but enhances the projective and immersive capacity that the system has to envelop us, 'kidnap us', in a dimension different from the real one in which we are enjoying a particular listening. Therefore, the 'presence concerns the form, which consists of unifying simulated sensory data and perceptual processing. This relationship can produce a coherent 'place' where one is 'in and where one can be 'present' and where one has the technological potential to act in that environment. Appreciating or not, on the other hand, the peculiarities of an environmental concern taste, interest, personal motivation, desire, the constructive quality of a given environment and therefore, the substance of what it represents; not the technical form of what constitutes it (Slater, 2003).

The debate on 'presence', 'immersion', 'virtual reality' often leads to epistemological problems that would seem to be marginal for the central theme of the fruition of content at a distance, in virtual environments for learning or through e-learning tools. However, the interpretative theme becomes fundamental to orient operational choices concerning educational design. Living the 'presence' within a virtual environment means to experience one's corporeity, one's rational mind, one's own experience, one's feelings in real union with the context that hosts this relationship and above all with the perception that there is a deep adherence between oneself and the context, so sensorially significant from being able to determine feelings of completeness and commitment to that environment.

Wearable Mobile Digital Technologies and the Hybrid Theory of the Mind

The 'salient' world is modeled from the environmental aggregate through a creative process. Each hypothesis is an imagination of a world in which we can be present; a model of sensory things in a salient world to which the mind gives form and meaning. Men act as if they are in the environment; they could not do it if that environment did not stimulate all its possibilities (Grimshaw-Aagaard, 2019). Hence, the need for active presence in the virtual environment and mobile use become the bases: a series of sensory biofeedback can in fact confirm and further stimulate the company within a given context. This would further define reality as a psychoactive and sense-active capacity of perception, presence, and immersion within a given virtual context, offering additional stimulation elements to ensure an adequate and vivid 'presence' of the benefits.

Therefore, an environmental design process intends to develop and stimulate all human perception components. However, this has nothing to do with motivations, tastes, aspirations or personal will which are cultural characteristics, but mainly with intrinsic components to a subject, regardless of its perceptive abilities. Indeed, as the development of models focused on enhancing the empathic, experiential, playful, and social nature of the topics, it also leads to a further emotional and motivational involvement necessary for an adequate learning processes, but it does not directly affect the 'presence' given mainly by a faithful sensory representation and appropriate stimulation.

The point of mediation between body and technology seems to be central also with the possibility of having digital technological objects with this function at more and more competitive costs. With mass industrial production, these tools become more and more portable, miniaturized, wearable, able to verify a series of body parameters with more excellent reliability. In a certain way, mobile devices have 'increased' the possibilities to have data in an autonomous, constant, and quantitatively intelligible way, without requiring the support of prominent technological instruments, significant economic investments, and above all, vital competencies to understand their functioning.

Wearable devices can read body parameters and highlight them through graphs, numbers, synthetic descriptions (think of electronic clocks, ranging from calories calculation to the pedometer, to the possibility of tracking the heart rate as a real portable electrocardiogram, or able to monitor and calculate your physical activity); in this case, we speak of reading coming from within the subject and so is somewhat extensively subjective. However, there is also an external function of which priority is mainly represented by communication in all its forms (phone calls, messaging of various kinds, infotainment), particularly as a feature of passing on/communicating throughout history. This represents who we are, our social status, our tastes, our socio-anthropological characteristics (Ju, 2016).

As in jewels over the millennia, which served indirectly to mediate their experiences and interactions in their implicit ability to communicate roles, rites of passage, elements of sentimental representation, nowadays wearable technology tries to represent, in addition to its nature of technological intermediation, a long series of social aspects for those who use them. It is no coincidence that often the fruition, use, social demand of certain technological products do not merely respond to a greater need for technology but above all, to an evident and growing need for self-representation which those individual technological 'jewels' can communicate already only in their nature as products and not as mediation technology.

The technological evolution that we would like to channel towards a primary function, as in training, has often been an element of the exaltation of the useless, of the unnecessary. We cannot fail to mention the role that the Sony Walkman has represented for entire generations, the possibility to enjoy subjectively, 'walking'— and above all, self-isolating through headphones—songs, music, the result of personal choices and not necessary to the unfolding of everyday life.

However, everything revolves around a fundamental principle that we have raised several times in this text: man is a being who expresses his true nature, is able to show his dynamic being, in movement, in continuous social and physical expression (Brinkmann, 2017). Brinkmann defines in his '*Hybrid Theory of the Mind*' the holistic expression with regard to the integration element that exists between brain, mind, and the body. He states that mental dimension cannot be eradicated from the possibility of intermediations occurring through the significant mediation of technologies (Brinkmann, 2012).

Hence, the mind is an element that exercises its substance through the brain and the body and becomes the object and importance of mediation. Intention is also mediated. In this mediation that belongs to the expressive tools, technical or technological, there is a relationship of continuous mediation between the mental self's expression and the essential support of the instrument that allows this mediation. This confirms that the technical and/or technological tool can never assume a neutral nature in what it can represent, precisely because it will allow the minds of those who use it to emerge. Inclusion of stimuli means creating environments that involve the subject sensorially and become technological, digital, immersive formulas of mediation for the mind. Tools allow the expression of what the mind can express. More are the possibilities, channels, senses, modes of expression, and sensory involvement; more are the opportunities for the mind to express its form.

A subject, who teaches and mediates between different cognitive dimensions by clarifying social contexts and cognitive meanings of a given culture, helps to identify himself as a mediator of competing intelligence. It is no coincidence that the structuring of training paths between social groups and different cultures in remote mode through e-learning works as if it was harmonized, structured, and 'adjusted' by the ethnological-cultural competence of those who can mediate between different cultures. Negotiating, evaluating, implementing, and observing in the field are the roles of those who exercise a link function in distance mediation processes (Fuchs et al., 2017).

The question of the relationship between education and technology collects further strength related to sustainable development. A system in which technology and education are enhanced serves as leverage to develop the world's resources more efficiently, thus using a smaller portion of its physical capacity (Amadei and Borgida, 2019). Motivation and acceptance of these systems by the students, who should benefit from them, remain open and significant for the operation and implementation of a training mediated by technology. A Chinese study, focusing on the level of acceptance of technology concerning the study motivation of the various respondents, discovered that the acceptance of technology and the technological self-efficacy of the students were strongly correlated with their attitude to technology-based self-learning (Pan, 2020).

Visual-Audio-Spatial Synchronization

The proof of a synthesis between spatial perception system of auditory type cannot undoubtedly be separate from the visual component, since the two neuronal characteristics, although coded differently in the brain and have a sensory concordance, must somehow be appropriately calibrated. One study (Zwiers et al., 2003) quantified the localization of human sound in response to spatially compressed vision (0.5 × lenses for two to three days that reduced the image to only parts of the entire field of view). Their results showed that spatial-scale vision induces systematic and adaptive changes in the localization of sound, and a reorganization of the visual system; in particular, a reduction in spatial visual gain induced by 0.5 × lenses. This means that to adequately adjust between visual-spatial perception (increasingly accurate as governed by the predominance of sight) and auditory spatial calibration, an active cross-modal experience is required (Zwiers et al., 2003).

This series of indications, deduced from a substantial amount of studies on the relationships between sound, space, action, and vision, clarify once again how perceptual processes and the organization of adequate responses to the needs of the subject in a real/virtual situation, can offer operational advantages if one can integrate stimuli and build them into a dense network of elements supporting learning processes. Body perception and action are also strongly and significantly modified by the presence of sound elements to interact with the subject during any motor activity. We have known for years *the doping effect* determined by the use of music with particular characteristics during some sports activities, such as marathons or competitions where strength and physical endurance are at stake. Music is considered as an element that alters the genuine functions of sports activity because it directly affects the perception of effort and the emphasis on performance (Terry et al., 2012; Karageorghis and Priest, 2008; Edensor and Larsen, 2018).

This alliance between sound, music, and physical activity has also been analyzed by Tajadura-Jiménez et al. (2019), who, in their work, demonstrated that altering walking sound changes people's perception as being more dynamic and light (Tajadura-Jiménez et al., 2019). The movement and completeness perception received from achieving an action intentionally set out, gives them a sense of wholeness and completeness of their bodily representation. However, how body awareness acts on the overall perception of neurocognitive processes is still shrouded in the mists of an articulated scientific debate. Burin et al. (2019) intended to verify if their body's perceived properties influenced the general motor performance through the use of an avatar in immersive virtual reality. They wanted to examine how much their own body's perception can modify and how much is consistent with their mental image with the actions that were being realized. The people involved in the study had first the opportunity to directly see and recognize themselves virtually in an avatar; they had to draw continuously straight vertical lines while seeing the virtual arm doing the same action (i.e., drawing lines) or differently, because of a change in the azimuth of the software (i.e., drawing ellipses).

The study showed that the subjects, when they perceived a movement of the avatar different from the task they had to perform, tended to slide towards the action of the drawing made by the avatar. Therefore, the body perception modulates the interference between assigned actions and visualized actions; in other words, we could say that the perception of the body self is always the result of mediation with the observed motor context (Burin et al., 2019).

Technology and Sound Production, VLE, Augmented Reality, and Performing Arts

Sound can also assume an *immaterial* value linked to the semantic construction of meaning within *anacoustic modes* (Seaback, 2020). The digital sound construction could be useful to us as a concept related to the sonorization of digital environments: the computer works as a simultaneous 'ontological' producer of information about what it has to materially produce (the tracks of a song, for example) and the real outcome of that performance, which is the sound production of that song. In this case, the sound becomes an actual physical substance that can be remodified, updated, reproduced, adjusted through organizational hierarchy that first sees the presence of structural computer elements, and then its effect. In this case, there is a clear difference concerning the sound expressed by an acoustic instrument, in which there is a relationship of absolute continuity between action on the device and the auditory effect of that action, to the point that the excellent story of a musician can be defined as real embodied knowledge, as already mentioned previously (Seaback, 2020).

On the other hand, the computer musician must continuously alternate a relationship between sound reality (already evident by the fact that we can listen to it) and the abstract bases of computer code, even when mediated by software. It produces a continuous feedback system between sound objectivity and machine language. This relationship takes over an objective data of non-predictability of the sound process related to the digital programming process; that is, there is a discrepancy to be filled continuously through cycles of continuous adjustments to achieve a meaningful and faithful relationship between the idea of the sound that one wants to produce and appropriate ways of programming to create them, adjust them, remove them (Seaback, 2020). It strongly appears that the integrations aimed at the interaction of different systems from the digital one to the sound system associated with corporeity, to the recalibration of immersive thought as the possibility to experience first-hand through augmented reality or virtual learning environments, some environments demand a series of complementary man/machine operations that require a high capacity of design, programming, and educational design.

Those who would seem to open the way to such elaborate interactions, express the possibilities of interaction between virtuality, digital environments, sound business, and the learning possibilities are the performing arts. Johannes Birringer (Birringer, 2018) offered an exciting documentary testimony on this subject and carefully described an artistic installation in which the spectator, the protagonist, is captured inside a magical and psychedelic world made of lights, sounds of nature, elements that reproduce meadows, and natural features and there they use their body as a mediator to modify projections, sounds, lights, through a system of motion-capture of body movement. Birringer's idea was to create a dance in which the dancer had the bodily ability to modify, through a series of instruments which could perceive movement, the very concept of theater, to create a consistently active and immersive performance in which the subject became both actor and user of the installation.

The main intention was to reach complete interaction with integration between tangible reality and virtual reality (VR) technology in 3D, so that the individual could allocate to virtual reality (mutilated of important sensory aspects) a series of objects and fundamental elements so that experience of the product combined a double dimension: the dreamy, unreal flight, the perception of fantasy elements, through virtual reality and real reality made of wind, natural sounds, physical objects through fundamental elements (Birringer, 2018).[40] We are therefore rapidly moving more and more towards real/virtual integration systems and within the constraints produced by this type of relationship, where a series of further mediators, including sound, are being progressively placed. These can both assign greater complexity to objects and environments but also give bio-sound feedback capable of absorbing the subjectivity acting, learning and create training processes in a more complex and sometimes more complete reality.

The sound represents a substantial element for a subject's spatial definition within a real though above all virtual learning environment (Grimshaw-Aagaard, 2019). However, this important outcome comes from a multimodal approach knowledge which offers substantial support to the realization of significant environmental experiences. This brings us to the hypotheses of Grimshaw-Aagaard (2019). The existence of an individual in an environment appears to be an interactive sum of the resonance of sound elements, which highlight the subject in the background. This aspect appears strongly in biofeedback systems, in immersive systems specially dedicated to games (Grimshaw-Aagaard, 2019).

Environmental Harmony and Sound: Isob and Esob

Rosenboom (2020) writes that all active processes, including reading, activate a series of multisensory experiences that directly affect the self-construction of images. For example, reading synthesizes endogenous ideas, produced precisely by reading a text, at times of memory and induced creativity that belong to the subject who reads. The creative production determined by reading is combined with the unconscious, implicit traits of personal memory also made of sounds, smells, situations similar to

---

[40] The fields of application of a mixed real/virtual reality and the use in this case of ultrasound seem to enter fully into the surgical-medical practice. Real-time ultrasound has become a crucial aspect of several image-driven interventions. One of the main constraints of this approach is the difficulty in interpreting the image's limited visual field. This problem has recently been addressed by using mixed reality, such as augmented reality and augmented virtuality (Ameri et al., 2018).

those experienced by the protagonist of the story. Rosenboom cites the experiment carried out by Luca Forcucci in Amazonia. Before proceeding to the recording of sounds coming from the environment around him, the author of the studio waited for time and for nature to 'harmonize' and perceive its presence. The idea of creating a relationship before proceeding to verify which sounds exist within a given natural environment serves to establish the disturbing elements within a stable balance in which man is simply a guest. Another element on which Forcucci's studies (Rosenboom, 2020) investigate is how sound environments, interference, and active presence can change the sound context itself from the sound's physical point of view.

However, it is particularly significant to understand that the dynamics associated with sound can take on various connotations. We cannot fail to consider the relationship between natural sounds (produced by natural elements and living beings), artificial sounds (made by natural and electronic instruments), and sound production. If it is true that there is a relationship between sound and objects and that this relationship is the result of an environmental interaction between the physical characteristics of this object, the action it undergoes as a result of natural physical forces could be defined as 'implicit sound of the objects' (Isob), which is different from the use and expression of a conscious, sought after, dynamic sound. This sound condition is the effect of all those natural relationships which are determined by the same composition of the objects as they are in nature. In turn, even natural elements, such as verbal, non-verbal, paraverbal expression produced by man or the sound expressions of animals are used for communication. These are implicit characteristics of natural sound. The opposite aspect to these conditions of the sound is the exact sound of the objects (explicit sound of the items—Esob), where we can resort to the instrumental and conscious use of the sound of natural objects for communicative, expressive, musical purposes, and the real sounds produced by instruments, both musical, acoustic and electronic produced by synthesizers to communicate, to create music, and to express organizational indications (think, for example, of the various alarm signals) (*see* Fig. 2).

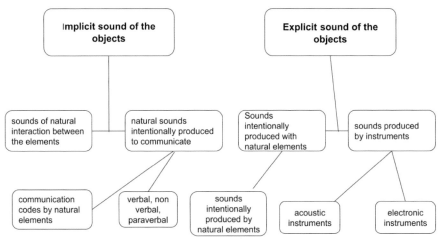

Fig. 2. Isob and Esob classification.

In both cases, sound interactions and rational expressions can occur either by environmental conditions external to the subject and determined by uncontrollable conditions (the most obvious example is the wind in the trees, the noise of rain, environmental noise, etc.) or through the voluntary act entrusted to the use, choice, willingness to produce sound. It should be pointed out that verbal, non-verbal, and paraverbal communication also belongs to implicit sound forms that disregard the 'explicit' and voluntary act of communicating. The phonic apparatus is a species-specific condition beyond the semantic act of communication it expresses. This is even more evident in animals' calls and sounds (Berent et al., 2016; Peyrin et al., 2012).

The relationship between sound/action, sound/expression, sound/communication seems very consolidated and, therefore, the result of an interaction assumes characteristics of strong interdependence and in some ways of complete overlap. The element of intense discrimination in the dimension that we have defined as Esob (External sound of the objects) is the voluntariness; this determines a great division between the implicit sound that presents characteristics of greater adherence to physical and biological principles (determined by the interaction of natural forces), and the explicit one of expressive, semantic, cultural, artistic, organizational.

In the explicit forms of sonorization or rational expression, there are numerous elements of cultural determination that can assume 'on-demand' characteristics that make the sound act become the expression of a cultural thought, or a communicative action. The two dimensions, Isob and Esob, can still be represented today in their constitutive form; however, the implicit aspects of sound can hardly express evolved sound forms, such as instrumental and electronic ones. This is because even the elaboration of the production of musical instruments or sound instruments (think of the archaic ways of a whistle obtained through the use of natural reeds) is represented as the evolution of two fundamental processes:

1. An expansion of the communicative acts that through the coding of sounds can determine the transmission of messages.
2. The artistic-musical expression that starting from ethnomusicological studies identifies a universal declaration belonging to man (Ciasullo, 2015, 2020).

A study which analyzed the function that sound had on the quality perceived by some users of iPhone, iPad, and iPad mini used for training purposes, discovered that, although the sound was of significant value in learning apps, there were no significant differences in the amount or quality attribution depending on the device used (Uther and Ylinen, 2019). This, according to our vision, does not demonstrate the ineffectiveness of the sound quality of the technological instrument but the fact that what is significant is the sound object of the App. This confirms that sound plays a fundamental role in the use of mobile devices and that:

- sound is not a secondary condition but central to the use of apps dedicated to training;
- high-definition audio quality is of little significance compared to general sound enjoyment until a higher-profile listening education is provided;
- you can start doing M-learning using sound as a mediator even if you don't have high-cost and high-definition tools available.

Audio Technologies in Mobile Devices

This accelerated and extraordinary progress, which mainly concerns the miniaturization of increasingly powerful computing components and sensors, actively involves also smartphones' audio sector. However, if for the external characteristics of smartphones, such as the size of the display, quality of the photographic compartment, fluidity of navigation and connection, the physiological predominance of the visual guarantees direct evidence of these characteristics, for the audio compartment, all this becomes much more complex to identify and discover.

The key to proper sound translation is the stereo speakers that can reproduce with a double speaker and with smartphone shells of quality materials, ranging from plastic, aluminum, and steel. One of the central elements in the definition of the excellent audio quality of the devices is the DAC (Digital to Analog Converter) chip dedicated to the conversion of sounds from digital to analog; its quality is able to ensure a better sound performance of the audio output from the smartphone. Then we have software conversion systems for the so-called 'compressed' audio formats, that is to say, to be transmitted more efficiently. Thus they undergo a reduction of their volume in terms of byte size and are 'decompressed' and restored to their natural size. The primary standards that can guarantee better sound performance are:

- the DTS:X surround 3D (able to create an 'immersive' sound);
- Hi-Res (high-resolution audio, with higher bit and frequency responses);
- LDAC (for high-resolution audio streaming);
- the Dolby Atmos, in addition to the classic Dolby Digital (Ziemer, 2020).

Essential for Bluetooth transmission of sound quality levels is the support of standards, such as aptX, EnhanXaptX, and aptX HD (also these are various systems of compression/decompression, encoding/decoding of audio data for wireless transmission made by the US company, Qualcomm) (Jariwala, 2020). It is crucial to emphasize that each type of audio information passes through a specific type of file of which differences mean better/worse audio quality. The most advanced and qualitatively better standards are AAC (advanced audio coding), FLAC (free lossless audio codec), and OggVorbis (variable bit-rate) files. All these algorithms have supplanted in terms of quality the Mp3 format (moving picture expert group-1/2 audio layer 3), especially those with low bit-rate (Winarski, 2020; Corrêa et al., 2020).

ASMR

One of the most exciting fields that we would like to include within remote, mobile, and virtual immersive training contexts is the field of ASMR (auto-sensory meridian response). However, a field not fully defined within the scientific debate but which is becoming more and more interesting is the study of some sound effects (mainly chills) resulting from listening to whispering sounds, the ticking of the rain, the movement of the hands, and some particular environmental sounds that would stimulate relaxation and sleep. Lately, various familiar domestic sounds and music have been introduced into the commercial environment, such as a whisper in a particular room, the crackling of a fireplace, the sound of a kitchen while preparing food (Chae et al., 2020).

Unlike other phenomena studied, such as the so-called 'aesthetic thrills' determined by certain types of music or particularly beautiful and involving scenarios, the basis of AMSR would not yet find justification from a neuroscientific and psychological point of view. However, two studies (Poerio et al., 2018), including a large-scale online experiment and a laboratory study have tried to verify the so-called ASMR response's emotional and physiological correlations. Both studies have shown that watching ASMR videos increases the chills produced by those listening or viewing only in people who have already experienced ASMR. This would indicate that there is undoubtedly a link between the production of the response in the subject and coordinated sound/video stimulus capable of stimulating it. The in-presence study has also shown that ASMR is associated with a reduction in heart rate and increased levels of skin conductance. Although not entirely justified from a scientific point of view, this would indicate that ASMR produces actual physiological phenomena capable of improving people's overall well-being when exposed to such visions or listening. Also, we noticed that those who find themselves having such an experience of sensory meridian stress are substantially related to specific categories of subjects, who tend to categorize in the same way all experiences they have had (Smith et al., 2020).

Numerous scientific studies, however, tend to analyze the critical trend that now accompanies the videos of ASMR and examines them in their sensitive nature. Considerable attention and strictly elaborative-psychological considerations are given in many studies in order to identify links between music and sound stimuli (Kovacevich and Huron, 2018). Included within the virtual learning environments in the use of contexts that reproduce places of teaching in the presence, or stimulating sound aimed at reducing the stress of those involved in the learning processes there could be a significant innovation in the structuring of sound mediators and sensory stimulation to affect the subject who understands. The aim would not be to push the subjects to find alternative ways of musical fruition but to build within a context aimed at learning appropriate sound contexts for the participants' active involvement. In this sense, it would be desirable to start a series of experiments to verify the opportunity for integration and implementation.

References

Aldhaban, F. (2012). Exploring the adoption of smartphone technology: literature review. pp. 2758–70. *In*: *The Proceedings of PICMET '12, Technology Management for Emerging Technologies.*

Ally, M. (2009). *Mobile Learning: Transforming the Delivery of Education and Training.* Athabasca University Press, Athabasca.

Amadei, C.A. and BaraldiBorgida, M. (2019). Education and technology as levers for sustainable change. pp. 61–72. *In*: Song, Y., Grippa, F., Gloor, P.A. and Leitão, J. (eds.). *Collaborative Innovation Networks: Latest Insights from Social Innovation, Education, and Emerging Technologies Research, Studies on Entrepreneurship, Structural Change and Industrial Dynamics.* Springer.

Ameri, G., Baxter, J.SH., Bainbridge, D., Peters, T.M. and Chen, E.C.S. (2018). Mixed reality ultrasound guidance system: a case study in system development and a cautionary tale. *International Journal of Computer Assisted Radiology and Surgery*, 13(4): 495–505.

Berent, I., Zhao, X., Balaban, E. and Galaburda, A. (2016). Phonology and phonetics dissociate in dyslexia: evidence from adult English speakers. *Language, Cognition and Neuroscience*, 31(9).

Birringer, J. (2018). Augmenting virtuality. *International Journal of Performance Arts and Digital Media*, 14(2): 224–28, London.

Bort-Roig, J., Gilson, N.D., Puig-Ribera, A., Contreras, R.S. and Trost, R.S. (2014). Measuring and influencing physical activity with smartphone technology: a systematic review. *Sports Medicine*, 44(5): 671–86.

Brinkmann, S. (2012). The mind as skills and dispositions: on normativity and mediation. *Integrative Psychological and Behavioral Science*, 46(1): 78–89.

Brinkmann, S. (2017). Persons and their minds: towards an integrative theory of the mediated mind. *Cultural Dynamics of Social Representation*, Routledge, London.

Burin, D., Kilteni, K., Rabuffetti, M., Slater, M. and Pia, L. (2019). Body ownership increases the interference between observed and executed movements. Public Library of Science: e0209899, *PloS One*, 14(1).

Casati, R. and Varzi, A.C. (1999). *Parts and Places: The Structures of Spatial Representation*. MIT Press, Boston.

Chae, H., Baek, M., Jang, H. and Sung, S. (2020). Storyscaping in fashion brand using commitment and nostalgia based on ASMR marketing. *Journal of Business Research*, March.

Ciasullo, A. (2015). Armonie bioeducative, *Scale e arpeggi pedagogici*, FrancoAngeli, Milano.

Ciasullo, A. (2020). *Opportunità degli ambienti virtuali tra sonorità, simbolizzazione e spazialità*. *In*: Panciroli, C. (ed.). *Animazione Digitale per la Didattica*. FrancoAngeli, Milano.

Cohn, A.G. and Hazarika, S.M. (2001). Qualitative spatial representation and reasoning: an overview. *FundamentaInformaticae*, 46(1-2): 1–29.

Corrêa, G., da Silva Stanisce, C., Pirk, R. and da Silva Pinho, M. (2020). A study on launch vehicle on board acoustic data compression. *Applied Acoustics*, 170: 107480.

DeNora, T. (2017). *Music-in-Action: Selected Essays in Sonic Ecology*. Routledge, London.

Edensor, T. and Larsen, J. (2018). Rhythm analyzing marathon running: 'a drama of rhythms'. *Environment and Planning A: Economy and Space*, 50(3): 730–46.

Elphick, M. (2018). The impact of embedded ipad use on student perceptions of their digital capabilities. *Education Sciences, Multidisciplinary Digital Publishing Institute*, 8(3): 102.

Fuchs, C., Snyder, B., Tung, B. and Han, Y.H. (2017). The multiple roles of the task design mediator in telecollaboration. *ReCALL*, 29(3): 239–56.

Godøy, R.I. and Leman, M. (2010). *Musical Gestures: Sound, Movement, and Meaning*. Routledge, London.

Godoy, R.I. and Jorgensen, H. (2012). *Musical Imagery*. Routledge, London.

Godøy, R.I., Song, M., Nymoen, K., Haugen, M.R. and Jensenius, A.R. (2016). Exploring sound-motion similarity in musical experience. *Journal of New Music Research*, Routledge, 45(3): 210–22.

Grimshaw-Aagaard, M. (2019). Sonic virtuality, environment, and presence. pp. 69–80. *In*: Braga, J. (ed.). *Conceiving Virtuality: From Art to Technology, Humanities—Arts and Humanities in Progress*. Springer.

Herdener, M., Esposito, F., Scheffler, K., Schneider, P., Logothetis, N.K., Uludag, K. and Kayser, C. (2013). Spatial representations of temporal and spectral sound cues in human auditory cortex. *Cortex*, 49(10): 2822–33.

Heylen, D.K.J., Nijholt, A. and Reidsma, D. (2006). Determining what people feel and think when interacting with humans and machines. pp. 1–6. *In*: Kreiner, J. and Putcha, C. (eds.). *First California Conference on Recent Advances in Engineering Mechanics*. California State University.

Jariwala, A. (2020). *Analysis and Test of Bluetooth Communication Systems*. Department of Electronics and Communication Engineering, Nirma University, Ahmedabad, 382–481.

Ju, A. (2016). Functionality in wearable tech: device, as jewelery, as body mediator. pp. 641–46. *In: The Proceedings of the TEI '16, Tenth International Conference on Tangible, Embedded, and Embodied Interaction, TEI '16, Association for Computing Machinery*. New York.

Karageorghis, C. and Priest, D.L. (2008). Music in sport and exercise: an update on research and application. *The Sport Journal*, 11(3).

Kovacevich, A. and Huron, D. (2018). Two studies of autonomous sensory meridian response (ASMR): The relationship between ASMR and music-induced frisson. *Empirical Musicology Review*, 13(1/2): 39–63.

Kukulska-Hulme, A. (2009). Will mobile learning change language learning? *ReCALL*, 21(2): 157–65.

Laitinen, M.V., Pihlajamäki, T., Erkut, C. and Pulkki, V. (2012). Parametric time-frequency representation of spatial sound in virtual worlds. *ACM Transactions on Applied Perception*, 9(2): 8:1–8:20.

Luangrungruang, T. and Kokaew, U. (2018). Applying universal design for learning in augmented reality education guidance for hearing impaired student. pp. 250–55. *In: 5th International Conference on Advanced Informatics: Concept Theory and Applications (ICAICTA)*.

Marsonet, M. (2020). *Logica e impegno ontologico, Saggio su StanislawLesniewski*. FrancoAngeli, Milano.

Norman, D.A. (2010). Natural user interfaces are not natural interactions. *ACM*, 17(3): 6–10.

Pan, X. (2020). Technology acceptance, technological self-efficacy, and attitude toward technology-based self-directed learning: learning motivation as a mediator. *Frontiers in Psychology*, 11.

Peyrin, C., Démonet, M.L., J.F., Pernet, C., Baciu, M., Le Bas, J.F. and Valdois, S. (2012). Neural dissociation of phonological and visual attention span disorders in developmental dyslexia: FMRI evidence from two case reports. *Brain and Language*, 120(3): 381–94.

Piaget. (1976). J. Piaget's theory. pp. 11–23. *In*: Inhelder, B., Chipman, H.H. and Zwingmann, C. (eds.). *Piaget and His School*. Springer, Berlin, Heidelberg.

Poerio, G.L., Blakey, E., Hostler, T.J. and Veltri, T. (2018). More than a feeling: autonomous sensory meridian response (ASMR) is characterized by reliable changes in affect and physiology. *PloS One*, Public Library of Science: e0196645, 13(6).

Rosenboom, D. (2020). Active imaginative reading … and listening. *Leonardo Music Journal*, September.

Schiavio, A., Gesbert, V., Reybrouck, M., Hauw and Richard Parncutt, D. (2019). Optimizing performative skills in social interaction: insights from embodied cognition, music education, and sport psychology. *Frontiers in Psychology*, 10(July).

Seaback, R. (2020). Anacoustic modes of sound construction and the semiotics of virtuality. *Organised Sound*, 25(1): 4–14.

Smith, S.D., Fredborg, B.K. and Kornelsen, J. (2020). Functional connectivity associated with five different categories of autonomous sensory meridian response (ASMR) triggers. *Consciousness and Cognition*, 85(October): 103021.

Tajadura-Jiménez, A., Newbold, J., Zhang, L., Rick, P. and Bianchi-Berthouze, N. (2019). As light as you aspire to be: changing body perception with sound to support physical activity. pp. 1–14. *In: The Proceedings of the 2019 CHI Conference on Human Factors in Computing Systems, CHI '19*, Association for Computing Machinery. New York.

Terry, P., Curran, M., MecozziSaha, A. and Bool, R. (2012). Effects of synchronous music among elite endurance athletes. *In: The Proceedings of the International Convention on Science, Education and Medicine in Sport (ICSEMIS 2012)*. London, Congrex, UK.

Toiviainen, P. and Keller, P.E. (2010). Special issue: spatiotemporal music cognition. *Music Perception*, 28(1): 1–1.

Traxler, J. (2005). Defining mobile learning. pp. 261–266. *In: IADIS International Conference Mobile Learning*.

Traxler, J. (2007). Defining, discussing and evaluating mobile learning: the moving finger writes and having writ…. *The International Review of Research in Open and Distributed Learning*, 8(2).

Uther, M. and Ylinen, S. (2019). The role of subjective quality judgements in user preferences for mobile learning apps. *Education Sciences*, Multidisciplinary Digital Publishing Institute, 9(1): 3.

Uther, M. (2020). Mobile learning—trends and practices. *Education Sciences*, Multidisciplinary Digital Publishing Institute, 9(1): 33.

Wan, N. and Nicholas, H. (2013). A framework for sustainable mobile learning in schools. *British Journal of Educational Technology*, 44(5).

Wigdor, D. and Wixon, D. (2011). *Brave NUI World: Designing Natural User Interfaces for Touch and Gesture*. Elsevier, Amsterdam.

Winarski, T.Y. (2020). Matched filter to selectively choose the optimal audio compression for a material exchange format file, April, *Google Patents*.

Xu, X., Akay, A., Wei, H., Wang, S., Pingguan-Murphy, B., Erlandsson, B. and Li, X. (2015). Advances in smartphone-based point-of-care diagnostics. *The Proceedings of the IEEE*, 103(2): 236–47.

Yurdagül, C. and Öz, S. (2018). Attitude towards mobile learning in English language education. *Education Sciences*, Multidisciplinary Digital Publishing Institute, 8(3): 142.

Ziemer, T. (2020). Conventional stereophonic sound. pp. 171–202. *In*: *Psychoacoustic Music Sound Field Synthesis*. Springer, Switzerland.

Zwiers, M.P.A., Van Opstal, J. and Paige, G.D. (2003). Plasticity in human sound localization induced by compressed spatial vision. *Nature Neuroscience*, 6(2): 175–81.

CHAPTER 8
Sound and Meaning in Specific Learning Disorders

Alessandro Ciasullo

> *The instruments are like the alphabet, and we must know how to manage them if we are to read nature; but as the book, which contains the revelation of the greatest thoughts of an author, uses in the alphabet the means of composing the external symbols or words, so nature, through the mechanism of the experiment, gives us an infinite series of revelations, unfolding for us her secrets.*
>
> —Montessori (1912)

Introduction

This chapter opens the field to the role that sound plays within the characteristics of learning specific problems, particularly in dyslexia. This leads us to focus on the structure of sound perception in identifying the disturbance, and because in the mobile devices that we are analyzing and trying to define in their theoretical and functional structure, a relevant role can be played by the mediation of sound. Already today, voice synthesis offers a valid role of support and compensation to overcome the difficulties imposed by the neurobiological problem of specific learning disorders. These technological peculiarities, precisely because we are in the presence of conditions which cannot be resolved through educational processes (if not in a strengthening of self-compensation capabilities), are essential and fundamental to widen the cognitive gap between people of the same age. The compensatory tools, therefore, and in particular the mobile technologies implemented with software for sound synthesis, become the cognitive prosthesis necessary to replace and bypass the functional deficiencies caused by the alteration of the mechanisms necessary to read, write, organize sentences, and to account.

Researcher of Education, University of Naples Federico II.
Email: alessandro.ciasullo@unina.it

The hypothesis of mediation between subject/object occurs through additional objects with physical nature, such as smartphones, tablets, notebooks, PCs, and immaterial and therefore, related to software development which, with the implementation of remote cloud systems, can develop higher-level assistive systems. These features become even more customizable because, thanks to the cloud, to the subsistence of the system beyond the material objects that allow its use, can be enjoyed everywhere, inside the car, the home, perhaps managed by home automation systems, the smartphone, the smartwatch, the smart TV.

All these technological objects are consistent with each other as they are linked to the same source to which to refer—semantic, mnemonic, technological, build coherent support, a smart environment in which the subject is no longer unknown or partial but integrated and assisted according to their physical, cognitive, social characteristics. In this cloud of continuous assistance, the subjects undergo real prosthetics which makes them experience a super-subjectivity that allows them to compensate for their difficulties and expand their cognitive peculiarities. The condition of subjective being, therefore, undergoes a process of real widening, of extension.

Specific Learning Disorders

The specific learning problems of dyslexia, dysgraphia, disorthography, and dyscalculia continue to be disorders, the complexities of which are not clarified by a prevailing etiology generally accepted by the scientific community. The problems associated with the primary aspects of learning, which are specifically reading, writing and numeracy, require medical, neuroscientific, psychological, and educational analyses, using strategies that have as yet no common ground as to the point of origin of the original problem/s. There are no objective epidemiological data since the difference between cultures, languages, education systems, and diagnoses do not help to clarify these disorders' overall incidence in the world. This is because the neurobiological data must be combined with the very significant cultural data.

However, it is believed that a world population, ranging from 5–17 per cent, is affected (Tasman et al., 2011). This does not make evident the characteristics of those affected from country to country, yet there are several main scientific options about the neurobiological nature involving those affected by this disorder. A study conducted by Alloway and Gathercole (2012) analyzed how working memory affected the performance of some children with dyslexia. The tests included *digit recall, word list matching, non-word list matching, word list recall, non-word list recall, and children's nonword repetition* test. The children were also subjected to four tests to verify their visuo-spatial performance through drawings.

What emerged from the study is that in children with dyslexia, there are significant variations, especially concerning the phonological function, which consists of neuronal attribution of sound, or sounds, to specific letters of their alphabet or ideographic system of some Asian language, including Chinese and Japanese. Investigation of the phonological function in the tests showed that in people with dyslexia, there was less ability to attribute the right sound to the corresponding graphic sign. This difficulty was repeated when unfamiliar phonological forms were presented to them and it became even worse than the reading characteristics they were believed to have achieved.

However, a particularly interesting finding in this study (Alloway and Gathercole, 2012, 25) is that spatial skills in the drawings were not significantly altered between the experimental and control groups. It is a basic observation as it would confirm that spatiality, spatial memory, spatial perception, and its representative derivations produced through drawings and graphic representations, not inherent to the meanings of writing, preserve their autonomy and in doing so, represent a dimension of a biologically 'primitive' cognition on which to act (Santoianni and Ciasullo, 2019; Casati and Varzi, 1999; Santoianni and Ciasullo, 2020). In this respect, the spatial perception can then represent an authentic element of the ability to manage the biological context of a subject that in turn, through spatiality, is in charge of the environment around him. On the other hand, it can also represent a possibility of mediation between 'biological intelligence' and the situated and cultural characteristics defined within the linguistic competence.

The linguistic component, given by the sound mediation of the word, although in some cases may be affected when attributing sound to the graphemic component of reading, does not appear disturbed in the typical language characteristics. Those who suffer from the interference of specific evolutionary disorders and particular learning disorders do not show verbal incapacity or language disorder, and when it happens, it affects a very low percentage of the population. In the mnemonic tasks (listening recall tasks) described by Alloway and Gathercole concerning the ability to remember verbally presented sequences, boys with dyslexia had no problem remembering them. While in *Counting Recall* and *Backward Digit Recall,* the performances between the experimental and control groups were significantly different because those with dyslexia had very low levels of ability to remember sequences. The same situation occurs for spatial memory related to matrix tests (graphic elements placed in a grid composed of shared spaces within which signs are placed in different positions). Also, in this case, the ability to remember or place the signs appropriately in their relative spaces presented was significantly impaired (*see* Fig. 3).

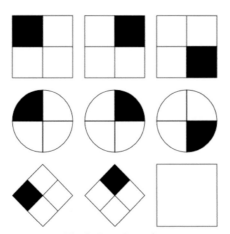

Fig. 3. Raven's matrix.

What appeared clear in the cognitive analysis of these children is a tendency for verbal regularity or, in some cases, a relatively mild language disorder. The reasons why the good memory of the verbal is not altered is not yet clear; however, as in spatial representation, this feature, unchanged or independently compensated, may represent an additional way to overcome, build and reconstruct paths suitable to bypass the difficulties caused by specific learning disorders. However, during visual exercises (reading), we assign great value to orthographic processing and, therefore, the words' conceptual organization within a given semantic context. This feature tells us a lot about the associative correlation between letters, words, and written language construction but does not adequately illustrate how the visual system can identify and decode the words and graphic signs that appear before his eyes.

A partial confirmation of an overall view of literal perception would seem to be the evidence that the left ventral occipitotemporal cortex seems mainly oriented to elaborate groups of letters and texts. We do not know enough about how the brain elaborates interpretative strategies when it has to implement processes that have nothing to do with the spelling organization of writing and, therefore, of the pre-orthographic organization, i.e., the processing of character strings, regardless of the type of character. The study of Lobier et al. (2012) used 14 expert adult readers who performed a visual categorization of multiple elements and a single element with alphanumeric (AN) and non-alphanumeric (nAN) characters under fMRI. The participants activated the posterior parietal cortex more strongly to process more elements than the single element. It was pointed out that some areas of the brain dedicated to decoding in reading were activated when there were groups of graphic symbols, regardless of the type of character; the parietal mechanisms are involved in the orthographic processing of character strings. This study shows that one crucial element in the visual recognition of words is activated, regardless of the meaning detected but always for large groups of words. This finding results in a further consideration, namely that the reading's decoding occurs primarily through a process of neurophysiological attention (Lobier et al., 2012).

Therefore, adequate attention levels would be a precondition in the ability to read and a condition that testifies an excellent attitude to levels of textual interpretation. This would lead to considering the elements typical of attention deficit hyperactivity disorder (ADHD) as related to learning disorders, both in a primary way (due to the low levels of attention compared to the visual classification of groups of letters) and in a secondary way because the reading difficulties would lead to a psychological state of frustration, avoidance and ultimately become the cause of early school dropout. In a study by Passolunghi et al. (2005), it was proven that even in the interpretation of a mathematical problem, children with attention deficit and hyperactivity disorder (ADHD), when they were asked in a mathematical problem to report the missing part to solve it (among a series of solutions that could be selected), they couldn't do it. Therefore, the problem appears to be related to the ability to link the contents, the meanings, that organic set of indications to textual sense characteristics to justify the possible resolutive choices (Passolunghi et al., 2005).

Much of the prevailing literature associates with the study of the etiology of specific learning disorders and in particular, dyslexia, which is often associated in comorbidity with other forms of specific learning disorders and seems to prove to be a very strong phonological processing disorder. Therefore it is a proper attribution to the graphic symbol, or the group of graphic symbols (Lobier et al., 2012) with which it should be associated. However, this hypothesis is not the only one. The progressive aspects resulting from our hypothesis is, as mentioned above, further strengthened because of the strong link between visual attention and the writing decoding process. Theoretical hypotheses support the impossibility of considering phonological processing in a completely autonomous sense and free from other neuronal collaboration forms. Moreover, it is not possible to clarify with absolute certainty whether phonological limits depend on other forms of deficit (Valdois et al., 2004).

These considerations open to further reflection on the role that the two functions—visual and phonological—have, making them equally responsible for specific learning disorders. It has been seen that learning disorders can both be associated with the perception of simple unitary elements because one cannot attribute verbal 'sound' to the graphic sign that is perceived through the sense of sight. However, they can also concern more complex elements, such as groups of words or the more complex semantics of more elaborate cognitive processes, such as mathematical learning. The so-called dyscalculia can have more complex forms that evolve into a simple form of a specific mathematical learning disorder; such difficulties are not connected to intellectual disability problems, physical or mental disorders.

In Morsanyi et al. (2018) study, a sample of 2,421 elementary school children on standardized math, IQ, and English tests, about 5.7 per cent of the sample with a gender balance between males and females had mathematical learning problems. More than half of that 5.7 per cent faced mathematical learning difficulties with language or communication difficulties; in some cases, ADHD and autism spectrum disorder profiles were present (Morsanyi et al., 2018). The most immediate reflection that emerges after this further data is that the process related to specific learning disorders cannot be attributed to the dysfunction of a single element and that in the analysis of the disturbance, the problem of visual crowding and phonological process must be considered in equal measure (Gori and Facoetti, 2015; Spinelli et al., 2002). In this difficulty of coordination between the ocular-visual system and sound attribution, there is a gap that affects the process of reading, writing, spelling organization, and calculation skills.

Therefore, according to our findings, specific learning disorders are a secondary process resulting from previous modification in the coordination itself. Even in those who are not affected by any form of disorder, the reading process is characterized by a significant and marked complexity associated with the encoding of alphabetical and numerical signs, their cataloging into sounds, and union in memory of graphic signs/sounds. The next process consists of identifying the graphic sign or number in a text, recovering the sign/sound union in memory, and assigning it the relative verbal sound to translate it into language. The element that blocks the whole process can be both primary, i.e., the one where you cannot attribute a mental sound to the graphic sign in the storage operation, and the second, when you have to retrieve the sign/sound union process from memory. According to this perspective, what is involved in encoding/decoding processes are no longer the cognitive components related to ocular/phonological coordination and, above all, the component of support, but the inscription of the entire process is memory. However, the memory would not be affected in different processes but precisely by the inscription of sign/sound encoding.

To partially confirm these hypotheses, or to be verified because we have not yet arrived at a unified vision of the problem, some findings have been demonstrated through some studies. Moll et al. (2014) found a very strong comorbidity between dyslexia and mathematical learning ability, and this is even when the cognitive domains of the two disorders (one reading, the other mathematical learning) are different. Initial considerations on this condition lead us to hypothesize exclusion of the cognitive characteristics of individual subjects from a disorder that, at this point, would be hierarchically superior to the particular cognitive conditions. However, there is some evidence in a study involving 99 elementary school children with reading and mathematical learning disorders, divided into four groups: in each of the two groups, there were children with only one of the two disorders; in another, both disorders were present in the control group. The study aimed to investigate, through tests, the speed of processing, time processing, and working memory (Moll et al., 2014).

One of the first factors that emerged is a particular co-existence in all those who had a specific learning disorder—the presence of a reduced level of attention. As we have seen previously, this could be attributed to primary causes of neurocognitive functioning or secondary causes of difficulties related to understanding specific processes. The attention processes analyzed through the transient analysis of covariance ANCOVA presented differences in it, depending on the disorder: although deficits in verbal memory were associated with both dyslexia and dyscalculia, the reduced processing speed was mainly related to people with dyslexia, less for dyscalculic; and the association with dyslexia was limited to the processing speed for familiar symbols. On the contrary, deficits in temporal processing and visuo-spatial memory were associated with dyscalculia, but not with people with dyslexia (Moll et al., 2014).

What emerges is a fundamental attention difficulty, often related to attention deficit hyperactivity disorder (ADHD) (Passolunghi et al., 2005); however, this cannot be considered the only 'reason' and only way to investigate in order to identify operational trajectories to compensate for the learning disorders produced.

Memory and the Role of Alpha, Beta, Delta, and Theta Waves

The theme of short-term working memory, which consists of manipulating information for short periods mentally and supporting daily activities, especially in early childhood, returns to central. This seems to be due to its ability and alteration in learning difficulties found within the primary cognitive areas to which the specific learning disorders refer. Nowadays procedures laid down to strengthen the short-term mnemonic component in children with disabilities or difficulties of various kinds (Down, Williams, specific language disorders, and ADHD) can identify procedures and technologies that can support these deficiencies (Alloway and Gathercole, 2012).

Adequate investigations have been carried out on the neurobiological nature of specific learning disorders and, in particular, about dyslexia. The real complexity is often evident in the diagnosis of SLD (specific learning disorders) resulting from the identification of whether the problems are associated with the emergence of reading problems coming from a low mechanism of compensation and a poor acquisition of the necessary reading skills or dependent on inadequate levels of neuronal communication at specific frequencies. Arns et al. (2007) tried to verify what happened in people with dyslexia concerning alpha, beta, delta, and theta waves through a QEEG (quantitative electroencephalography) and psychological tests.

What is commonly thought about brain waves is that they perform a very close function to the frequencies produced by music and that the functional balance of the brain is determined by 'harmonic' coherence, the balance that these ranges of brain waves can sustain. A study conducted by combining dyslexic children with a control group and carried out by the Brain Resource International Database (Luijtelaar et al., 2006) was to evaluate cognitive processes and their brain function according to EEG-related consistency parameters. What resulted was that the dyslexic group showed an increase in slow activity (delta and theta) in the right frontal and temporal regions of the brain with the attempt to compensate even for higher frequencies. This supports the hypothesis of a double deficit of dyslexia and shows that the differences between dyslexia and the control group could reflect compensation mechanisms (Arns et al., 2007).

Sound and Specific Learning Disorders. Memory, Phonetics, and Phonology

It seems to be quite clear now that we can no longer speak in a generic sense of dyslexia in the evolution of the diagnosis of specific learning disorders. As for other disorders, we are increasingly moving towards redefining the various and distinct differences within the diagnosis. The diagnosis model is increasingly aimed towards the endophenotypic hypothesis, accompanied by phonological awareness or magnocellular function that associate these characteristics of reading disorders with genetic and/or neurobiological characteristics (Galaburda, 2011).

It has been seen in some studies related to the induction of genetic abnormalities in murine models that these, otherwise normal, showed difficulties in processing some sounds. This has brought to the conclusion that similar anatomical abnormalities in humans could cause auditory processing deficits and predisposition to phonological deficits during and after language acquisition. Previous murine models have helped to understand the relationship between cortical developmental abnormalities and abnormal auditory processing. However, it is not entirely clear whether the presence of abnormal auditory processing during development is sufficient for dyslexia (Galaburda, 2018).

What is interesting for identifying a causal relationship between acoustic processes and short-term memory is that both the alteration of acoustic processes and the construction of appropriate mnemonic processes are involved in the hippocampus's functioning. In some mice's brain, where the manipulation of candidate genes produced anatomical abnormalities equivalent to those found in dyslexic brains, auditory and memory dysfunctions may occur. Therefore, the disturbance of neuronal migration to the cortex may be associated with an abnormal acoustic mapping in the cerebral cortex and which could be the crucial factor behind sound deficits and phonological processing in dyslexia (Galaburda, 2010).

There is a long and challenging possibility to recognize the etiological characteristics of dyslexia and the cultural and environmental conditions that play a direct role in understanding the phenomenon. It is essential to understand that even intellectual processes are in some way affected by cultural conditions. This leads us to understand that it is not entirely true that being involved in the diagnosis of dyslexia is not or cannot be just the intellectual peculiarities. For example, while in languages with opaque spelling, such as English, most dyslexics, especially young people, read slowly and make phonological errors; in languages like Finnish and Italian, where spelling is transparent, there is mainly slower reading but less full of errors accompanied by an inferior spelling ability.

In this work, it is essential to recognize a substantial difference that is emerging in the identification of sound discrimination and in the identification of learning disorders, especially dyslexia. It all revolves around the module of phonological recognition and decoding. In the studies of Berent et al. (2016), it was seen that it is necessary to proceed by applying a further differentiation between phonetic difficulties and phonology of grammatical structures. What emerged is that there is no strict link between the two functions—phonetics and phonology. Starting from the observation that in people with dyslexia, there was some form of slight alteration of language, sometimes linked to minimal and almost imperceptible traits—iterature attributes this characteristic to a problem of phonological processing. However, phonology is not the only process involved in the verbal action of speech, but also the result of inferior processes, which we would define as primary—as we did before—and which affect the auditory and phonetic system.

What emerges is that there are no reasons for homogenization of phonological and phonetic results because one concerns the recognition of grammatical structures, the other of sound discrimination, and phonetic identification. In all the experiments, the participants with dyslexia showed multiple phonetic difficulties, while their sensitivity to the phonological, grammatical structure was unchanged. These results demonstrate dissociation between the functioning of the phonetic and phonological systems in dyslexia. Contrary to the phonological hypothesis, phonological grammar seems to have been spared (Berent et al., 2016).

Intelligence and Specific Learning Disorders

However, what role does intelligence play in the self-compensation processes of a person with dyslexia? Intelligence, which remains an immaterial attitude with both objective and subjective characteristics, is influenced by attitudes, motivations, attention, and accumulated knowledge, which are associated with the social degree of the family to which it belongs, the opportunities, and the support obtained. In literate societies, dyslexia can interfere with knowledge acquisition as the prevalence of cultural acquisition passes through reading and writing; less so in societies, where knowledge passes mainly through imitation and narration. In this sense, it is not entirely possible to separate intellectual processes, as we measure them objectively, from a reading disorder. Even the parts of the intelligence tests that concern non-verbal skills are partly related to verbal skills because this is how the instructions for compilation, correction, and development are given (Galaburda, 1993; Galaburda et al., 1985; Galaburda, 2018).

A person with dyslexia with better mnemonic qualities will be helped to decode the text because he/she will be able to better guess words about his/her previous knowledge, even if he/she would stumble in reading a list of words or he/she cannot benefit from semantic, syntactic, and pragmatic cues. A person with dyslexia, on the other hand, equipped with well-developed attention and the organizational system, will be able to better manage information during the experience of acquisition and subsequent phases of recovery, to be less dependent on his phonological ability to draw meaning from the text. This does not mean that his phonological capabilities are more robust. Thus, the central system responsible for dyslexia, the phonological module, or access to it is independent of intelligence, as it can be strong or weak in intelligent and less intelligent children. However, a weak phonological module's action is brutal to ignore under the conditions induced by specific learning disorders (Galaburda et al., 1985; Galaburda, 1993; Galaburda, 2018).

Cognitive Assistants, Speech Synthesis, Accessibility, and Teacher Training

All inclusion processes go through the use and implementation of technological systems; in this sense, we can say that mobile technology and inclusive processes are closely related. When we refer to inclusive processes, we refer to an extended range of environmental, organizational, and technological possibilities to offer support and compensation to those in difficult situations. However, as we know, if a system can improve the learning of those with various problems, it will also involve and offer solutions that are valid even for those who do not have any altered functioning. It is no coincidence that the origin from which we move with the cardinal principle of inclusion precisely tends to enhance the differences, subjective, individual. In this field, mobile technologies become protagonists. They are easy to implement and, above all, they enter fully into our daily lives. This means that our learning dynamics are already modified informally, without the systems being used in formal contexts, such as the school.

In the previous chapters, what are strongly modified are the places of fruition and the times in an educational process that from a prerogative of the so-called 'time-school' becomes a continuous dynamic act. Technological trends related to support and compensation for both special education and traditional curricular education have been linked for a long time mainly to studying the role of visual dictionaries, sign language, vocal exercises, or history books. All these studies have often emphasized the results after using these tools but are minimal about the processes (Genc et al., 2019). However, few studies describe the entire process of technological integration with a holistic approach and look at the use of mobile technologies and M-learning. The study of Genc et al. (2019) through a survey of the literature produced by consulting the Scopus catalog with the keyword '*mobile learning and special education*' reveals in the analysis of the studies published on the subject that even we give great importance to the numerous reports referring to individual instruments used; this does not present an organic research on the entire topic.

M-Learning Accessibility

Interventions in education, but more generally related to the use of so-called assistive technology for cognition (ATC), have always aimed at the support and compensation of functional activities that require multiple skills like intricate attention, executive reasoning, perspective memory, self-monitoring for the improvement or inhibition of specific behaviors, and sequential processing. They also have an essential function for subjectivity about information-processing problems that may affect visual, auditory, and linguistic abilities or that may depend on the alteration of these primary forms of perception (Lopresti et al., 2004).

The theme of M-learning, related to accessibility remains, however, a poorly explored topic (Pieri, 2011), but is beginning to receive increasing interest from the scientific community. What seems particularly strange about the lack of attention on studies related to *mobile learning accessibility* could, however, be justified by the fact that mobile devices and, in particular, smartphones are 'constitutively' accessibility-oriented tools. Pieri (2011), in his survey of the literature on *mobile learning accessibility,* identifies two main strands: one concerning the study of technologies and their potential applied and the other referring to holistic evaluations of the pedagogical principle. What emerges is that many studies are mainly oriented to verify the technological characteristics applied to M-learning with an evident prevalence of studies related to the description of technologies and applications.

In this sense, this research tries to overcome this dichotomy mediating between technological instances, investigating in our case the role of audio and sound in learning processes through technologies on the one hand while, on the other, offering an overall view on the relationship of technology/education looking at theoretical educational aspects. So, clearly even in the elaboration, updating, and improvement of technological tools for learning, they should be considered in cases of altered physical conditions, as instruments to implement the needs of a special pedagogy. In particular this happens especially with smartphones, which are undergoing the most critical evolutionary process also because they are the mobile devices par excellence. As stated by Pieri (2011, 52–53) in the case of auditory sensory disability, the use of only audible alerts about calls or the content sought, together with a considerable operational difficulty determined by a complex operating language, can be a closing element rather than a compensatory possibility or support to the subject. Therefore, mobile learning usability should represent as a guide to a strong possibility of interchangeability of stimuli. We mean to say that the real strength of M-learning, and in this case of M-learning accessibility, in addition to the function developed by select apps (real simplified software that performs essential functions through quick, intuitive interfaces), must be to replace sound stimuli with visual ones and vice versa, visual ones with appropriate voice synthesis tools for screen reading (Apple voiceover, TalkBack Android).

Today's mobile devices' centrality allows them to provide everyone with good possibilities and learning opportunities, overcoming the difficulties imposed by their physical, mental, and sensory conditions. It is clear that mobile devices have social tension aimed at relationships, both in their primary function associated with telephone communications, and the use of cultural and social content through consultation of sites for information and entertainment and the ability to enjoy customizable music content. However, between possibility given by web access to and its continuous web interrelation lies its substantial weakness/educational strength. In fact, what stands out about their use is the lack of adequate management tools, use, and a real education in mobile media. In this sense, an overall pedagogical vision could generate a responsible approach but keeping in mind the cases regarding mobile learning, to be aware of both technical management of the object and the training-oriented use.

As in all training processes, it is not a question of having to award licenses, certifications, or certificates but investing in practices of responsible use of these technologies, and this can happen only if there is willingness of an integrated teacher/student training on technologies starting from elementary school. In a 2010 Israeli research by Goldfus and Goteman on the reading and comprehension skills of students of English courses at the University of Tel Aviv, it was found that the possibility of compensating for their difficulties occurs through programs capable of reproducing the sound of the so-called text-to-speech words; of recognizing vocal sound production, and translating it into the speech-to-text text; or writing programs with spell-checker and word prediction. The areas of technological intervention used could be summarized as follows:

- decoding (sound out words) reading comprehension by text-to-speech programs;
- handwriting directionality by speech-to-text programs;
- expressing words in written form by word processors and word prediction programs; encoding (spelling) proofreading programs spell checkers;
- organization by outlining/brainstorming programs (Goldfus and Gotesman, 2010, 21).

With the study of apps and software for learning, some applications already present and scientifically studied in various case studies relating to the use of OCR (optical character recognition) in children with dyslexia have given excellent results in terms of compensation (Fälth and Svensson, 2015). Fälth and Svensson's study (2015) aimed to see if a multifunctional application (Prizmo) for iPhone/iPad could provide an adequate level of compensation for some boys with dyslexia. A small group of 5th grade and middle-school students and their teachers could use their smartphones and tablets for about six weeks with the Prismo application, which is an OCR combined with a text-to-speech program. The results showed that the ability to decode words increased for many students and they found the application useful even after the end of the study; but, above all, the prolonged and familiar use of this program limited the expected effect on its use. It was noted that the implementation of such software and technology tools 'significantly' implemented the understanding, productivity, and cognitive success of these students. In this case, the research of 2010, the great technological evolutions that we have achieved subsequently with the evolution of various portable devices, began to define an integrated and organic approach mediated by the same category as mentioned above of *M-learning accessibility*.

Comparing the experimentation and use of iPad for mobile learning as assistive technology and also as a technological object to support learning among children with disabilities of various kinds, from sensory to psychic, showed that the contribution that tablets can make is far better than other less mobile technologies, such as PCs and notebooks (Chmiliar and Anton, 2015). In this pilot study, the researchers' questions about the *participatory action research* set up were:

1. How did each student use the iPad?
2. Which apps did the students choose to use and why?
3. Which apps did the students choose not to use and why?
4. Did the iPad, the course materials on the iPad, and the apps help support the students in their course work?
5. What supports did the students require to use the iPad effectively in their studies?

A series of questions to which they tried to answer by carefully observing the use that could be made of these technologies made sure that the advantages obtained were far more significant than the difficulties. However, the difficulties had all been described as easily overcoming precisely due to the versatile tools and therefore, easily modifiable, modular, expandable and reprogrammable. What emerged in the study of Chmiliar and Anton was (2015) that:

- there is a need for adequate preparation for the conscious and 'total' use of the tool mediated by adequate training because if the use of content, such as social media, e-mail, video viewing was familiar, the rest of the object's potential was less easy to manage. There is also a problem associated with the expected time for one's cognitive needs, that is, if it takes too long to search for the right app, finding the most suitable program and so to organize the iPad might reduce attractiveness and success of the object itself significantly;
- the tool is all the more effective, the more it can offer applications (apps) that allow the user to realize precisely what he has in mind and what he needs. The fact that there are too many apps whose actual function is known only after downloading them causes a sense of frustration that is not good for the training process;
- students need support in using the most suitable apps for their purposes. It is not enough to illustrate the presence of some specific apps to get immediately aware of use and support immediately valid; it has been noticed that some students with poor mnemonic skills but with the right visual aptitude to guide them towards the use of applications for the construction of mind maps contributed enormously to their educational success;
- the use of certain apps compared to others is strongly linked to the personality and specific tastes of each student; this must lead us to consider that the internal organization of these devices must also be strongly customized;
- the use of iPad not owned by the user does not help the right structure processes of customization of the tool's use (Chmiliar and Anton, 2015, 133).

From these considerations, some indications emerge that can be interesting for the organization's conscious realization and the use of some devices for mobile learning. It highlights that for the intended use of such tools, an adequate strategy of training, support, and guidance is necessary before their use. Relying on their easy use is to represent an autonomous process of growth. It is not enough if they do not first develop real user guides. Students need apps in the tool at their disposal to be clear and intelligible in their function right away, without the need to download them to understand their usefulness. This tells us that it is advantageous to draw up a list of suitable applications with a detailed description of their function and efficiency.

Mobile learning devices perform their function adequately if the subject who uses them can perceive it as her/his object and therefore finds it customizable, modifiable, customizable, and we would add 'decorative' and therefore it is perceived as her/his own. Pads not proprietary but entrusted only for research or educational purposes in the classroom cannot perform their function of assistance and support adequately because they cannot guarantee the process of subjectification, customization, and individualization to which such a tool should respond.

This last consideration opens up to other considerations of an economic nature: can mobile learning tools, considered at various levels increasingly important, especially in the presence of disabilities of various kinds, be considered objects whose purchase must be considered by schools? Is it the responsibility of the state as a direct educational guarantee to support those in educational difficulties? Or should it be a cost borne by those in difficulty? We would be inclined to the second hypothesis, namely that the state should guarantee these mobile devices as fundamental for the educational processes of those who have disorders or disabilities and should therefore finance them totally on a par with other forms of activities and materials supported by national welfare.

The so-called assistive technology (AT) has unquestionable potential and should not be understood as exclusively benefiting the disadvantaged situations caused by disability but should be understood as a system for the implementation of teaching. This clarification is necessary to avoid another problem of discrimination against those who use mobile devices. Often in the school experience, the possibility offered to individual students to use tablets or smartphones to support and sustain their learning activity is interpreted, especially in elementary schools, as an advantage in favor of the individual student and not as a good help. This could lead to the paradox of non-inclusive inclusion: while we try through the use of some tools to create a climate of inclusion, the presence of tools activated and used for a single child could create a visible climate of exclusion and differentiation.

In the use of learning technologies and in particular, in those that we imagine can be represented in *M-learning accessibility,* the questions that should be referred to are those imagined by Bryant and Bryant (2011): '*what is it, how is it made and how is it used?*' The first question refers to what types of support we imagine and what relationship is possible between technology and software in our possession; the second refers to the tool used if it can be used in its functions alone or must be implemented with additional special external peripherals suitable for the individual types of assistive needed; the third has to do with the use we make of it and refers to the needs for which it was designed, structured, organized, customized, and individualized and therefore the use for which it is considered necessary (Bryant and Bryant, 2011). Dell et al. (2012) have divided the development of assistive technologies for education into three macro-categories—low-tech, mid-tech, and high-tech technologies. Among the last ones, we find desktop computers; laptop computers; tablet devices; mobile devices; software; smartphones, augmentative communication devices (Dell et al., 2012).

Siri, Alexa, Cortana, Google Assistant: Can You Help Me?

A further transformation towards technologies' possibilities is indeed represented lately by what are called 'virtual assistants': Apple Siri, Alexa Amazon, Cortana, Microsoft, Google assistant. A series of systems that, through AI and the advantages of cloud computing (i.e., within servers external to individual devices), are increasingly used to manage information, respond, implement actions, search for specific news, entertain, organize personal data, remember events, implement personal reminders, and translate (Hoy, 2018).

All this happens through verbal interaction with a synthetic interlocutor who can recognize the voice of the individual user, to adapt to the semantics of the latter, to suggest anticipating routine needs, and to interact with home automation systems for the implementation of actions in the control mechanisms of their home, their communication tools, and their entertainment preferences, both music and television, both on social networks and on the web. Lopez et al. (2018), in their systematic analysis of these four main virtual assistants about some areas of applications, such as access to music services, agenda, news, weather, To-Do lists, and maps with directions, place them within the specialized area of what is called natural user interfaces (NUI). Natural interfaces occur through a tactile, verbal, immediate visual relationship and without further mediation or other peripheral elements.

This ease of use has meant that the technology industry has invested heavily in developing instruments suitable for their use. These are further defined *speech-based natural interfaces* within the so-called *human-computer interaction* (López et al., 2018). In 2019, Daley and Pennington (2020) stated that in the world about 100 million virtual assistants have been sold, which the authors call real 'cognitive assistants' and for about 10 years also, Siri (first commercial son) has helped to support and ensure its voice presence thanks to the use of many devices at Apple home.

What appears increasingly relevant within the growth prospects in the use of AI for training processes is that these tools are cheap, can be easily implemented within their mobile devices, can report and manage a long series of elements in streaming mode, and tend to be highly customizable to create an assistive cloud around each person. The implementation experiment of the object called *Echo* that supports the vocal assistant Alexa from Amazon within some classes in rural Idaho (Dousay and Hall, 2018) has shown how the lesson, thanks to what we could define as a real virtual assistant, has become strongly enriched with stimuli, with elements necessary in the immediacy of the lesson, with timers for the structuring of exercises. This has allowed them to overcome, on the one hand, the limit imposed by the use of IWB (interactive whiteboard) as compared to multimedia content requires more time between the activation of the blackboard, the web search of the topic or song, and its performance. *Echo,* on the other, allows immediate activation through the pronunciation of simple verbal requests to which it responds.

These possibilities benefit the integration of the frontal lesson through virtual assistants' use to integrate to the lesson a series of additional information to the advantage of a more significant offer of cognitive objects. The advantages, however, also exist in private use because it has been seen that, especially in the function of the repeater of verbal stimuli, that the tool could generate feedback to implement and strengthen the pronunciation characteristics of certain words in those who faced language difficulties. This happened because the input words, if not adequately expressed in their specific verbal characteristics, could not activate the virtual assistant and this involved an obstacle to overcome through a series of trials and errors (Medvedev Gennady, 2019).

Integration of Technological Objects, Cloud, Customization, and Individualization

The role of technologies is increasingly moving towards an integrative dimension that tends to combine the tools, such as smartphones, tablets, notebooks, and the latest speech-based-natural interfaces in a holistic system that also involves the so-called wearable devices (smartwatch, in particular). The integration of these various elements produces a cloud of service that exists and works beyond the individual devices; this means that virtual assistance, the interactive interface, is no longer represented simply by the technological tool, but by the intangible services that are behind and continue to evolve thanks to a real bionic semantics to regulate itself. This feature of continuous updating produced by the ability of the central systems to update to the characteristics of the single user, his verbal expressions, timbre characteristics, prevailing requests, can be supported only if the ability to store this complex amount of contents takes place in a series of external central servers accessed through personal credentials. This allows access to the same content from various devices, without having to fall on the memories of the individual technological devices used.

In both theoretical and practical terms, how external ontologies can be integrated into the new frontiers of *digital humanities* is addressed (Meghini et al., 2019). This long series of technological possibilities available for those with disabilities, disorders of various kinds, or anyone involved in training processes creates a new and articulated approach to training that guarantees two main elements fundamental to activate good inclusive possibilities and ensure quality cognitive processes: personalization and individualization.

Imagining unique, personalized, individualized paths with traditional didactics in the presence guaranteed by individual teachers which is the furthest that can exist from the reality of genuinely-inclusive training processes! Inclusion is realized by guaranteeing individual training needs so that it can be realized and not even if we simply offer the possibility to use a smartphone or tablet. There is a need to realize an intelligent, modular, mobile, functional, and customizable technological transition so that appropriate forms of support for education, training, and learning, in general, can be realized. To do this, we need choices guided by clear organizational principles which are changeable, implementable, revisable, and above all, focused on enhancing the individual in his need to train, express his skills, and be autonomous in managing the world around him.

One of the possibilities to organize the environments, the equipments, the tools inclusively are represented by the universal design for learning, indicating a way both to real contexts and to virtual and/or artificial ones.

Teacher Training

The real challenge, writes Qahmash in one of his essays (2018), is no longer so much in the observation of the advantages offered by the many and increasingly advanced features that high-tech technologies for education offer, but preferably in the way in which these technologies can become part of school contexts. In this context, analysis of school systems as an expression of national and supranational states as well as the role of respective ministries and offices for education, representing the needs of individual schools within the various territories of which they are an expression, teacher training plays a crucial and fundamental role.

The sensitivity to the problems related to special education, the technical ability to manage new technological systems, and the competence in flexibly structuring of alternative forms of training are all characteristics that refer to individual teachers' scientific skills. There can be no real development and growth or progress except through a structured and robust training of the teaching class (Qahmash, 2018). From this series of considerations, it is possible to understand that the current mobile technology potential for learning is beyond all possible limits, especially among those suffering from disorders or disabilities. However, the conditions for this potential to be well expressed are linked to some fundamental point of view:

- train and train yourself in the in-depth use of the instrument you want to use and the type of specific need of who will be the end-user;
- select the type of apps and peripherals suitable for the specific needs of the user, avoiding unnecessary dispersion in the test of apps whose characteristics are not known in depth;
- have highly trained lecturers on new learning technologies to interface with the scientific and academic world about discoveries in the field of technology;
- realize a dynamic training pact between the subject in training, families, teachers, schools, the world of university research for education and computer research, in order to create more and more apps, software, and technologies useful for inclusion.

Universal Design for Learning and Universal Instructional Design

The universal design for learning (UDL) consists of three guiding principles: representation, action/expression, and commitment. Therefore, its nature is to imagine an educational design system capable of reaching everyone and allowing everyone to express their functional characteristics in the best possible way. The strength of these principles lies in its ability to be in continuous movement—a process that does not stop and therefore can create flexible solutions for today, and also to organize future training contexts, always starting from a central element: the subject and its specific training needs (Courtad, 2019).

The idea behind the UDL is based on the observation that the individual has multiple possibilities to represent what he knows or may know (principle one), that there are different ways to express what one has understood and that one may know, and therefore to demonstrate that one has understood (principle two) and so there are different possibilities to involve subjects in the training processes (principle three) (CAST, 2018). In the guides of CAST (center for applied special technology, which is the body in charge of updating and drafting the internal indications of the UDL), the first point talks about *engagement,* which answers the question of why we learn in a certain way. A series of rhetorical questions are asked, which require a strong reflection on how we approach the structuring of learning environments and the realization of lessons. This helps us to understand how in the educational reality, there is not a single way to learn and therefore there cannot exist a single methodological mode that can be good for everyone.

For example, you could structure a lesson in a real/virtual learning environment that is stimulating, new, fascinating, and able to exalt the principle of discovery, but this for some could be something extremely positive and for others it could represent a further cognitive effort even if reassured by their routines or environments they are familiar with. In this sense, the 'why' a particular person learns in a certain way is fundamental in order to understand and realize the methods, the educational actions, and the tools that allow them to learn in the most suitable way so that all are the stimulus to discovery, or that induces them to a more incredible feeling of serenity in a context more familiar to him.

Creating multiple ways of *representation* is providing anyone with the individual differences, despite limitations imposed by their disability, despite the disorders that afflict their learning, a way or means to achieve 'what' is the best way to gain knowledge. However, this UDL component is only feasible strictly through the possibilities offered by technologies, certainly the mobile ones. The following indications, such as providing multiple ways to encourage perception through the ability to modify and customize the vision of certain information as well as to access an audio content and modify it according to particular needs, to have the opportunity to view a visual content according to different strategies, and to allow us to understand that the modularity, flexibility, and interactivity guaranteed by M-learning allows adaptation of the contents to the needs of each individual in training.

In the construction of one's cognitive processes, a significant force within the *representation* can offer options for language and symbols (including letters). That is to offer a series of vocabularies, graphic representations, alternative meanings, linguistic and mathematical decoding supports, representations of deep meanings through graphic representations can offer the subject in training access to cognitive contents, and immediately modify their form. This, once again, requires the support of mobile technology as it can give an immediate response to these needs.

A further step associated with *representation* would be offering to those who have to learn, depending on their difficulties, multiple options for understanding. Not everyone learns in the same way (Santoianni, 2014). This means that even the organization of cognitive content must proceed according to organic representations that have different settings and meet everyone's needs. Through the activation of previous knowledge, necessary integrations with the most recent learning can take place since knowledge does not proceed through a continuous thread and never proceeds by total jumps. Understanding in an acceptable way can also pass through the opportunity to have available schemes, relationships between elements, and a guide that helps in the processing, visualization, and management of information. However, all this requires the subject to have the ability to communicate adequately what he has understood, how he has understood it, and the motivations that has led him to understand. This is why it is essential that he can apply generalizations and proceed to identify causal links between the elements at his disposal.

The further macro category on which to act to encourage the forms of learning expressed in the UDL is the one that refers to the 'what' to learn in order to realize and make multiple possibilities of *action and expression,* in those that it comes defined like the learning's strategic ambit—to offer possibilities of physical activity through various methods for action and navigation, and optimize access to assistive tools and technologies; to increase the possibilities of expression and communication through various media and tools for communication, with the use of multiple tools for construction and composition to build skills to increasing degrees to practice and achieve good performance adequately; to create opportunities for executive functions that consist of guiding the subject to select the appropriate objectives for his or her needs and possibilities, helping him/her to develop plans and strategies, facilitating his or her ability to manage information and resources, and above all, strengthening his/her ability to monitor progress (CAST, 2018).

This long series of information taken directly from the indications of the UDL help us to understand how in reality, the management of learning processes, the awareness of one's cognitive levels, the need to organize better a system around the subject that learns and grows and evolves in society, involves a capacity for a reorganization of environments, learning styles, how one goes to school and above all, to elaborate universal strategies for an inclusive system.

Accessibility, collaboration, and community are material and immaterial pivots on which the UDL is built. That is why its realization can find reasons for implementation in tools that are not intended merely as assistive technology but as a real contiguous reality in constant support of the learning subject (Rogers-Shaw et al., 2018). Technology that offers alternative modes of representation, fruition, and action to the present can make the educational process inclusive, constant, fluid, self-organized, and total.

In this position of systematic construction of processes of alternative management of educational dynamics, M-learning could represent a frontier to satisfy and stimulate the interests of the subject, to support and strengthen the personality of those in difficulty. Its persistence can encourage self-regulation, activate adequate perceptual processes, and compensate for difficulties in order to strengthen linguistic and symbolic skills by offering the opportunity to modify symbolic systems directly through the graphical and visual interface and allowing the subject to express himself in all those ways that overcome their limits.

M-learning, the VLE (virtual learning environment) systems, maybe integrated into an all-inclusive system made of sound, spatiality, and above all, modification possibilities, which can be the basis for developing a subjectivity 'beyond the obstacles'. Therefore, the guidelines of universal design for learning (UDL) include flexible technology integration. What should be clear is that these processes go through intense and structured teacher training. The model of technology integration in teacher education (TPCK) for technology integration and use of mobile devices in teacher training appears to be a training path which can include in its strategic and operational choices the development of integration strategies between mobile technology within the parameters of the UDL (Shambaugh and Floyd, 2018).

Universal Instructional Design and M-Learning

The transition to the application of mobile technological tools for learning now seems to us more sustainable as we have overcome what Nielson in 2009 defined as a series of possible limits imposed by technologies of the time. However, it is good to remember that even today, in many places on the globe, even in the peripheral areas of advanced nations, we still have to deal with the issue of the digital divide and the difference in the connection between well-managed areas served by an excellent quality of communication networks and others with great difficulties of connection and in some cases, zero.

The significant variability and great differences between mobile tools only compared to the operating systems of individual manufacturers; slow downloads and limited Internet access; screens not large enough with low resolution, color, and contrast; difficulties in inserting text on smartphones; limited memories—these seem to be the difficulties of about a decade ago (Nielsen, 2009). Now they seem to have been overcome in some cases, but not completely.

The recommendations of UID (universal instructional design) are based on eights principles: (1) equitable use, (2) flexible use, (3) simple and intuitive, (4) perceptible information, (5) tolerance for error, (6) low physical and technical effort, (7) community of learners and support, and (8) instructional climate. Elias (2011), in a precise systematization of UID to be applied to M-learning, identified eight principles to refer to a useful, flexible, and inclusive organization of mobile learning systems:

1. Providing fair and accessible use to people with different characteristics and located in different places to do this requires that the characteristics of such communications are effectively simple.

2. Flexible use and organization to ensure a wide range of individual skills, time preferences, and connectivity levels.

3. Simplicity and intuitiveness of the system allow everyone quick intelligibility, avoiding unnecessary complexity that would only interrupt the cognitive flow and the educational progress that M-learning should guarantee to everyone.

4. Easily perceptible information by adding captions, descriptors, and transcripts prevents many things written or not correctly perceived from misuse of their meaning.

5. Tolerance for error. UID principles reduce and minimize the dangers and negative consequences caused by errors in software operation by designing learning environments with error tolerance and resilience. In this case, in the design of apps and learning environments, it would be useful to verify beforehand its management and any technical limitations regarding its usability.

6. Low physical and technical effort. As with online learning, M-learning should be developed with little physical effort, thanks to the NUI tactile that can perceive input quickly and effortlessly. This category should also include video and audio features made so as not to overload the subject's brain activity in training.

7. Student community and support. As in other forms of learning, the learning community's support has an extraordinary and irreplaceable function. That is why the organization, support, and internal communication should be facilitated through group development and appropriate internal communication tools, beyond the immediate M-learning environment.

8. Instructive climate. In this case, technological mediators or teachers play a significant role, or teachers can mediate with the end-users to clarify aspects related to the use of the content and technical problems within the platforms themselves. This should not be confused with the role played by the course designer.

At the end of this articulated path, what emerges as significant is that training systems focused on M-learning are made of elements which must harmonize in a relationship of interdependent functioning: the individual, learning communities, platforms or online content, technological tools such as smartphones and tablets in the process of exchange entrusted to communication networks, structures able to develop appropriate training projects and especially the construction of learning environments that consider the primary stimuli, including the extraordinary mediation of sounds as natural/synthetic elements to include everyone in their diversity.

The next path of this natural tendency towards technology that crowds our societies and reduces its cost as it goes along, constitutes a reasonable basis for further democratization of educational processes, provided that everyone, according to their specific needs, their diversity, their disorders, their sensitivity, their emotionality, has equal access to the future of education already widely rooted in our human community.

References

Alloway, T.P. and Gathercole, S.E. (2012). *Working Memory and Neurodevelopmental Disorders*. Psychology Press, Taylor & Francis, London.

Arns, M., Peters, S., Breteler, R. and Verhoeven, L. (2007). Different brain activation patterns in dyslexic children: evidence from EEG power and coherence patterns for the double-deficit theory of dyslexia. *Journal of Integrative Neuroscience*, 6(01): 175–190.

Berent, I., Zhao, X., Balaban, E. and Galaburda, A. (2016). Phonology and phonetics dissociate in dyslexia: evidence from adult English speakers. *Language, Cognition and Neuroscience*, 31(9): 1178–1192.

Bryant, D.P. and Bryant, B.R. (2011). *Assistive Technology for People with Disabilities*. Pearson Higher Education, London.

Casati, R. and Varzi, A.C. (1999). *Parts and Places: The Structures of Spatial Representation*. MIT Press, Boston.

CAST. (2018). *UDL: The UDL Guidelines*.

Chmiliar, L. and Anton, C. (2015). The ipad as a mobile assistive technology device. *Journal of Assistive Technologies*, 9(3): 127–135.

Courtad, C.A. (2019). Making your classroom smart: universal design for learning and technology. pp. 501–510. *In*: Uskov, V.L., Howlett, R.J. and Jain, L.C. (eds.). *Smart Education and E-Learning. Smart Innovation, Systems and Technologies*. Springer, Singapore.

Daley, S. and Pennington, J. (2020). Alexa the teacher's pet? A review of research on virtual assistants in education. pp. 138–146. *In: Association for the Advancement of Computing in Education (AACE)*.

Dell, A.G., Newton, D.A. and Petroff, J.G. (2012). *Assistive Technology in the Classroom: Enhancing the School Experiences of Students with Disabilities*. Pearson, Boston.

Dousay, T.A. and Hall, C. (2018). Alexa, tell me about using a virtual assistant in the classroom. pp. 1413–1419. *In: Association for the Advancement of Computing in Education (AACE)*.

Elias, T. (2011). 71. Universal instructional design principles for mobile learning. *International Review of Research in Open and Distributed Learning*, 12(2): 143–156.

Fälth, L. and Svensson, I. (2015). An app as 'reading glasses'—a study of the interaction between individual and assistive technology for students with a dyslexic profile. *International Journal of Teaching and Education*, 3(1): 1–12.

Galaburda, A.M., Sherman, G.F., Rosen, G.D., Aboitiz, F. and Geschwind, N. (1985). Developmental dyslexia: four consecutive patients with cortical anomalies. *Annals of Neurology*, 18(2): 222–233.

Galaburda, A.M. (1993). *Dyslexia and Development: Neurobiological Aspects of Extra-Ordinary Brains*. Harvard University Press, Boston.

Galaburda, A.M. (2010). Neuroscience, education, and learning disabilities. *Human Neuroplasticity and Education*, 27: 151.

Galaburda, A.M. (2018). The role of rodent models in dyslexia research: understanding the brain, sex differences, lateralization, and behavior. pp. 83–102. *In*: Lachmann, T. and Weis, T. (eds.). *Reading and Dyslexia: From Basic Functions to Higher Order Cognition*. Literacy Studies, Springer.

Genc, Z., Masalimova, A., Platonova, R., Sizova, Z. and Popova, O. (2019). Analysis of documents published in scopus database on special education learning through mobile learning: a content analysis. *International Journal of Emerging Technologies in Learning (IJET)*, 14(22): 192–203.

Goldfus, C. and Gotesman, E. (2010). The impact of assistive technologies on the reading outcomes of college students with dyslexia. *Educational Technology*, Educational Technology Publications, 50(3): 21–25.

Gori, S. and Facoetti, A. (2015). How the visual aspects can be crucial in reading acquisition: the intriguing case of crowding and developmental dyslexia. *Journal of Vision*, The Association for Research in Vision and Ophthalmology, 15(1): 8–8.

Hoy, M.B. (2018). Alexa, siri, cortana, and more: an introduction to voice assistants. *Medical Reference Services Quarterly*, 37(1): 81–88.

Lobier, M., Peyrin, C., Le Bas, J.F. and Valdois, S. (2012). Pre-orthographic character string processing and parietal cortex: a role for visual attention in reading? *Neuropsychologia*, 50(9): 2195–2204.

López, G., Quesada, L., Guerrero, L.A. and Alexa vs. Siri vs. Cortana vs. Google Assistant (2018). A comparison of speech-based natural user interfaces. pp. 241–250. *In*: Nunes, I.L. (ed.). *Advances in Human Factors and Systems Interaction. Advances in Intelligent Systems and Computing*. Springer.

Lopresti, F.E., Mihailidis, A. and Kirsch, N. (2004). Assistive technology for cognitive rehabilitation: state of the art. *Neuropsychological Rehabilitation*, 14(1-2): 5–39.

Luijtelaar, G.V., Sumich, A., Hamilton, R. and Gordon, E. (2006). *Brain Resource International Database*.

Medvedev, G. (2019). Ask alexa: language practice with amazon echo. pp. 187–190. *In: Society and Languages in the Third Millennium*. Communication. Education.

Meghini, C., Bartalesi, V., Benedetti, F. and Metilli, D. (2019). *Sistemi basati su ontologie per le Digital Humanities, CNR*. ISTI.

Moll, K., Göbel, S.M., Gooch, D., Landerl, K. and Snowling, M.J. (2014). Cognitive risk factors for specific learning disorder: processing speed, temporal processing, and working memory. *Journal of Learning Disabilities*, 49(3): 272–281.

Montessori, M. (2013). *The Montessori Method*. Transaction Publishers.

Morsanyi, K., van Bers, B.M.C.W., McCormack, T. and McGourty, J. (2018). The prevalence of specific learning disorder in mathematics and comorbidity with other developmental disorders in primary school-age children. *British Journal of Psychology*, 109(4): 917–940.

Nielsen, J. (2009). Mobile Usability. *First Findings*, Nielsen Norman Group, California.

Passolunghi, M.C., Marzocchi, G.M. and Fiorillo, F. (2005). Selective effect of inhibition of literal or numerical irrelevant information in children with attention deficit hyperactivity disorder (ADHD) or arithmetic learning disorder (ALD). *Developmental Neuropsychology*, Routledge, 28(3): 731–753.

Pieri, M. (2011). *L'accessibilità del mobile learning. Italian Journal of Educational Technology*, 19(1): 49–56.

Qahmash, A.I.M. (2018). The potentials of using mobile technology in teaching individuals with learning disabilities: a review of special education technology literature. *TechTrends*, 62(6): 647–653.

Rogers-Shaw, C., Carr-Chellman, D.J. and Choi, J. (2018). Universal design for learning: guidelines for accessible online instruction. *Adult Learning*, 29(1): 20–31.

Santoianni, F. and Ciasullo, A. (2019). Digital and spatial education intertwining in the evolution of technology resources for educational curriculum reshaping and skills enhancement. *In: Virtual Reality in Education: Breakthroughs in Research and Practice*. IGI Global, Pennsylvania.

Santoianni, F. and Ciasullo, A. (2020). Teacher technology education for spatial learning in digital immersive virtual environments. *Examining the Roles of Teachers and Students in Mastering New Technologies*. IGI Global, Pennsylvania.

Shambaugh, N. and Floyd, K.K. (2018). Universal design for learning (UDL) guidelines for mobile devices and technology integration in teacher education. *In: Handbook of Research on Digital Content, Mobile Learning, and Technology Integration Models in Teacher Education*. IGI Global, Pennsylvania.

Spinelli, D., De Luca, M., Judica, A. and Zoccolotti, P. (2002). Crowding effects on word identification in developmental dyslexia. *Cortex*, 38(2): 179–200.

Tasman, A., Kay, J., Lieberman, J.A., First, M.B. and Maj, M. (2011). *Psychiatry*. Wiley, Hoboken.

Valdois, S., Bosse, M.-L. and Tainturier, M.-J. (2004). The cognitive deficits responsible for developmental dyslexia: review of evidence for a selective visual attentional disorder. *Dyslexia*, 10(4): 339–363.

Part IV
Daniele Agostini

CHAPTER 9
Mixed Reality Mobile Technologies and Tools in Education and Training

Daniele Agostini

Introduction

Augmented reality (AR) is a term coined by the Boeing researchers, T.P. Caudell and D.W. Mizell. They created a heads-up (HUD), see-through, head-mounted display (HMD), enabled for head-position sensing and real-world registration. Using it, a worker could have her or his field of view augmented (Caudell and Mizell, 1992). This technology was thought to add a visual level over and above the user's sight by means of HMD. Nowadays, researchers often define AR in the same way as just an augmentation of the sense of sight, with the difference that now it can be used with glasses, headsets or mobile device displays used as HUD. Nevertheless, one can agree with the work of Schraffenberger and van der Heide (2016), who argue that AR is ordinarily multimodal and involves many other senses than that of sight.

Thus, AR is a technology which heightens our sensory perception of reality through the superimposition of a computer-generated layer to one or more of our senses. Another basic characteristic of this technology is that it implies an anchor with reality or, in other words, something that links the computer-generated layer with reality. It is most commonly used to heighten the sense of sight, providing the user with contextual information, three-dimensional images or models which interact with the environment and other real objects (Azuma et al., 2001). This feature is usually achieved by sensors. In most of the appliances, the sensors are GPS and the camera. One does not ignore that one of the most effective forms of augmented reality refers to the sense of hearing. Audio-guides are a form of AR that offer additional and contextual knowledge to our sense of hearing.

Postdoctoral Researcher in Educational Technology and Digital Interpretation, University of Padua.
Email: daniele.agostini@unipd.it

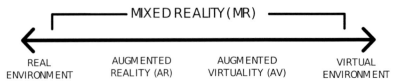

Fig. 1. Reality-virtuality continuum by Milgram and Kishino, 1994.

Augmented reality acts within a continuum with two polarities: the real environment and the virtual environment (Milgram and Kishino, 1994) (*see* Fig. 1). The areas which lie between these two poles are part of a so-called 'mixed' reality. AR acts within the mixed reality interval, which is closer to the real environment, whilst augmented virtuality is closer to the virtual environment.

As a medium that connects the virtual with reality, the potential of augmented reality in the field of education is increasingly studied by researchers who consider it as one of the next-generation medium with a prominent role in future learning best practices (Dede, 2008).

Mixed reality has been used since the beginning to broadly refer to technologies which are in continuum but cannot be classified as 'just' AR or VR (virtual reality). In fact, one of the earliest definitions of MR, from the 1st International Symposium on Mixed Reality, is '*the overlaying of virtual objects on the real world*' (Billinghurst and Kato, 1999; Pan et al., 2006) which is very similar to the current AR definition. It seems that the fact of having 'virtual objects' superimposed on a real environment, instead of plain textual information, was the discriminant to classify it as MR. Being in a continuum, one cannot talk of boundaries and strict classifications, but with the latest AR technologies, such as Apple AR Kit and Google AR Core, these categories become, if possible, more blurred than ever.

The Range of AR and MR Technology

Since the advancement of mobile technology, there is now a large array of affordable handheld devices with significant processing power and many sensors that can be used for virtual, augmented, and mixed reality.

Initially, there were two kinds of distinctions: smartphones and tablets are best suited to growing usage of AR/MR, and big manufacturers of smartphones and tablets are coming up with increasingly productive ways of using AR/MR. Some of these tools, due to their built-in features, can be used alone (Google Glasses and Microsoft Hololens, for example). In other cases, smartphones or similar devices are used as computers to handle these glasses, thereby making smartphones and similar devices more affordable (e.g., Sony SmartEyeglass, Epson Moverio).

As a result of the wide variety of technological solutions available, mobile operative systems are becoming more scalable and richer in AR and MR apps. Some of these apps are 'general purpose', enabling us to use AR/MR with 3D images or models when needed: HP Reveal, BlippAR and ZappAR, for example. The second kind of programme should be found within a particular context, such as a variety of well-known historical sites or catalogues (such as Ikea's).

This kind of classification, though useful for a first glance at what is available, is not so helpful for a better understanding of, and theoretical work on, the technology.

For this reason, it is possible to propose an alternative classification of AR/MR, not making a distinction between hardware and software, but based on context and hardware-software technology as a whole. This can be called a classification of AR/MR experiences. There are attempts to create taxonomies for this kind of technology, but it is preferable to use the word 'classification' instead. In fact, while for some macro category one can easily distinguish a hierarchy (for example, distinguishing between fixed and mobile or types of technology), for others it is hard to identify sub-categories and hierarchies that are not merely related to hardware. Therefore, it is not a taxonomic style tree like that of Darwinian evolutionary theory, but rather of an intricate bush-style diagram, more like the theory of punctuated equilibriums of Eldredge and Gould.

Mixed Reality Experiences Classification

Within AR/MR technologies can be identified a variety of characteristics or dimensions, the combination of which delineates technologies, software and, above all, application contexts. The first dimension is the above mentioned virtuality continuum, but not all the others are polarities of a continuum; several are part of a discrete succession:

- Portability dimension:

  Static ◄———portable———mobile———► ubiquitous/wearable/pervasive

Traxler (2005) made this dimension by trying to classify e-learning and M-learning devices. This dimension is directly related to the hardware required to run a specific AR/MR appliance. An example of AR/MR fixed installations is the use of 3D mapping projectors which project layers of information and images on a real surface. They can be interactive or not.

Another reason to have a fixed or a moveable device might be the need for great computing power to have a real-time rendering of a 3D scene, for example. In that case, one could use a powerful desktop or a powerful laptop, which is portable but not exactly pocket-sized. On the other polarity, the most common kind of AR/MR experiences involves smartphones, smart glasses, and other wearables.

- Sensory range:

Range:	Visual	Auditive	Haptic	Olfactory	Gustative	Vestibular

AR/MR technology can add an informative computer-generated layer over one or more of our senses. Traditionally we reckon to have five senses: sight, hearing, taste, smell and touch. Each sense is enabled by a sense organ or sensor. In order to augment our senses, the AR/MR device needs a component to do that and, often, a sensor that lets it acquire the same kind of information as our senses to give the right information at the right moment and in the right place. For example, one of the most common AR/MR involves the sense of sight. To add a computer-generated layer to our vision, the device needs to let us see that layer by using a screen, a projector or a light emitter. At the same time, the device needs to know what we are seeing or where we are, so it needs a sensor, like a camera or the GPS sensor, to give us the right information.

An example of visual AR/MR are the apps like Peak Finder that shows us through the screen of the phone or specific headsets the same panorama that we are looking at, but with all the names of the mountains around us. Pokémon Go is another example of visual AR/MR. To have an example of an auditive AR/MR, one can look at all that apps in the market which provide information through sounds or speech in relation with the place where one is or where one is heading to. Good examples are audio-guides, satnavs and soundwalks. Is it also possible to use 3D sounds, thus adding the dimension of space to the sound. Another example of an audio AR is Shazam: it is an app that can listen to the music one is hearing and tell the name, artist and album. It can also provide links to listen to that song.

There are other applications which use haptic feedback[41] to give us more information. It is mostly used in the fields of simulation and telesurgery, but one can find widespread uses, such as the Google Maps navigator which provides tactile feedback while navigating in 'on foot' mode to tell when to turn. Other examples are haptic clocks or an app that helps in taking level photos through haptic feedback. In regard to the other two traditional senses of smell and taste, there are some experiments and even commercial products. To enable that augmentation, it would be necessary to add a taste and smell peripheral to the device (Sardo et al., 2018). One of the latest examples is the iPhone Duo, which is a device connected to the smartphone that, through a specific app, allows one to send pictures with a primary and secondary note of scent.

All the capabilities mentioned above are directly connected to the sensors which the AR/MR device contains.

- Sensors range:

  A sensor is the electronic equivalent of a sensorial organ. To register information from the environment, an electronic device needs sensors. Every sensor is a possible link with the reality. In Table 1, one can see a list of sensors with which a modern high-end smartphone is equipped. Each one of it is sensitive to one particular physic aspect of the reality and can collect different types of information. For example, the accelerometer is sensitive to acceleration (even the gravitational one) and can collect data on the spatial movement of the phone. The magnetometer is sensitive to magnetic fields and is used to detect them, but, mostly, to detect the magnetic North for spatial orientation. From a combination of these sensors, the smartphone's operating system (OS) is capable of being aware of more complex situations.

If the augmented reality needs anchors to connect the virtual world with the real one, the sensors are these anchors. What is a gyroscope/posture sensor?

---

[41] Haptic feedback includes tactile feedback and kinesthetic feedback, the former being what you can sense on the surface of your skin, like touch, texture, pressure or vibration. The latter is given from sensors in joints, tendons and muscles and lets you feel the approximate weight, size, and the relative position to your body (Minamizawa et al., 2010).

Table 1. Sensors on a high-end smartphone: they are sensitive to one particular physic aspect of the reality and can collect different types of data.

Smartphone Sensor	Environment Feature								Data Types					
	Light/infra red	Sound	Acceleration	Eletro/Magnetic Field	Radio frequency	Temperature	Pressure	Humidity	Visual	Auditive	Spatial	Environmental/Athmospherical	Communications/security	Biometrical
3D Facial	Yes								Yes (3D)				Yes	Yes
Accelerometer (Gravity)			Yes								Yes			
Altimeter				Yes			Yes				Yes	Yes		
Ambient light	Yes								Yes			Yes		
Barometer							Yes					Yes		
Bluetooth					Yes						Yes		Yes	
Camera (2 or more)	Yes								Yes				Yes	Yes
Fingerprint				Yes									Yes	Yes
Gesture	Yes										Yes		Yes	
GNSS (2 or more)					Yes						Yes	Yes	Yes	
GSM/3G/4G/5G					Yes								Yes	
Gyroscope (Posture)			Yes (3D)								Yes			
Hall effect				Yes							Yes		Yes	
Heart rate	Yes													Yes
Hygrometer								Yes				Yes		
Magnetometer (Compass)				Yes							Yes	Yes		
Microphone (2 or more)		Yes								Yes		Yes	Yes
NFC											Yes		Yes	
Oximeter	Yes													Yes
Proximity	Yes										Yes		Yes	

Table 1 Contd. ....

...Table 1 Contd.

Smartphone Sensor	Environment Feature								Data Types					
	Light/ infra red	Sound	Acceleration	Eletro/ Magnetic Field	Radio frequency	Temperature	Pressure	Humidity	Visual	Auditive	Spatial	Environmental/ Athmospherical	Communications/ security	Biometrical
RGB Colours	Yes								Yes			Yes		
Step counter			Yes								Yes			Yes
Thermometer						Yes						Yes		
Touchscreen				Yes									Yes	
WiFi					Yes						Yes		Yes	

The gyroscope is a crucial sensor for most mixed reality applications. A gyroscope is a device that maintains orientation and angular velocity. It consists of a rotor (a spinning wheel) mounted on a gimbal (*see* Fig. 2). The rotor is free to assume any orientation; however, while rotating, it is not affected by any of the gimbal tilting or rotation due to the conservation of angular momentum.[42] People in ancient civilizations, such as Greece and Rome, knew this principle, but the German Johann Bohnenberger made the first known instrument of this kind in the modern era. Léon Foucault used it in an experiment to 'see' the Earth's rotation movement. Hence, he was the first to name it gyro-scope, from the Greek *gyros*, which means 'circle' or 'rotation', and *skopeein*, which means 'to see'. Since the 19th century, it was used as an aid to navigation, mining, flight, and ballistics, often associated with a compass ('Gyroscope', 2021).

Nowadays, the most common version is the micro-electromechanical one (MEMS), which can be found in many smartphones and uses vibrating micro elements to function. These sensors can detect—with very high accuracy—any movement that diverts from the initial position in any direction. This means that, with this sensor feedback, a software program can keep track of them and calculate, with very high accuracy, any direction the device is pointing towards on three or six axes. A similar effect, but with a much lower accuracy, can be caused by combining the feedback of the accelerometer and compass sensors.

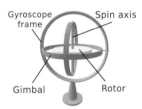

Fig. 2. A classical gyroscope. The spinning wheel has freedom of rotation in all three axes and will maintain its spin axis direction regardless of the orientation of the outer frame.

Sensors, Landscape and Heads-up Attitude

In practical terms, the gyroscope sensor enables a smartphone app to understand where the device is pointing into a hypothetical sphere around it (*see* Fig. 3).

On the other hand, it cannot understand where one is physically located on the surface of the Earth or in relation to other objects. Therefore, we used other sensors as well. After the gyroscope, the second most important tool was the A-GPS sensor. In the open-air, it can pin down a person's location on the surface of the Earth with a precision of roughly 4 metres. It can be used in order to trigger outdoor AR, MR and VR experiences. Hypothetically, the third sensor of importance would have been the compass. Its work should have been to align the gyroscope-guided AR, MR and VR systems with the real landscape—in other words, to align the virtual North with the real one. Depending on the quality of the compass sensor, one can be happy or disappointed to discover that the compass sensor is too imprecise to do such a job.

---

[42] The total angular momentum of a system remains constant unless acted on by an external torque.

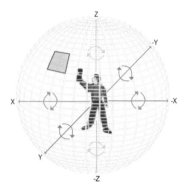

Fig. 3. The gyroscope sensor axes. The gyroscope sensor also detects yaw and rotation.

Fig. 4. Heads-up vs. heads-down attitude. A heads-up attitude integrates the screen content with the landscape, while a heads-down one decontextualises it.

Nonetheless, the most important advantage of using the gyroscope is to allow a heads-up attitude in using the app and exploring the environment. In a heads-down attitude, which is the usual one people assume when looking for information on a smartphone or even a booklet on the move, one controls the device with one hand, possibly using the other one to hold it. Users' eyes see the device screen, and all the rest in the field of view, which most likely consists in the user's own legs, feet, and the floor and which remain blurred and unimportant (*see* Fig. 4). Eventually, users look up to catch up with the environment around them and to compare the information received from the app with the real environment. By contrast, using the AR with the gyroscope allows users to look at the landscape around them while also looking at the device screen (*see* Fig. 4), which displays the same landscape with a layer of contextual information on it. The typical pose of people using this modality is to hold the device with one hand at eye level towards the direction they are looking at. This use of the device permits the brain do the job of fusing reality and AR (Caudell and Mizell, 1992), triggers cognitive dissonances, and allows users to concentrate on exploring the environment rather than in using the device.

Augmented and Mixed Reality Type Classification

Thus, the AR/MR experiences can be classified, depending on which sensor or cluster of sensors the application uses. The two most commonly used are GPS[43] and the camera. The former is used in all the applications that trigger events based on one's position, as an example Google Maps or the Google Assistant. The latter is adopted from the kind of apps with a vision-based AR/MR that shows pictures or 3D models on the camera feedback. Examples include the Ikea catalogue app, the ZappAR which is used mainly for commercial communications, and SnapChat's famous AR/MR function that modifies people's faces. Pokémon GO, on the other hand, uses both of these sensors to deliver the AR/MR experience. The GPS triggers the encounter with the Pokémon that one would be able to see through the camera.

More advanced AR/MR framework, like Apple ARKit and Google ARCore, use a combination of sensors. They use camera, gyroscope, and accelerometer together with machine learning algorithms to deliver the best possible AR experience. A possible sub-division of the sensor options is the type of anchor, or marker that the AR/MR uses in order to link the real and the virtual.

Location-triggered AR and MR

These can be categorised as *location-triggered* AR/MR experiences based on location sensors, but there is more than one possibility to achieve that:

- GNSS based uses as the link the location data provided from the GNSS sensor (mainly latitude, longitude and altitude).

- A-GPS based stands for assisted GPS, and relies on more than one sensor to improve the accuracy of the GNSS location. It cross-references the data from GNSS sensor, cellular network when available (GSM/UMTS/LTE), visible Wi-Fi networks and, where available, barometer—used to give an approximate altitude—to provide quicker and more accurate GNSS position. This is the most common locative system for outdoor AR/MR.

- Location services based provide data from above-mentioned sensors, notably triangulating WiFi network signals and can also give an approximate position in the absence of a GNSS sensor or/in indoor situations.

Proximity-triggered AR and MR

A second category is *proximity-triggered AR/MR*, which is a technique based on sensors that need an electronic tag which they can recognize:

- Bluetooth Beacons based technology delivers a very accurate position system based on Bluetooth beacons, which are little devices that emit a Bluetooth wireless signal and need a power supply. They are used mainly in indoor

---

[43] GPS actually means a generic GNSS (global navigation satellite system). In fact, current smartphones use not just the US Navstar global positioning system (GPS), but also the Russian global navigation satellite system (GLONASS), the Chinese BeiDou navigation satellite system (BDS), the Japanese quasi-Zenith satellite system (QZSS)—which is technically a regional satellite navigation system which augments the performance of the GPS in Asia and Oceania—and the European Galileo GNSS.

settings where the GNSS signal cannot be received. Its maximum range is about 70 metres for regular beacons and 450 metres for long-range beacons.

- Near field communication (NFC) based is the same wireless technology that is used for Apple, Google, and Samsung Pay contactless platforms. Although it uses a specific NFC sensor, in AR experiences it can be used more or less like a more sophisticated and reliable QR (quick response) code (Miglino et al., 2014). To use it, it is sufficient to bring the device close to the NFC tag, which is a very thin and inexpensive piece of circuitry. Since it uses the principle of electromagnetic induction, the tag does not need a power source. NFC will also recognize other NFC sensors. The range of the most common sensors is usually less than 20 cm.

Vision-triggered AR/MR

It can be categorised as *vision-triggered AR/MR* for all the experiences where one has to point the camera sensor towards the surrounding environment to trigger an augmentation of it. Here, too, there is more than one approach:

- QR code-based approach which uses the camera sensor with software capable of reading and decoding QR codes, which are bi-dimensional barcodes. They contain data embedded and, usually, a link to a webpage, a web app or a smartphone function. They are common in both outdoor and indoor experiences to deliver highly contextualized information. An example is the QuizerRo experience where they are used for location-based games to trigger riddles and other contents that will bring the user to the next stop-over (Erenli, 2013).
- Marker-based is the most used type of vision-triggered AR. It is based on a so-called fiduciary marker, that is a picture or a pattern already stored in the system. The computer vision algorithm looks for that same marker in the camera feedback and, when it finds it, the AR is triggered. One education-oriented example worth mentioning is Mirage Make AR ('Mirage - *Réalité Augmentée et Virtuelle Pour l'enseignement*' n.d.).
- Marker-less is the rarest kind of AR/MR technology because it requires the computer vision to recognize the unknown features following models and categories rather than a well-known marker. This approach is however becoming more common lately, thanks to machine learning and artificial intelligence (Feng Zhou et al., 2008) and widespread framework as Google AR Core and Apple AR Kit. Between the best educational, general purpose software which uses this approach are Cospaces Edu and Metaverse ('CoSpaces Edu for Kid-Friendly 3D Creation and Coding' n.d.; 'Metaverse Studio' n.d.).

Virtuality Classification

- The 'virtual' dimension:

  It is the most used and generally accepted amongst the scientific community. It can be used as one of the dimensions of our classification, but, since we use the mixed reality concept, here we will be more specific and pinpoint significant kinds of MR in the continuum. In particular, can be found two different and conventional ways of implementing AR: direct and indirect (Wither et al., 2011). These techniques refer to a visual AR, but the same principle could be used for an auditive AR. In direct AR, the reality is captured and augmented in real-time from the smartphone sensors. As an example, applications like Ikea's and BlippAR use direct AR because they add a computer-generated layer on the real-time feed from the camera. On the other hand, indirect AR does not use real-time feed but information that has been previously stored in the correct format inside the device. Most of the open-air AR/MR applications use that technique. It provides a series of technical advantages in respect of alignment issues, and, in particular, because of the difficulties in linking the virtual layer with the real one in contexts on which there is no control over contrast and light conditions (Wither et al., 2011). Other sensors, like GPS, gyroscope, and accelerometer link the indirect AR to reality.

One may say that indirect AR, from the user-experience point of view, is very similar to a traditional AR experience, but, from a technical point of view, it is a space that is more virtual than real. For this reason, it can be seen as one of the most typical examples of MR. This classification could be used in literature review to analyze all the different experiences of mixed reality mobile learning (MRML).

Mixed Reality Mobile Learning Applied to Outdoor and Heritage Education

A field that could serve as a brilliant example of how technology and paradigms evolved in MRML is the outdoor and heritage education. The reason is that these matters cannot be effectively covered in the classroom environment due to the fact that they are inherently artefact-based and context-based. One needs to show landscapes (in particular, the land and the action of humans on it—reification and organization) and artefacts (e.g., utensils, jewellery, buildings, remains, etc.) that most of the time cannot be seen, de-contextualized, or even moved from their original places. At the same time, there is a need to give precise and contextualized information about these artefacts. For these reasons, the outdoor and heritage education was one of the fields where MRML has evolved since its beginning.

The starting point was the creation of virtual environments. The idea was to immerse the student (or the researcher or even the tourist) in the original context, thanks to the virtual recreation of it. It can be called 'mobility through immersion' which solved—and solves nowadays—a wide variety of technological and logistical problems. In the meanwhile, mobile learning tools, platforms, and methodologies were improving and growing, paving the way for the fusion of virtual, 3D, immersive content with mobile learning which gave birth to MRML.

The Starting Point is Virtual

Looking back over the last 20 years, one can recall many examples of applications that use virtual reality to show and recreate ancient artefacts and sites that nowadays exist in a completely different form or which are now totally non-existent. Some of earliest, which are discussed later in this chapter, may be listed here: the Virtual Hagia Sophia (Foni et al., 2002), Virtual Campeche (Zara et al., 2004), the Ancient Malacca Project (Sunar et al., 2008), Virtual Pompeii (Jacobson and Vadnal, 2005), and the Virtual Prior Park reconstruction in Bath (Tredinnick and Harney, 2009). This kind of software bears in mind specific aims (Noh et al., 2009):

- To document construction of an historical object in order to reconstruct it in case of destruction.
- To create resources for the promotion of cultural and historical studies.
- To reconstruct historical monuments or parts which no longer exist.
- To visualize scenes from difficult or practically impossible angles.
- To interact with objects without the risk of damage.
- To promote tourism and virtual exhibitions.

Today virtual reproductions of historical sites are often available based on software, such as Open Virtual Worlds, which allow the creation of environments that permit a virtual interaction with other users and the guide, and interesting educational outcomes, for example, the virtual reconstruction of St. Andrew's Cathedral in Scotland (Kennedy et al., 2013). This idea of virtual guided visits has been very important during the Covid-19 pandemic. Many institutions have set up virtual reality visits to let people be in contact with cultural heritage even during lockdowns and museal-visit suspensions. Some had an already existing infrastructure, while others started at the moment, relying on robust and easy-to-use platforms. One must not forget these interesting initiatives that seek to engage the audience in game and virtual world environments, such as Second Life, Minecraft and Animal Crossing. Some examples are the Museum of Angers (France) that prepared a virtual exhibition inside the game 'Animal Crossing: New Horizons'; the Smithsonian American Art Museum (USA) that, since 2019, is virtually accessible with the software 'Beyond the Wall'—downloadable on Steam, the famous videogame download platform; the Louvre Museum has two different virtual experiences: the virtual visit of the whole museum and a virtual 3D experience of Leonardo's masterpiece 'Mona Lisa: Beyond the Glass'. There are many other experiences in this sense and many platforms are used; the most used emerge as Google Arts and Culture, Matterport, and VIVE Art.

But the Evolution is Mixed

In the first two decades of the new millennium there is not the same quantity of examples as far as mobile augmented and mixed reality for cultural heritage are concerned. However, as mentioned previously, there have been great advances in this field in recent years. This software has similar aims as those which use virtual reality, but its use is best seen in training, educational and didactic situations because of the affordance aspect of AR/MR mentioned beforehand.

Let us now move on to examine some particularly significant examples. Archeoguide, at the beginning of the new millennium, was one of the most ambitious projects in this field (Gleue and Dähne, 2001; Vlahakis et al., 2002). This used to be a client-server application. The server aspect contained a series of information on three-dimensional sites and models linked to a specific geographical place. The client aspect was made up of a laptop along with a specific software installation, a GPS device, and a head mounted display with a specially mounted camera in front. Thanks to the GPS data, the client could download this contextual information, including the 3D models. These models featured the structures as they would have appeared soon after completion and could be accurately placed on real-life images taken by a camera which, combined to AR, could then be presented to the user by means of the head-mounted display. This portable system, which seems cumbersome today, was necessary because, in 2001, devices, such as the present smartphones endowed with the necessary calculation potential, were non-existent. Its total weight varied from 6,8 to 7,3 kg, depending on the type of display.

With a slightly lighter set of hardware to carry in a backpack as well, Dow et al. (2005) created the mixed reality tour 'The Voices of Oakland'. It was designed to let visitors discover the histories of the Oakland Cemetery and of the people that have been buried there. There are many stories coming to life since the park is connected with centuries of history and especially with the American Civil War. This mixed reality is not visual but auditive, which means that the perception of being augmented is not the sight, but the hearing. As a unique case in all the review, it uses the Wizard of Oz (WOz) technique to deliver the experience, which consists in a human operator acting behind a system that is believed to be autonomous (Hanington and Martin, 2019).

One of the first educational outdoor mobile AR/MR heritage experiences was conceived in 2005 by Gorospe et al. (2006) with the name 'Lurquest'. In this project, high-school students had two introductory lessons before the visit. During the visit, they used PDA devices equipped with GPS in order to collect data about the site of Santa Maria la Real de Zarautz, in the Basque country. Results of this experience with 52 students seems to confirm the validity of this teaching methodology and technology used to promote high motivation to learn, learning autonomy as well as students' and teachers' satisfaction.

The second one took place in 2006, thanks to Squire and Klopfer (2007) and it was aimed at K12 students. It used Pocket PC with embedded touchscreen and GPS sensor, which is essentially comparable in weight with current smartphones, although the screen is very small and to be precise, one needs to use a little pen. In this learning experience, children had the opportunity to survey their environment for the presence of toxins in the water with the aid of a map,[44] contextualized informations, and instructions. The experience was not individual but required collaboration between children that covered different roles.

The third experience was held in the Carnuntum archaeological site (Austria) in 2007 by Lohr and Wallinger (2008) under the name of project 'Collage'. As the project Lurquest, it used PDAs equipped with GPS to augment an out-of-classroom

---

[44] Maps have always been very important, and sometimes central, parts of AR mobile apps. Usually, they would show our position as well as points of interest around us. The map is usually alternative (or, sometimes, parallel) to a menu system and provides more contextualized information. For example, to access information of a monument from a map that also shows our position, carries more information and meaning than selecting the same monument from a list.

activity for secondary school students and it had the same collaborative and role-play elements. The tasks that pupils had to complete related with school subjects, like Latin, History and Physics. One peculiar characteristic of this project was that teachers were watching while monitoring and communicating with the students through the devices. During the activity, 12 PDAs for the students and one laptop for the teachers were used. They had 33 participants and one device was used for every three pupils. Experts were interviewed, and they remarked engagement and collaboration in the teams, the pedagogical significance of that kind of game-based learning, and the power of PDA as a mediation tool (Lohr and Wallinger, 2008).

These were the only three projects selected for this review in the first decade of the 21st century. There have been others, but these are in small number of those that have been tested in a proper user experience. Of course, other similar projects were devised, like augmented reality in cultural heritage (ARICH) project (Mourkoussis et al., 2002), the project PRISMA (Fritz et al., 2005) which started with the aim of '*design, develop and implement a new 3D visualization device based on AR technologies*' (Fritz et al., 2005, p. 2) and 'Ancient Pompeii' project (Papagiannakis et al., 2005). Because of the early stage of the technology, they all had to face the problem of reduced portability and inadequate mobile operating systems and hardware. For the same reason, most of them were aimed at the technical side of the research more than at heritage education and interpretation. Another issue with such research was that, because of the equipment involved, it was hard to have experiences with many testers, or testers different than the researchers at all.

AR and MR Projects in the Last Ten Years

Moving a step in the second decade of the 21st century, the number of outdoor MR/AR experiences for heritage education grows dramatically. Experiences with users rely in most cases on the affordances of the new smartphones and tablets: many sensors embedded, new operating systems, powerful CPUs (central processing unit), and GPUs (graphic processing unit), more storage and working memory, bigger screens, easy interaction through the touchscreen, and high-resolution cameras.

Narrowing the review simply at the level of strictly educational experiences, or, in other words, the experiences that have the heritage education as the first aim, can be found in 10 projects, of which eight had students as the target audience and two as the general public. Just three of them have been created by technical sciences researchers (Angelopoulou et al., 2012; Erenli, 2013; Chang et al., 2015), while the others come from psycho-pedagogical field. On the other hand, if one does not consider just the education-aimed projects, overall, technology researchers have created half the experiences (in a selection where every project is about heritage education and have been tested in a heritage-education experience). That is very significant about the not often well-coordinated effort of technology and education experts to deliver respectively new technological and new methodological tools.

On the subjects of tools, in the last decade, all the education-oriented experiences rely on smartphones and/or tablets with few of them using VR headsets. In fact, the advent of smartphones, with the iPhone as the precursor, and tablets, again with an Apple product, the iPad, as the pioneer, disrupted the market of educational technologies. With every other producer copying those two models, in a few years the market was saturated with these devices which now are very affordable even in educational contexts (Sarwar and Soomro, 2013). Analyzing the context dimensions, one can notice that three experiences have been formal, four non-formal, one informal, and one both formal and non-formal. This data shows on the one hand that these technologies are inherently cross-contextual, while, on the other hand, it is difficult to develop experiences for true informal learning. This is not necessarily a downside, since formal and non-formal learning often result in a collaborative and shared experience, while informal learning is often an individual one. This is because it needs a very complex system to bring informal and casual learning in the frame of a collaborative work. Most of the times, the collaborative work requires to have staff to direct and organize people in the context, shifting the experience to a non-formal one.

All of these are some of the experiences that have inspired us, as they have fascinating and innovative elements in their design. One of the projects more in tune with our principles is the one of Chang et al. (2015) on the sense of place (SOP). With SOP, the authors intend the combination of feelings of attachment, dependence, concern, identity, and belonging that people develop regarding a place. Their study is based on the synergy between the framework of the human–computer–context–interaction (HCCI) (Greeno et al., 1996) and the strategy of historical–ceo–context–embedded–visiting (HGCEV) to conduct the visitor to reach the higher level of SOP through the following steps which are included in the app design and content—to find out the past geographical and historical information about the heritage site; to establish its geographical and historical context; when visitors visit the heritage site, the context allows them to feel interested in and interact with the heritage site; and further to establish the interaction among visitors, the heritage site itself, and the geographical and historical context of the heritage site.

A second exciting experience is the one by Smørdal et al. (2016) because of the kind of MR that they use, which is indirect (Wither et al., 2011) and based on a situated simulation approach. That is an on-site augmented reality showing how the place was in the past, how it would be in the future or how it could have been in an alternate reality given certain conditions (like global warming). They involved a 9th year science class in an experience divided into two hours of classroom preparation on the topic of climate change, one hour-and-a-half of situated simulation (field trip), two hours' work and construction of knowledge after the field trip, and finally one hour of presentations in the classroom (5 to 10 minutes per group). The situated simulation represented the place of the Oslo Opera House and surroundings in the year 2222, in a possible future where the climatic changes raised the level of the water more than 2 metres higher than today. The simulation provided also links to information and material as well as clues of what could have happened. Results of this experience underline how powerful the method of the situated simulation is for a situated learning and experiential knowledge. Students were able to make relevant connections between different school disciplines and use external sources to implement that knowledge. Also, they have been able to ponder causes and effects providing likely and original ideas about what might have created that situation (Smørdal et al., 2016).

Other AR/MR Outdoor Heritage Apps

Other notable apps are in the same line of the already mentioned projects but are thought for a completely informal context. They are, or were, available for download on the various app stores and have been used, some more, some less, from the general public. There is no research data for them; nonetheless, they are expressions of the same wave of interest and enthusiasm for the use of AR and MR technology for education and interpretation of heritage.

One of the most interesting was developed in the year 2011 and presented from the Region of Apulia and available for Android and iOS under the name of 'Puglia Reality+'. This application relied on operative systems, sensors, and the power of the new smartphones to provide an AR experience at various levels. Visiting various cities in Apulia, one had at his disposal an AR which, taking advantage of the smartphone's camera and GPS, managed to place virtual labels on real images in an AR visible on screen. The labels were interactive and, when selected, gave access to photographs and information on the monument or the structure selected. If one visited one of the archaeological sites where this option was available, the application was able to superimpose 3D models on the real things. This method allowed the visitors to see the structure as it was originally intended, thus giving him a tour of the mixed reality presented to him on the screen. A very similar app was iTTP, which guided the visitors along the touristic routes in Turin and surroundings, and Tuscany+, but that did not support reconstructions of the past. Both of them were developed only for iOS.

The Italian Ministry of the Cultural Properties, Activities and Tourism (MIBACT), in 2011, has created one of the most advanced applications of this kind to date: 'i-MiBAC Voyager', developed only for iOS, allows one to see how the site of the Imperial Forums in Rome looked like in Roman times. It includes an audio guide and can be used both at home and on site. Thanks to GPS/Compass/Accelerometer synergy, it was one of the first software that allowed us to look at the environment in a heads-up attitude, through a smartphone or a tablet as it was a window on the past (Bonacini, 2014).

The French company, GMT Éditions, developed in 2014, Izzyguide 3D (Bideran and Fraysse, 2015), which uses the same kind of technology as Puglia Reality+, but is more advanced for it allows a more interactive experience for the user and a richer media and content. From Izzyguide, they evolved the software with Poitiers 3D and Avignon 3D: applications that allow one to follow a guided tour to the respective cities, displaying the evolution of the same place through the centuries by the use of maps (without geolocation) and through mixed reality. These applications, in addition to the information accessible from the menu, allow one to view interviews with experts and listen to audio-guide style information within the virtual tour. Only the 3D Avignon application, the most advanced of the two, also incorporates small interactive games. Of the same series, there are also the apps 'Perpignan 3D' and 'Saint-Crespin-sur-Moine'.

Another particular example has the same functionalities of the above-mentioned apps, but it uses Epson Moverio AR glasses instead of a Smartphone. Its name is Art-Glass. Thanks to that different approach, one can have the superimposition of the information and of the reconstruction directly in her or his field of view (FOV). Still she/he will be able to see through the glasses and see the real environment. It is a very immersive experience, with hands free and one uses a pointer at the center of her or his FOV to select contents. It works also as an audio guide with the narrator speaking to the user as well as virtual characters that could appear, thanks to the AR technology. It was also used in outdoor environment for the Roman archaeological site of the *capitolium* and the Roman theatre in Brescia (Italy) and now at the James Monroe's Highlands at Charlottesville (Virginia).

There are apps that are maybe more interesting from a teacher's point of view, mainly because they are frameworks that let one develop their own AR experiences; in particular, it is very easy to create scavenger hunts and other experiences involving storytelling and places. To pick two of the best, the focus is on FreshAiR and Huntzz, which equire installing their own app and which allow one to use the trial that one has developed and to look at all the trails developed from other users. One cannot have one's own stand-alone app. FreshAiR is a very easy to use and flexible framework that has been developed after specific design principles for AR learning (Dunleavy, 2014). It allows the use of GPS to trigger events; it embeds a refined events logic,[45] a map, and an AR viewer in order to see the points of interest (POIs) in the landscape. It allows the use of rich media elements, including 360-degree videos. It can be used as a collaborative way, thanks to the creations of different roles and interactions through objects. Huntzz has not as many options, but it is a well-established platform with many heritage trails and especially developed for scavenger hunts. Both of them are available for iOS and Android.

An Example of the Use of the Proposed Classification: Summary of AR/MR Outdoor Experiences

In Table 2, the review of the outdoor mobile augmented and mixed reality for heritage experiences is summarized. In this section, the word 'experience' takes the place of the word 'app' because it will be used as the criteria explained in the paragraph 'The Range of AR and MR Technologies' to list selected experiences which took place in the last 20 years. It does not limit our view at the app, but considers the research of which it is a part (if that is the case), the context, and the technology. It is worth noting that only actual experiences and apps tried on the field were counted. Thus, all the papers of research which just designed an app and/or tried it in a lab have been excluded from the list. Looking at this tabular summary, it is interesting to note how, on 25 experiences, six (the highlighted ones) have as target audience of primary or lower secondary schools. All six studies were developed by, or in collaboration with, education departments, except one. Five of them have therefore a strong educational design and objective. Two of these experiences predate ours while the other two came subsequently. Comparing them, one can notice that two of them were run in formal settings and all of them are thought to be collaborative experiences.

---

[45] That means one can describe conditions to trigger events with a granular logic control, for example, one can decide to play a given sound or show information only if the user is within 5 metres from a certain position, has a given object in its inventory, and has already visited another place.

Table 2. Mobile augmented and mixed reality for heritage experiences classification table. Only studies with user experience were reviewed.

Authors	Research				Context				Technology					
	Main Objective	Target audience	Department of research	Subject of publication	Learning Dim.	Collaborative Dim.	Kinetic Dim.	Setting	Portability Dim.	Virtuality Dim.	Sensory Range	Sensors Range	Interaction	Device
Amato, Venticinque and Di Martino 2013	Enhanced visit experience	General public	Information Engineering	Mobile Computing and Multimedia	Informal	Individual	Mobile	Outdoor, monuments	Ubiquitous	Augmented Reality	Visual	GPS, RFID, Camera	Touch-screen	Smart-phone
Angelopoulou et al. 2012	Educational	11–16 y.o. children	Computer Science	Mobile Wireless Computing	Informal, Non-formal	Individual, Competitive, Collaborative	Mobile	Indoor, Outdoor, monuments, museum	Ubiquitous	Augmented Reality	Visual	Camera, GPS	Touch-screen	Smart-phone
Boyer and Marcus 2011	Enhanced visit experience	General public	Digital Humanities	Digital Humanities	Informal	Individual	Mobile	Outdoor, streets	Ubiquitous	Augmented Reality	Visual	GPS, Camera	Touch-screen	Smart-phone
Caggianese, Neroni and Gallo 2014	Enhanced visit experience	General public	Computer Science	Augmented and Virtual Reality	Informal	Individual	Mobile	Outdoor, buildings, monuments	Wearable	Augmented Reality	Visual	Camera, Depth sensor, GPS, Accellerometer, Magnetometer, Gyroscope	Touch-screen	Custom Headset
Cavallo, Rhodes and Forbes 2016	Enhanced visit experience	General public	Computer Science	Augmented and Mixed Reality	Informal	Individual	Mobile	Outdoor, streets, buildings	Ubiquitous	Augmented Reality	Visual	A-GPS, Gyroscope, Camera	Touch-screen	Smart-phone
Chang et al. 2015	Educational	1st Year University students (~19 y.o.)	Tourism, Technology, Geography, Educational Psychology	Educational Technology	Non-Formal	Individual	Mobile	Indoor, Outdoor, buildings	Ubiquitous	Augmented Reality	Visual, Auditive	Camera	Touch-screen	Tablet

Reference														
Gorospe, Ibañez Etxeberria and Jiménez 2006	Educational	High school students	Education	Teaching social sciences	Informal	Collaborative	Mobile	Outdoor, archaeological site	Ubiquitous	Augmented Reality	Visual	GPS	Touch-screen	PDA
D'Auria et al. 2015	Enhanced visit experience	General public	Information Technology, Physics	Digital Information Management	Informal	Individual	Mobile	Outdoor, monuments, buildings	Ubiquitous, Wearable	Augmented Reality	Auditive	gyroscope, accelerometer, magnetometer, GPS, microphone	Touch-screen, Voice	Custom headset
Dow et al. 2005	Enhanced visit experience	General public	Computing, Literature, Communication and Culture, Interactive Media Technology	Computer Entertainment Technology	Non-Formal	Individual	Mobile	Outdoor, monuments	Mobile	Mixed Reality	Auditive	GPS, head-orientation	Controller	Custom Headset + laptop + controller
Erenli 2013	Educational, Training	Schools, Organisations	Applied Sciences	Corporate Learning	Formal	Collaborative	Mobile	Outdoor, monuments, streets	Ubiquitous	Augmented Reality	Visual	Camera (QR Code reader), GPS	Touch-screen	Smart-phone
Georgiou and Kyza 2017	Educational	Middle and High School Students	Media, Cognition and Learning	Mobile and Contextual Learning	Formal	Collaborative	Mobile	Outdoor, natural environment	Ubiquitous	Augmented Reality	Visual	GPS	Touch-screen	Smart-phone
Guimarães, Figueiredo and Rodrigues 2015	Enhanced visit experience	General public	Art and Communication	Digital Heritage	Informal	Individual	Mobile	Outdoor, gardens	Ubiquitous	Augmented Reality	Visual	Camera	Touch-screen	Smart-phone
Haugstvedt and Krogstie 2012	Enhanced visit experience and exhibit	General public	Computer Science	Mixed and Augmented Reality	Informal	Individual	Mobile	Outdoor, streets	Ubiquitous	Augmented Reality	Visual	Camera, GPS	Touch-screen	Tablet

Table 2 Contd. ...

...Table 2 Contd.

Authors	Research				Context				Technology					
	Main Objective	Target audience	Department of research	Subject of publication	Learning Dim.	Collaborative Dim.	Kinetic Dim.	Setting	Portability Dim.	Virtuality Dim.	Sensory Range	Sensors Range	Interaction	Device
Kamarainen et al. 2013	Educational	Primary school (6th year)	Education	Computers and Education	Non-Formal	Collaborative	Mobile	Outdoor, monuments	Ubiquitous	Augmented Reality	Visual, Auditive	Camera, GPS	Touch-screen	Smart-phone
Kang 2013	Heritage awareness	General public	Cinematic Content	Wireless personal Communications	Informal	Individual	Mobile	Outdoor, buildings, streets	Ubiquitous	Augmented Reality	Visual	Camera, GPS	Touch-screen	Smart-phone
Squire and Klopfer 2007	Educational	K12 Students	Education	Education Tech	Formal	Collaborative	Mobile	Outdoor, monuments	Ubiquitous	Augmented Reality	Visual	GPS	Touch-screen	Pocket PC
Lee et al. 2012	Enhanced visit experience	General public	Human Interface Technology	Mixed and Augmented Reality	Informal	Individual	Mobile	Outdoor, streets, buildings, monuments	Ubiquitous	Augmented Reality	Visual	GPS, Compass, Accellerometer	Touch-screen	Tablet, Smart-phone
Liestal 2014	Educational	General public	Media and Communication	Cultural Heritage	Informal	Individual	Mobile	Outdoor, monuments, streets, buildings	Ubiquitous	Mixed Reality	Visual, Auditive	GPS, Camera, Compass, Gyroscope	Touch-screen	Smart-phone
Lohr and Wallinger 2008	Educational	3rd and 7th grade school students (13 and 17 y.o.)	Educational technology	Wireless, mobile and ubiquitous technology in Education	Formal	Collaborative	Mobile	Outdoor, archaeological site	Ubiquitous	Augmented Reality	Visual	GPS	Keyboard	PDA
Pacheco et al. 2015	Enhanced visit experience	General public	Synthetic, Perceptive, Emotive and Cognitive Systems	Digital Heritage	Informal	Individual	Mobile	Outdoor, memorial site, buildings	Ubiquitous	Mixed Reality	Visual	GPS, Compass, Gyroscope	Touch-screen	Tablet
Petrucco and Agostini 2016	Educational	Primary schools students (5th year)	Education	e-Learning and Knowledge society	Non-formal	Individual, Interactive (TRI-AR)	Mobile	Outdoor, monuments, streets	Ubiquitous	Mixed Reality	Visual	A-GPS, Gyroscope	Touch-screen/ Virtual pointer	Smart-phone, Tablet, VR Headset

Reference													
Pintus et al. 2005	Educational	Schools	Education	Mobile Learning	Formal	Individual	Mobile	Outdoor, archaeological site	Ubiquitous	Augmented Reality	Visual	GPS	Touch-screen	PDA
Kortabitarte et al. 2018	Educational	Secondary students	Education	Education	Formal	Individual	Mobile	Indoor, classroom	Ubiquitous	Augmented Reality	Visual	Camera	Touch-screen	Smart-phone
Pombo and Marques 2017	Educational	Primary and Secondary students (9-11 y.o. and 13-14 y.o.)	Education and Psycology	Computers in Education	Non-formal	Collaborative	Mobile	Outdoor, parks	Ubiquitous	Augmented Reality	Visual	Camera	Touch-screen	Smart-phone
Smordal, Liestol and Erstad 2016	Educational	9th grade students	Media, Communication and Education	Technology, Culture and Education	Formal	Collaborative	Mobile	Outdoor, environment, buildings	Ubiquitous	Mixed Reality	Visual	GPS, Camera, Compass, Gyroscope	Touch-screen	Smart-phone
Vlahakis et al. 2002	Enhanced visit experience	General Public	Computer Graphic	Computer Graphic in Art History and Archaeology	Informal	Individual	Mobile	Outdoor, monuments, archaeological sites	Mobile	Augmented Reality	Visual, Auditive	DGPS, Camera, Compass	Touch-screen	Laptop + HMD + Sensors devices/ Pen Tablet

References

Amato, A., Venticinque, S. and Di Martino, B. (2013). Image recognition and augmented reality in cultural heritage using OpenCV. pp. 53–62. *In: The Proceedings of International Conference on Advances in Mobile Computing & Multimedia MoMM '13*. New York. Association for Computing Machinery.

Angelopoulou, A., Economou, D., Bouki, V., Psarrou, A., Jin, L., Pritchard, C. and Kolyda, F. (2012). Mobile augmented reality for cultural heritage. pp. 15–22. *In: The 4th International ICST Conference, Mobile Wireless Middleware, Operating Systems and Applications*. June 2011, London. Springer.

Azuma, R., Baillot, Y., Behringer, R., Feiner, S., Julier, S. and MacIntyre, B. (2001). Recent advances in augmented reality. *IEEE Computer Graphics and Applications*, 21(6): 34–47.

Bideran, J. and Fraysse, P. (2015). *Guide Numérique et Mise en Scène du Territoire, entre Médiation Patrimoniale et Stratégie de Communication Touristique*, Études de Communication, *Langages, Information, Médiations*, 45(December): 77–96.

Billinghurst, M. and Kato, H. (1999). Collaborative mixed reality. pp. 261–84. *In*: Ohta, Y. and Tamura, H. (eds.). *Mixed Reality*. Springer, Berlin Heidelberg.

Bonacini, E. (2014). *La realtà aumentata e le app culturali in Italia: storie da un matrimonio in mobilità, Il Capitale Culturale*. Studies on the Value of Cultural Heritage, 9: 89–121.

Boyer, D. and Marcus, J. (2011). Implementing mobile augmented reality applications for cultural institutions. *In: MW*. Philadelphia, PA.

Caggianese, G., Neroni, P. and Gallo, L. (2014). Natural interaction and wearable augmented reality for the enjoyment of the cultural heritage in outdoor conditions. pp. 267–82. *In*: De Paolis, L.T. and Mongelli, A. (eds.). *Augmented and Virtual Reality, Lecture Notes in Computer Science*. Springer, Cham.

Caudell, T. and Mizell, D.W. (1992). Augmented reality: an application of heads-up display technology to manual manufacturing processes. *In: The Proceedings of the Twenty-Fifth Hawaii International Conference on System Sciences*.

Cavallo, M., Rhodes, G.A. and Forbes, A.G. (2016). Riverwalk: incorporating historical photographs. Public outdoor augmented reality experiences. pp. 160–65. *In: IEEE International Symposium on Mixed and Augmented Reality*.

Chang, Y.L., Hou, H.T., Pan, C.Y., Sung, Y.T. and Chang, K.E. (2015). Apply an augmented reality in a mobile guidance to increase sense of place for heritage places. *Journal of Educational Technology & Society*, 18(2): 166–78.

CoSpaces Edu for Kid-friendly 3D Creation and Coding, accessed 13 February 2021.

D'Auria, D., Di Mauro, D., Calandra, D.M. and Cutugno, F. (2015). A 3D audio augmented reality system for a cultural heritage management and fruition. *Journal of Digital Information Management*, 13(4): 203–9.

Dede, C. (2008). Theoretical perspectives influencing the use of information technology in teaching and learning. pp. 43–62. *In*: Voogt, J. and Knezek, G. (eds.). *International Handbook of Information Technology in Primary and Secondary Education*. Springer, Boston.

Dow, S., Lee, J., Oezbek, C., MacIntyre, B., Bolter, J.D. and Gandy, M. (2005). Exploring spatial narratives and mixed reality experiences in Oakland cemetery. pp. 51–60. *In: The Proceedings of the 2005 ACM SIGCHI International Conference on Advances in Computer Entertainment Technology ACE 05*. New York, Association for Computing Machinery.

Dunleavy, M. (2014). Design principles for augmented reality learning. *TechTrends*, 58(1): 28–34.

Erenli, K. (2013). Gamify your teaching using location-based games for educational purposes. *International Journal of Advanced Corporate Learning*, 6(2): 22–27.

Feng Zhou, H., Duh, B. and Billinghurst, M. (2008). Trends in augmented reality tracking, interaction and display: a review of ten years of ISMAR. pp. 193–202. *In: 2008 7th IEEE/ACM International Symposium on Mixed and Augmented Reality*.

Foni, A.E., Papagiannakis, G. and Magnenat-Thalmann, N. (2002). Virtual hagia sophia: restitution, visualization and virtual life simulation. pp. 2–7. *In: UNESCO World Heritage Congress*.

Fritz, F., Susperregui, A. and Linaza, M.T. (2005). Enhancing cultural tourism experiences with augmented reality technologies. *In: 6th International Symposium on Virtual Reality, Archaeology Cultural Heritage*. Pisa.

Georgiou, Y. and Kyza, E.A. (2017). A design-based approach to augmented reality location-based activities: investigating immersion in relation to student learning. pp. 1–8. *In: The Proceedings of* the 16th World Conference on Mobile and Contextual Learning, M-Learn. New York, Association for Computing Machinery.

Gleue, T. and Dähne, P. (2001). Design and implementation of a mobile device for outdoor augmented reality. pp. 161–68. *In: The Proceedings of the 2001 Conference on Virtual Reality, Archaeology, and Cultural Heritage, VAST 01*. Archeoguide Project, New York, Association for Computing Machinery.

Gorospe, J., Ibañez Etxeberria, A. and Jiménez, E. (2006). *Lurquest: Aplicación de Tecnología m-Learning al Aprendizaje Del Patrimonio, Iber, Didactica de Las Ciencias Sociales*, 50(October): 109–23.

Greeno, J.G., Collins, A.M. and Resnick, L.B. (1996). Cognition and learning. pp. 15–46. *In*: Berliner, D.C. and Calfee, R.C. (eds.). *Handbook of Educational Psychology*. Prentice Hall, New York.

Guimarães, F., Figueiredo, M. and Rodrigues, J. (2015). Augmented reality and storytelling in heritage application in public gardens: caloust gulbenkian foundation garden. *Digital Heritage*, 1: 317–20.

Gyroscope. (2021). *In*: Wikipedia. https://en.wikipedia.org/w/index.php?title=Gyroscope&oldid =998344971.

Hanington, B. and Bella, M. (2019). *Universal Methods of Design Expanded and Revised: 125 Ways to Research Complex Problems, Develop Innovative Ideas, and Design Effective Solutions*. Rockport Publishers.

Haugstvedt, A. and Krogstie, J. (2012). Mobile augmented reality for cultural heritage: a technology acceptance study. pp. 247–55. *In: 2012 IEEE International Symposium on Mixed and Augmented Reality*.

Jacobson, J. and Vadnal, J. (2005). The virtual pompeii project. pp. 1644–49. *In: Association for the Advancement of Computing in Education (AACE)*.

Kamarainen, A.M., Metcalf, S., Grotzer, T., Browne, A., Mazzuca, D., Tutwiler, M.S. and Dede, C. (2013). EcoMOBILE: integrating augmented reality and probeware with environmental education field trips. *Computers & Education*, 68(October): 545–56.

Kang, J. (2013). AR teleport: digital reconstruction of historical and cultural-heritage sites for mobile phones via movement-based interactions. *Wireless Personal Communications*, 70(4): 1443–62.

Kennedy, S., Fawcett, R., Miller, A., Dow, L., Sweetman, R., Field, A., Campbell, A., Oliver, I., McCaffery, J. and Allison, C. (2013). Exploring canons cathedrals with open virtual worlds: the recreation of St. Andrews Cathedral, St Andrews Day, 1318: 273–80. *In: 2013 Digital Heritage International Congress (DigitalHeritage)*, 2.

Kortabitarte, A., Gillate, I., Luna, U. and Ibáñez-Etxeberría, A. (2018). *Las Aplicaciones Móviles como Recursos de Apoyo en el Aula de Ciencias Sociales: Estudio Exploratorio con el App 'Architecture Gothique/Romane' en Educación Secundaria, Ensayos, Revista de la Facultad de Educación de Albacete*, 33(1): 65–79.

Lee, G.A., Dünser, A., Seungwon, K. and Billinghurst, M. (2012). CityViewAR: a mobile outdoor ar application for city visualization. pp. 57–64. *In: 2012 IEEE International Symposium on Mixed and Augmented Reality, Arts, Media, and Humanities*.

Liestøl, G. (2014). Along the appian way. Storytelling and memory across time and space in mobile augmented reality. pp. 248–57. *In*: Ioannides, M., Magnenat-Thalmann, N., Fink, E., Žarnić, R., Yen, A.Y. and Quak, E. (eds.). *Digital Heritage. Progress in Cultural Heritage: Documentation, Preservation, and Protection, Lecture Notes in Computer Science*. Springer, Cham.

Lohr, M. and Wallinger, E. (2008). Collage. The carnuntum scenario. pp. 161–63. *In: Fifth IEEE International Conference on Wireless, Mobile, and Ubiquitous Technology in Education*.

Metaverse Studio, accessed on 13 February 2021. https://studio.gometa.io.

Miglino, O., Di Ferdinando, A., Di Fuccio, R., Rega, A. and Ricci, C. (2014). Bridging digital and physical educational games using RFID/NFC technologies. *Journal of E-Learning and Knowledge Society*, 10(3).

Milgram, P. and Kishino, F. (1994). *A Taxonomy of Mixed Reality Visual Displays*.

Minamizawa, K., Prattichizzo, D. and Tachi, S. (2010). Simplified design of haptic display by extending one-point kinesthetic feedback to multipoint tactile feedback. pp. 257–60. *In: 2010 IEEE Haptics Symposium*.

Mirage. Réalité Augmentée et Virtuelle pour l'Enseignement, accessed on 13 February 2021. https:// mirage.ticedu.fr/.

Mourkoussis, N., Liarokapis, F., Darcy, J., Pettersson, M., Petridis, P., Lister, P.F. and White, M. (2002). Virtual and augmented reality applied to educational and cultural heritage domains. *In: Business*

Applications of Virtual Reality, Workshop in Conjunction with 5th International Conference on Business Information Systems. BIS.

Noh, Z., Sunar, M.S. and Pan, Z. (2009). A review on augmented reality for virtual heritage system. pp. 50–61. *In*: Chang, M., Kuo, R., Kinshuk, Chen, G.D. and Hirose, M. (eds.). *Learning by Playing, Game-based Education System Design and Development, Lecture Notes in Computer Science.* Springer, Berlin, Heidelberg.

Pacheco, D., Wierenga, S., Omedas, P., Oliva, L.S., Wilbricht, S., Billib, S., Knoch, H. and Verschure, P.F.M.J. (2015). A location-based augmented reality system for the spatial interaction with historical datasets. *Digital Heritage*, 1: 393–96.

Pan, Z., Cheok, A.D., Yang, H., Zhu, J. and Shi, J. (2006). Virtual reality and mixed reality for virtual learning environments. *Computers & Graphics*, 30(1): 20–28.

Papagiannakis, G., Schertenleib, S., O'Kennedy, B., Arevalo-Poizat, M., Magnenat-Thalmann, N., Stoddart, A. and Thalmann, D. (2005). Mixing virtual and real scenes in the site of ancient pompeii. *Computer Animation and Virtual Worlds*, 16(1): 11–24.

Petrucco, C. and Agostini, D. (2016). Teaching cultural heritage using mobile augmented reality. *Journal of E-Learning and Knowledge Society*, 12(3).

Pintus, A., Carboni, D., Paddeu, G., Piras, A. and Sanna, S. (2005). Mobile lessons: concept and applications for "on-site" geo-referenced lessons. pp. 163–65. *In*: Attewell, J. and Savill-Smith, C. (eds.). *Mobile Learning Anytime Everywhere: A Book of Papers from MLEARN 2004.* Learning and Skills Development Agency, London.

Pombo, L. and Marques, M.M. (2017). Marker-based augmented reality application for mobile learning in an Urban park: steps to make it real under the EduPARK project. pp. 1–5. *In*: *2017 International Symposium on Computers in Education (SIIE).*

Sardo, J.D.P., Semião, J., Monteiro, J.M., João, A.R., Pereira, M., de Freitas, A.G., Esteves, E. and Rodrigues João, M.F. (2018). Portable device for touch, taste and smell sensations in augmented reality experiences. pp. 305–20. *In*: Mortal, A., Aníbal, J., Monteiro, J., Sequeira, C., Semião, J., Moreira da Silva, M. and Oliveira, M. (eds.). *INCREaSE*. Springer, Cham.

Sarwar, M. and Soomro, T.R. (2013). Impact of smartphones on society. *European Journal of Scientific Research*, 98(2): 11.

Schraffenberger, H. and van der Heide, E. (2016). Multimodal augmented reality: the norm rather than the exception. pp. 1–6. *In*: *The Proceedings of the 2016 Workshop on Multimodal Virtual and Augmented Reality*. MVAR 16, New York, Association for Computing Machinery.

Smørdal, O., Liestøl, G. and Erstad, O. (2016). Exploring situated knowledge building using mobile augmented reality. *Qwerty Open and Interdisciplinary Journal of Technology, Culture and Education,* 11(1): 26–43.

Squire, K. and Klopfer, E. (2007). Augmented reality simulations on handheld computers. *Journal of the Learning Sciences*, 16(3): 371–413.

Sunar, M.S., Zin, A.M. and Sembok, T.M.T. (2008). Improved view frustum culling technique for real-time virtual heritage application. *The International Journal of Virtual Reality*, 7(3): 43–48.

Traxler, J. (2005). Defining mobile learning. pp. 261–66. *In*: *IADIS International Conference Mobile Learning.*

Tredinnick, J. and Harney, M. (2009). An interactive tool for the exploration of contextual architecture: case study: 18th century prior park, bath. pp. 623–630. *In*: *Computation: The New Realm of Architectural Design, 27th ECAADe Conference Proceedings.* Istanbul (Turkey), 16–19 September, Cumincad.

Vlahakis, V., Ioannidis, M., Karigiannis, J., Tsotros, M., Gounaris, M., Stricker, D., Gleue, T., Daehne, P. and Almeida, L. (2002). Archeoguide: an augmented reality guide for archaeological sites. *IEEE Computer Graphics and Applications*, 22(5): 52–60.

Wither, J., Tsai, Y.T. and Azuma, R. (2011). Indirect augmented reality. *Computers & Graphics, Semantic 3D Media and Content*, 35(4): 810–22.

Zara, J., Benes, B. and Rodarte, R.R. (2004). Virtual campeche: a web-based virtual three-dimensional tour. pp. 133–40. *In*: *Proceedings of the Fifth Mexican International Conference in Computer Science*, ENC.

CHAPTER 10
Teaching and Learning Interactions with Mobile Mixed Reality Tools

Daniele Agostini

IIIIIIIIIIIIIIIIIIIIIIIIIIIIIIIIIIIIIIIIIIIIIIIIIIIIIIIIIIIIIIIIIIIIIIIIIIIIIIIIIIIIIIIIIIIIIIIIIIIIIIIIIIIIIIIIIIIIIIIIIIIIIIIIIIIIIIIIIIIIIIIIIIIIIIIIIIIIIIIIIIIII

Introduction

The use of mobile mixed reality tools for education is an exciting way to engage students and to make them understand concepts that would have been difficult, or even impossible, to convey with other media. The devices needed to use such technology are inherently disruptive in teaching and learning practices, settings, and contexts. This technology moves the educational word's axis from a transmissive model of teaching to a co-constructive one, from a static attitude to a mobile and active one, form a formal context to an informal one, from a place-centered model to a student-centered one. This chapter aspires to provide some theoretical and methodological instruments, as well as examples, to understand, handle, and better exploit all the resources that these new tools have to offer.

Some Activity Theory Basis

Activity Theory is a powerful instrument, especially suited for analyzing studies based on technological tools and human-computer interaction (HCI). Here is briefly presented the evolution of the concept of Activity Theory from its beginning. Activity Theory designates a conceptual framework to answer better the classic questions *'who is doing what, why and how'* (Hasan and Kazlauskas, 2014: 9). It provides powerful instruments for the interpretation of human activities, and in its history, one can pinpoint three generations. The first one, dated back to the first half of the 20th century was further developed until the eighties having Vygotsky as forefather;

Postdoctoral Researcher in Educational Technology and Digital Interpretation, University of Padua.
Email: daniele.agostini@unipd.it

the second and the third were developed respectively at the end of the eighties and the nineties. In 1993, Yrjö Engeström described the Activity Theory (AT) as 'the best-kept secret of academia'. Although not very well known outside the social, pedagogical, and human-technology interaction sciences, nowadays, it is no longer such a well-kept secret. In fact, since the beginning of the nineties, the growth of research about, or using, the AT has been exponential (Roth and Lee, 2007). AT is a very sophisticated tool to enable a better understanding of the complexity of human activity and to ask meaningful questions related to every action (Kaptelinin et al., 1999). Vygotsky[46] and his colleagues (notably, Aleksej Leont'ev e Aleksandr Lurija) created it because, unlike machines and animals, human activities are guided by a meaningful purpose and carried out by employing tools. They developed the triangle of artefact mediation, of which the current AT model is an expansion. First, Leont'ev and then Engeström worked on that expansion. Their works are similar in many aspects but not identical. Engeström, having worked at it subsequently, incorporated some of the work of Leont'ev in a new model. In the last 20 years, other authors decided to work on the Leont'ev model rather than on the Engeström one. Among them, the most influential are arguably Bonnie Nardi, Kari Kuutti and Victor Kaptelinin, who refined the AT to better apply to information systems.

The Beginnings

The first-generation model is very similar to the Vygotskij concept. The unit of analysis is mediated action. It can be visualized as a triangle with one vertex, the subject, on the other one the mediational means and the third one is the object. The subject can be an individual, a dyad or a group; the mediational means are tools, such as writing, speaking, technology, etc.; the object is the motive that leads to an outcome. This triangle represents how Vygotskij brought together cultural artefacts with human actions to avoid the dualism between individuals and society (*see* Fig. 5). During that period, the focus of the theory was on individuals. The fundamental concept is that tools or signs, which are culturally defined or created, mediate every human activity. The subject interacts with a tool to achieve an outcome. During this external interaction with the tool, the individual's internal mind is transformed (Aboulafia et al., 1995).

This is a crucial concept for AT and mixed reality experiences, and the reason is that the tools that have been used in human activity are themselves the result of a long process of cultural and evolutionary development.[47] Since they are not neutral,

---

[46] Lev Semyonovich Vygotsky (1896–1934) was a Russian psychologist. He is considered the father of what is now called the 'cultural-historical' school of psychology. The peculiarity of this school is to indissolubly bind the human mind with the society and culture it belongs. He greatly influenced the world's psychology and pedagogy. In Russia, his students, Luria and Leont'ev, developed the Cultural-Historical Activity Theory (CHAT) (Roth and Lee, 2007) while in Europe and US, from the seventies on, his theories have influenced social-constructivist theories, complementing Piaget's constructivism (Daniels, 2010).

[47] In fact, the concept of mixed reality can be found back since the 17th century and even before, only with different technology available. Some examples are the 'Claude Glass', the anamorphosis and *trompe d'oeil* techniques.

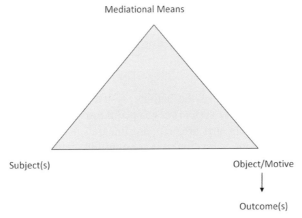

Fig. 5. Representation of the first generation of activity theory.

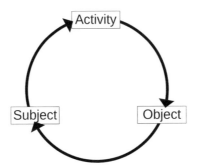

Fig. 6. Leont'ev's ringstruktur.

the subject will be influenced by interaction with them, as will the object. Leont'ev called this phenomenon *ringstruktur* (ring structure) (*see* Fig. 6).

Subject, activity and object are on the same level, and the object closes the circle, influencing the subject. In Leont'ev's thought, the activities are ordered in a hierarchical system (*see* Fig. 7) in which an activity comprises a series of actions and an action has a series of operations. Take the example of the basic skill of executing a mathematical addition. The activity, in that case, is training in mathematical addition. An action is to solve an addition, and an operation is to sum one and one. Nevertheless, the activity has a motive, in this case, to give to children literacy in maths; the action

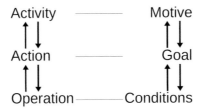

Fig. 7. Hierarchical levels of activity.

has a goal, which is to get the result; and the operation has conditions which are to add the numbers correctly and to obtain a correct result for the numbers given. It is essential to highlight that the operation always takes place in the subject's mind.

The Following Generations

From Leont'ev's conception of the activity system, Engeström (2015: 78: The structure of a human activity system) developed the so-called 'second generation'. It is based on the concept that artefacts are integral, inseparable components of the human being, with the mediation as a constant link between components of an activity system.

To advance the design of Activity Theory, Engeström extended the original triangular representation (*see* Fig. 8) to make it possible to analyze systems of activity at the macro level of the group and the community rather than micro-level focus on the particular participant or agent working with tools. The aim is to emphasize the social and collective aspects/elements of the activity structure as they play a crucial role in influencing an individual's (subject's) transformation. The extension has been rendered by incorporating community, rules, and division of labour, and the importance of discussing their relationships with each other. The 'community' includes the non-subjects or people in the system. Rules may be formal or informal, overt or implied, and provide the system with instructions during its operation. The division of labour has two dimensions—vertical and horizontal. The vertical balances between power and status, and the horizontal separates what people do (tasks) based on their role in the community.

In this second generation of AT, where the object is represented as an oval, the object-oriented actions are often, directly or indirectly, marked by complexity, surprise, perception, sense-making, and the possibility for improvement. As is visible in Fig. 9, the six elements peculiarly interact with each other. These interactions produce 'tensions' or contradictions[48] within the activity systems which are the

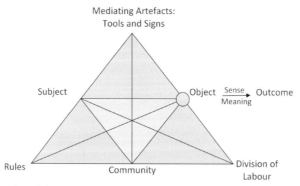

Fig. 8. Representation of the second-generation activity theory. The process of attribution of sense and meaning to the object leads to the outcome.

---

[48] Collins et al. (2002) in their paper 'Activity Theory and System Design: A View from the Trenches' call them 'tensions', finding the word 'contradiction' too strong.

driving force of change and thus, development (Engeström et al., 1999). Usually, many participants' feedbacks articulate contradictions inside and between the elements. To investigate, these tensions or inconsistencies may offer perspective on system dynamics, and provide insights into how the system progresses (Collins et al., 2002). Tensions or contradictions between two elements of the triangle can evidence problems in a system. It is important to note that the contradictions represent a fundamental structural component of the system that affects all the elements; for example, the educational use of smartphones and tablet in classrooms can be considered useful, and is needed in order to teach with MR technology. This is a great educational choice, but it can lead to contradictions between rules and subject (students) and tools because smartphones are also powerful devices that are sources of distraction.

Inside every AT model, there are four higher-order functions originating from the relationship between the three vertexes of each triangle (Holt and Morris, 1993; Nardi, 1998) which represent different aspects of human activity:

- Production is the creation of the object needed to reach the aims of the system.
- Distribution divides the work in the community following the social laws.
- Exchange records the social interaction produced by the activity.
- Consumption is the function that comes after the others, which realizes the prefixed aims of the subject and the community.

Activity Theory and Human-Computer Interaction (HCI)

Since Leont'ev, and, in particular, after the first Engeström postulation, AT has been increasingly adopted in researches and notably in those fields of study where it is necessary to analyze dynamics between participants, the activities are mediated by artefacts and are object-oriented. In the following paragraphs, one can find a selection of AT-based research concerning augmented reality, mobile learning, and computer-mediated activities.

Since the beginning, human-computer interaction and computer-mediated activity have been amongst the main subjects of research that used AT as a framework. In the nineties of the last century, several authors worked to expose, systematize, and give theoretical support to the growing corpus of research applying AT to HCI (Nardi, 1996; Kaptelinin, 1996b; Kaptelinin et al., 1995). Finally, there was the proposal of five principles and a four-step checklist to develop and analyze computer-mediated experiences through AT (Kaptelinin et al., 1999) (Table 3). This research's latest evolution is represented by a new, comprehensive, activity model specifically thought for HCI and comprehensive of updated AT checklists for developing and analysing AT in HCI projects (Kaptelinin and Nardi, 2006, 2012).

Other reflections came from researchers who based their work on Engenstrom's and Kuutti's, intending to define and design constructivist learning environments (CLEs) (Jonassen, 2013). A CLE should be the direct consequence of adopting a constructivist approach, with the focus on meaningful learning, and should promote exploration, experimentation, construction, collaboration, and reflection. It can be

Table 3. Activity theory checklist.

Principle	Checklist
Object-orientedness →	Means and ends
Hierarchical structure of activity →	Social and physical aspects of the environment
Internalisation and externalisation →	Learning, cognition and articulation
Mediation →	(Present throughout the checklist)
Development →	Development

technology-based. A six-step, AT-based, model was created to guide practitioners in the process of creation and analysis of CLEs (Jonassen and Rohrer-Murphy, 1999):

1. Clarify the purpose of activity system
2. Analyze the activity system
3. Analyze the activity structure
4. Analyze mediators
5. Analyze the context
6. Analyze activity system dynamics

Each one of these steps has sub-steps to be followed with meaningful questions to help better understanding of what to analyze.

Applications of AT in Mobile Learning and Mixed Reality Learning

Since the first decade of the new century, AT has been used as the theoretical framework for developing an augmented reality-based system to enhance group work (Fjeld et al., 2002). On their research towards a task model for mobile learning, Taylor et al. (2006) regard AT as a powerful tool to analyze activity systems as classroom, workspaces, and learning communities. They chose it as a foundation for their model, in parallel with Conversation Theory (Pask, 1976). Papadimitriou et al. (2007) asked students to collect information in a museum, using PDAs. They reported that AT was handy for the researchers, enabling them to see how operations informed actions and to see the role of the facilitators as well as students and devices. They concluded by saying that AT seems the ideal conceptual tool to use in the context of a technology-enhanced museum visit as it lets you see beyond outcomes, tool, and context.

Lorda Uden, in 2006, acknowledged the work done with AT in HCI. She used it in the design of mobile-learning experiences and proposed a framework ad hoc. She found three main limitations on three AT's main strengths: the researcher's requirement to really understand the activity system he is studying, the difficulty in unravelling activity systems, and the difficulty in distinguishing between the levels of activity, actions and operations. The benefits outweigh the limitations though, providing a view of the whole learning system and describing all the interacting elements and their relationships. Another significant advantage is that AT looks at the activity system as dynamic and evolving regarding conflicts, breakdown, and discontinuities as vital dynamics (Uden, 2006).

In 2010, Walker used the principles in Table 1 in order to analyze the activity of visitors to museums. They were asked to construct trails by means of mobile technology to understand how people made meanings in such a context. Since he found AT ideal for investigating tool mediation but lacking a comprehensive description of the museum meaning-making context, he built a conceptual model for the design and analysis of trails. This draws a theoretical basis from the contextual model of learning (Falk, 1991; Falk and Dierking, 2018) and uses AT methodology.

In mobile learning activities, the guide's role or of the instructor, has always been something that is not easy to define and difficult to analyze. With the aid of AT, Cowan and Butler (2013) tackled the issue, finally proposing a change in the triangular Engeström model. Examining the four higher functions and their interrelations, they found tensions and imbalances between elements of AT, which affected the learning process negatively. They modified the AT, providing a three-dimensional representation and thus adding the teacher in the very center of it with the role of control and balance between elements that are necessary for effective learning. To understand the effectiveness of WhatsApp in mobile learning, Barhoumi (2015) takes advantage of AT in a quantitative study underlining how the three levels, called community, individual, and technological influence, in fact, adequately describe online participation.

The interest in using Activity Theory to analyze and design learning experiences with mobile technologies' mediation is growing. *Mobile Learning Design: Theories and Application* is a book published in 2015, where the editors (Churchill et al., 2015) selected papers on this subject. AT is used as a framework, or as one of the foundations, for the theoretical model in five of the 24 articles presented:

1. Churchill, Fox and King use it in their RASE (resource, activity, support, evaluation) learning design framework that aims to get an advantage from the multiple affordances of the mobile learning technologies.

2. Burden and Kearney conceptualize authentic mobile learning by providing a model of it. AT is present with its concepts of boundary and boundary objects in order to better understand the continuity between home and school, formal and informal, physical and virtual.

3. Rozario, Ortlieb and Rennie employ the six-step AT (Jonassen and Rohrer-Murphy, 1999) as the primary tool, along with a case study design, to understand how and if the different pedagogies, professional learnings, and mobile technologies support teachers to foster a learner-centered and interactive approach. They affirm that using AT as a lens provides an ideal position to better understand the relations between context, mobile technologies (both hardware and software), and collaborative and interactive learning.

4. Cook and Santos describe three phases of mobile learning and they push forward the research with a project aimed at developing a mobile platform for help-seeking for healthcare in the UK. They use Vygotsky's cultural-historical approach in the logic engine of the social semantic server that relates people to data, people with people, and data with data.

5. Khoo uses an app in order to let pre-school children view and represent addition and subtraction skills. It enabled them to acquire new strategies to learn and understand these operations. He employed AT as the primary framework of this research to analyze and to categorize four dimensions: subject-tool-object, subject-community-object, subject-division of labour-object, and subject-rules-object.

Examining the studies reviewed above, one may argue that Activity Theory literature provides one of the best frameworks for MR learning and teaching. An experience of MR learning may involve different institutions with different subjects, communities, and rules, which, nonetheless, may have partially shared objects. These objects crystallize in the shared object, which is a mixed-reality experience. In fact, Jonassen, with reference to the Engeström work, highlights the nested nature of Activity Theory dynamics. So, a learner group could be the subject of educational activity. However, it could also have been the result (object) of a previous activity aimed at the group's constitution.

Table 4 shows the above-mentioned studies, together with other relevant applications of the AT in the field of mobile learning and augmented and mixed reality mobile learning. It specifies the AT 'school', in other words, to which concept of AT the authors referred; the AT practice type, meaning what model or practical execution of the AT the authors chose; the elements of AT to which the authors focussed; and finally the kind of technology used in their studies.

Three Key Concepts of Mixed Reality Mobile Learning Experiences

When thinking of designing, executing or analyzing a mixed reality mobile learning experience, one needs a serious toolkit of instruments to channel and interpret the vast amount of processes and interactions that are going to happen. Three of them are particularly useful to accomplish these tasks. Even if not made explicit in every study, they are always in the background. Taking them into account one can secure the right perspective for design, action, analysis or assessment of these complex experiences.

Role of MR Technology in Reaching the Goals and Relating with the Object of the Activity

The concept of 'functional organ' could help us to understand the kind of interaction that takes place between a student and an MR device. A functional organ is defined as the result of the temporary fusion of internal and external resources, human capabilities, and tool properties to attain goals that could not be attained otherwise (Ukhtomsky, 1978; Leont'ev, 1981; Zinchenko, 1996; Kaptelinin, 1996b). A functional organ is inherently aimed at a specific task, in the context of which it has its specific function. In our context, it can be defined as a temporary, goal-oriented alliance between the human and the machine (the subject and the mediating artefact in AT terms).

Table 4. Studies that employed the activity theory in mobile learning and augmented/mixed reality mobile learning.

Title	Authors and year	AT school	AT practice type	Focal AT elements	Technology
Physical and Virtual Tools: Activity Theory Applied to the Design of Groupware	Fjeld et al. (2002)	Leont'ev, Engeström	Bødker (1996), Kaptelinin (1996a)	Tools, exteriorisation	AR, static/ fixed (non-mobile).
Towards a Task Model for Mobile Learning: A Dialectical Approach	Taylor et al. (2006)	Engeström	Waycott (2004), this paper: *Task Model*	Context, tools, appropriation	PDA, mobile
Analysis of an Informal Mobile Learning Activity based on Activity Theory	Papadimitriou et al. (2007)	Kuutti	Avouris et al. (2004); Ergazaki et al. (2007)	Actions, tools	PDA, mobile
Activity Theory for Designing Mobile Learning	Uden (2006)	Engeström	Jonassen and Rohrer-Murphy (1999)	Context, tools, actions, conflicts	Mobile (*theoretical*)
Designing for Meaning Making in Museums	Walker (2010)	Leont'ev	Kaptelinin and Nardi (2006)	Context, tools, visitor (subject), artefact	Mobile phones/ recorders, mobile
Using Activity Theory to Problematise the Role of the Teacher during Mobile Learning	Cowan and Butler (2013)	Engeström	Daniels and Warmington (2007), this paper: 3D Activity System Model	Consumption, exchange, production and distribution	AR, PDA (with GPS), mobile
The Effectiveness of WhatsApp Mobile Learning Activities Guided by Activity Theory on Students' Knowledge Management	Barhoumi (2015)	Engeström	Engeström (1993)	Levels: Technological, individual, community	Smartphone, mobile
Framework for Designing Mobile Learning Environments	Churchill, Fox, and King (2015)	Engeström	This paper: *RASE (Resources, activity, support, evaluation)*	Activity, support, evaluation	Smartphone/ tablet, mobile
Conceptualizing Authentic Mobile Learning	Burden and Kearney (2015)	Engeström	Engeström et al. (1995)	Boundary crossing, boundary object	Mobile (*theoretical*)

Table 4 Contd. ...

...Table 4 Contd.

Title	Authors and year	AT school	AT practice type	Focal AT elements	Technology
Interactivity and Mobile Technologies: An Activity Theory Perspective	Rozario et al. (2015)	Engeström	Jonassen and Rohrer-Murphy (1999)	Interactivity, tension	Smartphone/ tablet, mobile
Three Phases of Mobile Learning State-of-the-Art and Case of Mobile Help Seeking Tool for the Health Care Sector	Cook and Santos (2015)	Vygotsky	Cook (2010)	Social interaction, scaffolding, mediating tool, context	Smartphone, mobile
Enacting App-based Learning Activities with Viewing and representing skills in preschool mathematics lessons	Khoo (2015)	Engeström	Engeström (1999)	Subject, objectives, tool	Smartphone, mobile
Mobile collaborative informal learning Design: Study of collaborative effectiveness using Activity Theory	Baloch et al. (2012)	Vygotsky, Brown	Brown et al. (1989)	Community, collaboration, task	Smartphone, mobile
PDAs as Lifelong Learning Tools: An Activity Theory-based Analysis	Waycott et al. (2005)	Engeström	Engeström (1993)	Subject, tool	PDA, mobile
Reinterpreting Mobile Learning: An Activity: Theoretic Analysis of the Use of Portable Devices in Higher Education	Wali (2008)	Engeström	Engeström (1993)	All the main elements of the activity system	Laptop, portable
Experimentation of an Augmented Reality App to Communicate Cultural Heritage: The Hestercombe Gardens Augmented Visit	Agostini and Petrucco (2018)	Engeström	Agostini and Petrucco (2020), *TRI-AR*	Subject, human mediator, technological mediator	MR, smartphone/ tablet, mobile

MR technology enables the use of the smartphone or another device to create a virtual visual (or auditory or haptic) overlay that appears to be overlaid atop the perceived real world. Thus, MR technology emerges as a perfect example of a functional organ. They can help students to attain goals that are difficult—or impossible—to attain without them. But which goals are they? Which kind of interaction or understanding comes from the use of these new MR functional organs?

A non-exhaustive list comprehends:

- The capacity to imagine very different and distant contexts or epochs: thanks to MR technologies, one can see and be immersed in a faraway country, on the top of Mount Everest or in Pompeii in AD 79.

- The opportunity to understand non-existent or non-explorable complex systems: one can explore ancient cities and see how the monuments and remains we see today were part of the city system, create a molecule using his or her own hands, wander between the organelles of a cell, and fly around a DNA helix.

- Surrogate an actual travel experience or training: for example, using aircraft simulators or evacuation and job security simulators.

- The possibility to visualize and manipulate theoretical and abstract concepts: for example, to see and interact with geometry, numbers, letters, and words for educational purposes.

- The ability to manipulate unavailable or extremely fragile or hazardous artefacts: such as paintings, ancient manufacts, or modern weaponry.

In addition, the functional organ experience often contributes to the engagement of the students, and of the learners in general, creating internal conditions conducive to learning. MR forced them to work with their internal resources to normalize the cognitive dissonance between reality and mixed and augmented reality through a process of reinterpretation of reality by means of their newly acquired knowledge.

Affordance in HCI

Another construct that can help in the definition of the interactions between students and MR technology is the concept of affordance. It is a concept of great importance in the field of educational technology (Osborne, 2014). It was coined in 1966 by the American psychologist, James J. Gibson with reference to the complementarity between animals and the environment and what the environment can offer to the animal (Gibson, 2014). It was then implemented and finally applied by Donald A. Norman, an American cognitive scientist, in the field of human-machine interaction, in 1988. His definition of affordance is, therefore, a property or a function of an object that can be inferred from external features such as, for example, shape, size, weight and, in general, the design (Norman, 1999). Norman later changed the name with a more specific one: 'perceived affordance'.

The main idea is that the use of a tool rather than another affects the cognitive processes that are activated and the relationship with the environment. Kaplan (2017), Maslow (1966), Tomkins and Messick (1963) and Vygotsky (1980) support the thesis that the tool is not neutral. It substantially influences the way of thinking and acting. The following 'hammer' example is iconic and used in different situations by each one of them (except Vygotsky) to exemplify this concept: 'I suppose it is tempting, if the only tool you have is a hammer, to treat everything as if it were a nail (Maslow, 1966: 12)'. This concept from Abraham H. Maslow is often referred to as 'Maslow's Hammer', but the American psychologist was not the first to express it. In fact, the American philosopher, Abraham Kaplan, expressed this concept in 1964, giving it the name 'the law of the instrument': "I call it the law of the instrument, and it may be formulated as follows: Give a small boy a hammer, and he will find that everything he encounters needs pounding" (Kaplan, 2017: 28).

Applying the same idea to educational technology, one can remember how often the technologists, teachers, and education practitioners decided to use a particular instrument to do many things; even things that would have been better accomplished with other tools. One can think about how often the 'educational' technology is developed by technicians and engineers, who know very well the technology but very little of education. Lastly, one can resolve to use the best educational tool to reach a specific educational goal (tools and 'technology' do not need to be digital).

On the other side, if one is enquiring about the possible use of a specific tool, he/she must be very open and ready to acknowledge that the tool might not be the best for his/her chosen field (and, maybe, there could be an unexpected discovery).

Mobile devices and MR apps, for example, have different affordances than traditional tools, like booklets, and will change the learning and cognitive processes. They will also change the relationship between subject and tool, subject and object, and object with the tool. So, what are the affordances of the mixed reality mobile technologies and how they work with the educational field? This will be the focus of the next chapter. However, I would anticipate just the main affordances of the applied technology and their advantages when compared with paper mediums and other multimedia technologies:

1. MR creates a situation where the learner could detect a cognitive dissonance between the real element and the virtual element thanks to its unique way of visualizing and aggregating information. This causes a strong motivational impulse to reduce the dissonance and thus modify existing knowledge or integrate it with other elements to give coherence to one's understanding.

2. MR in teaching fascinates and engages students (Luckin and Stanton, 2011). In our study this technology made learning fun, increased the desire to participate, and made the experience memorable—all these factors being of vital importance to meaningful learning.

3. MR improve the remembering of the content, mostly thanks to its AR content, but one has to be careful in designing the software to avoid cognitive overload (Sweller, 2011) and to make the interface as transparent as possible.

4. MR can show superimposed three-dimensional objects and imagery, representing also abstract concepts. Therefore, the quality of internal models that the students make up in their minds is higher.

5. MR ability to plunge the user into mixed and virtual environments is crucial for a better understanding of the connection between various objects in a complex system.

6. The usage of MR tools enables a whole new level of interactions—not just between the student and the tool, but also between student and teacher, and student and environment through the tool. The main reason for this is the affordances of the MR tool. In particular, the fact that, unlike a book or a notebook, it is capable of interaction and is responsive to the actions of students. This interaction raises interest and motivation. It enhances understanding and remembering as well.

7. The correct use of MR technology encourages students to see themselves as problem-solvers and independent researchers through first-person experience and interaction.

8. MR mobile technology frees students from a static context and allows each individual to personalize their own learning context or space, thanks to mobile apparatus. Thanks to sensors and networks, this allows not only a situated means of learning but also a collaborative means. It also allows immediate sharing and modification of content, which promotes an authentic co-construction of knowledge.

9. Thanks to these sensors and actuators present in the devices, learning takes place not only due to augmented sight but also through augmented hearing and haptic feedback. This characteristic sheds new light on the uses of this technology in various contexts: in situations, for example, when one cannot look at a screen or where there are users with visual impairment.

Finally, there might be negative affordances, or, in other words, affordances that, in certain contexts, can act as obstacles or distractors for the student. One of these may be MR devices of being less transparent to the pupils than more traditional media, such as books or illustrations. The same affordance can be positive as it excites and engages but, being too much on the pupil's centre of attention, seems to be a not-so-desirable peculiarity of a mediating tool in an educational context.

Awareness and Interference Caused by Mixed Reality Tools

During research experimentations with MR tools, most of the time the experimental group's subjects—using MR technology—were more aware of the surrounding environment than the control group's colleagues, who used traditional media, such as booklets and pictures. For example, in an experiment conducted with a group of fifth-grade primary school students while visiting the Roman remains in Verona, the experimental group representations included more depictions of the monument on which the students were focussing as well as more elements of the surrounding landscape. Furthermore, most of the experimental group's drawings depicted a foreground, the monument, and a background, whereas the control group tended to pay attention—and draw—only the monument itself, without a clear context. The experimental group elements were often spatial references, such as streets, plants, other monuments, and people, contributing to an understanding of the relationship between the monument and the landscape.

As mentioned, using a tool instead of another always changes the interaction between the subject and the tool, and the mediation of the tool between the subject and the object of an activity. Taking, as an example, the experimentation mentioned above, one can pinpoint the different interactions just looking at the drawings produced by students of the two groups. Control group drawings do not show any sign of mediation between them and the cultural artefact except for, rarely, the guide and the teacher. No drawing directly represents the booklet, although some drawings seem to have been inspired by the booklet pictures. The booklet appears to function as a 'transparent' mediating tool, even if no tool is ever entirely 'transparent' as it always carries along with its own mediating means and effects. The experimental group's drawings indicate that the AR device (i.e., the smartphone or the tablet) was not transparent for everyone. In some cases, on the contrary, it might have distracted the students. The list below describes the four possible levels of technology interference and their frequency:

1. Fixed focus on technology (5 per cent): The student attention is wholly attracted by the mediating tool and misses to focus on the object. This is the kind of interference that one hopes to avoid. It seldom occurs when using MR technology but might show up. In Fig. 9-(3), the drawing only represents the technology (i.e., the device) without any information about the heritage or the context.

2. Focus on heritage through technology (12 per cent): The student attention is on the object as it is presented through the technology mediation. In Fig. 9-(2) the students represented the heritage in the frame of the device and the app (i.e., the pupils drew the device on paper in the first place as a frame for the heritage, then where there should have been the screen of the device, they represented the heritage).

Fig. 9. Examples of drawings of the first three categories of interference.

3. Focus on heritage and technology interaction (30 per cent): Basically, there is a first- and third-person view of the interaction with technology. The student is able to (*see* Fig. 9-(1)) represent how the heritage and context are in reality, integrating it with elements of knowledge acquired through the technology. He or she can represent himself in the learning context with references to the use of the technology.

4. Focus on augmented heritage (53 per cent): This kind of interference is a positive one and worth something to pursue. It is a synthesis of real and virtual information. These drawings represent the past or the present augmented by visual imagery or written information where the technology is transparent and invisible, as illustrated in Fig. 11. One variant of this kind of interference involves use of the drawing sheet as an organized space that mimics the app's function and categories (e.g., keeping the space for a map on the top-right corner).

In summary, although being aware of mediating tools can help students at the level of meta-cognition in their learning process, in this kind of experience, ensuring the transparency of those tools would help to prevent them from acting as an overly powerful distractor. They should deliver information while remaining as unnoticed as possible, for example, designing an educational MR app, one should avoid menus and complicated interactions. Rather, it should be possible to use it as a simple 'frame' to look at reality in the most transparent way possible.

It is interesting to gather intelligence on the use of the technology captured in drawings. Clues as to the attribution of meanings, importance, and emotions were founded. One of the first aspects that emerged from some drawings relates to emotions raised by the technology. While nothing was found about the use of AR technology, there were clues about the use of VR technology. In Fig. 9-(1), the girl using the headset is looking around at a virtual landscape, and is smiling; elsewhere (*see* Fig. 10), children have been represented in a 'jaw-dropping' expression of amazement while using the Google Cardboard headsets and seeing in the mixed reality what the guide is telling them. In addition, some pupils paid attention to the device's brand, its shape, and the position of software buttons. Some of them also remembered the interface of the MR app in minute detail. This seems to indicate a particular interest in the tool itself and its working principles, as well as an acquired competence in the use of the MR app. The view that the use of a tool rather than others affects the cognitive processes and the relationship with the environment gets confirmed.

Fig. 10. Details of drawing with the guide explaining and pupils looking around with and without cardboard VR.

Fig. 11. The ideal level of interference with the MR tool.

Four-X: An Activity Theory-based Teaching Model for Mobile MR Technology

To develop a strategy of didactic design and practice for the use of augmented and mixed reality mobile technologies for heritage education in non-formal learning contexts, such as educational visits to cultural sites, the Four-X method—formerly called Tri-AR—was implemented (Agostini and Petrucco, 2020), based on the second generation of the Activity Theory (Engeström, 2015). The process involved seven fifth primary school classes in Italy and two in England: a total of 194 pupils aged between nine and 11. The experimentation took place in Verona, during visits to the Roman Verona and Hestercombe, England, during visits to the historic landscape garden.

Four-X: Why and How

Here the challenge was to design and experiment with an app with augmented and mixed reality (AR/MR) technologies and, at the same time, build an educational model that could take into account:

- The context of the monument (in Verona) or the landscape view (in the Hestercombe Landscape Garden).
- The AR/MR content.
- The interactions between the subjects involved (e.g., a guide and students).

Furthermore, the model had to allow an evaluation of primary school classes to find if this approach helped students to understand concepts related to historical, cultural, and artistic content. Four-X is the model expressly developed for this purpose, based on the more general one of the historical-cultural Activity Theory (AT) (Leont'ev, 1981; Vygotsky, 1980). The name is given by its derivation from the Activity Theory triangle of Vygotsky and then of Engeström. A fourth vertex makes it tri-dimensional and it is proposed as divided into four steps of interaction with the mixed reality educational technologies. In Fig. 12, one can see the interactions predicted in our model. These four elements include:

- The subject of the action, which in our case is the student.
- The app or, in general, the mediating tool.
- The human mediator, who can be the guide or the teacher. This is not present in the classic model of AT, being part of the larger community.
- The setting/environment, so called to emphasize its spatial characteristic; in this case, it is the cultural heritage intended as the cultural object with which one interacts and the landscape. In this particular research, cultural heritage is represented by its tangible and visible artefacts in the form of *chronotopes*, or physical crystallizations of a culture and its adaptation to a specific place (Agostini and Piva, 2018; Bertoncin, 2004).

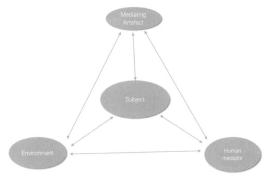

Fig. 12. The Four-X model.

This approach is less complete but more specific than the classic activity triangle contemplated by second- and third-generation AT (Engeström et al., 1999). Here the subject (in our experimentation, the student), the human mediator (the guide) and the technological mediator (the MR device with the app) are connected to each other by a network of interactions that are translated into a modular experience (visit) format through a series of rules. These rules define a specific sequence of interactions and are communicated to the subject before starting the experience. The paradigm of reference is that of Vygotsky (1980): human beings interact and learn, thanks to the mediation of tools and artefacts that expand the 'zone of proximal development' (ZPD) but also need interactions with people, or with the community. The human mediator (e.g., teacher or guide) is crucial in mobile augmented and mixed reality experiences. The model was used both to count the number of interactions between the elements and to highlight the processes of internalization and externalization from the point of view of the subject.

Internalization and Externalization Processes during the Visit

Internalization and externalization processes take place at two different levels of the activity: at the level of mental processes/visible behavior and at the inter-psychological/intra-psychological level (Vygotsky, 2012). The inter-psychological level represents the mental processes externalized and shared with the community, while the intra-psychological level represents the processes internal to the single mind. During an activity, these processes took place continuously; however, when the Four-X procedure is in place, they become more evident, thanks to the given rules.

The Example of a Visit to a Cultural Heritage Site

Always keeping the students as the subject, it follows a description of the traces of internalization and externalization identified through the observation of four visits to Roman Verona and two to Hestercombe, UK. Testing it in different contexts allowed us to verify that the methodology worked in different contexts with different types of heritage and artefacts.

Four-X application to educational practice consists of a four-steps model (*see* Table 5): first explanation, second exploration, third expression, fourth exchange.

Table 5. Synoptic table of activated processes.

Four-X phase	Subject	Technological mediator	Human mediator	Environment
1st: Explanation	Internalizes information and meanings simultaneously from the human mediator (HM) and through the environment.	Standby	Introduces and describes the environment.	It is mediated in its interpretation by the human mediator (HM).
2nd: Exploration	Actively acts, using the technological mediator (TM) with to explore the environment. By looking for correspondences between what he/she knows, what he/she has internalized in phase 1, and the information acquired through the TM, students internalize information and meanings.	Active	Encourages subject to use the technological mediator (TM) to explore and discover, to pay attention to details already explained through interactions with TM and ask for feedback by asking specific questions.	It is mediated in its interpretation by the TM and the HM.
3rd: Expression	Gives feedback (externalizes) and further and freely explores the surrounding environment through the app. Also, freely asks questions from HM and interacts with peers.	Active	Answers subject questions using TM, if needed.	It is observed in the elements of interest through the mediation of TM and peers.
4th: Exchange	Interacts freely with HM and peers, referring directly to the environment, and/or through TM when useful.	Active	Interacts with students and makes sure all the important elements have been internalized.	It is again observed, mediated or not by TM, HM and peers in the elements of interest.

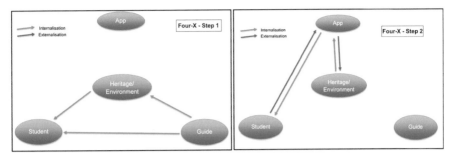

Fig. 13. Phases 1 and 2 of the Four-X model used in the Verona and Hestercombe visits.

In the explanation phase (*see* Fig. 13 on the left), the guide gives an introductory description of the place or monument, explains its history, and its use at the time of construction and at various moments in the past. Furthermore, it underlines the differences of the place as it appears today when compared to how it appeared in past eras. In this phase, the students, who are always the subjects of the action, internalize information and meanings simultaneously from the two sources—on the one hand, directly from the guide and, on the other, through the heritage that they have in front of and around them. The heritage, in turn, is mediated and mediates the words of the guide.

In the exploration phase (see Fig. 13 on the right), the guide encourages students to use the app to discover nearby details already explained in the presentation of the place. The guide asks students to discover content through AR and MR interactions and at the same time asks for feedback by asking specific questions.

In this phase, students are asked to actively respond by using the app as a mediator with the assets. This process allows students to externalize, at the level of actions, the information they have just gained from the guide. By looking for correspondences between what they have just learned and information in the app, students internalize information and meanings.

During the expression phase (*see* Fig. 14 on the left), students provide feedback and explore the surrounding environment further and freely through the app. They also freely put their questions to the guide. In this case, the externalisation process is also at the inter-psychological level. The pupils, in fact, share their observations and questions with their classmates and guide so that the community can help them solve them. Once the community has found the answer, it can be internalized on an intra-psychological level.

In the exchange phase (*see* Fig. 14 on the right), the guide answers students' questions by using the app if necessary. Students interact with the guide by referring directly to the monuments and the place, or by using the app when they think it is useful. At this point, all the previous interactions are possible, including the more classic ones, which do without the AR and MR mediator. Therefore, the processes of internalization and externalization are possible, both at the mental/behavioral level and at the inter/intra-psychological level according to the level of initiative of the student. Table 3 presents the same model, highlighting the phases, roles, and processes.

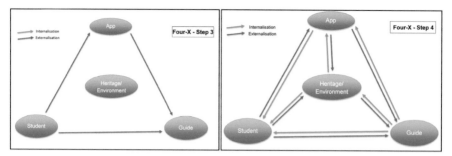

Fig. 14. Phases 3 and 4 of the Four-X model used in the visits to Verona and Hestercombe.

Outcomes of Four-X Application

The first tangible result was already noticed during the design phase of the two experiences. The process of creating the scripts of the two visits to Italy and England was facilitated and standardized, thus gaining a directly comparable structure.

While carrying out the experiences, the four steps were presented to students and teachers at the beginning of the visit, specifying that they would be repeated at each stage or stopover. This helped the guide (who was already aware of it) and students to manage timing, questions, and device usage. In terms of observation, it also made it possible to accurately place the observations of pupils' behaviors and interactions at specific moments of the stage. For example, only during the exchange phase the pupils felt free to interact with each other, indicating to each other what they discovered, thanks to the app.

Finally, during the elaboration of data, this Four-X structure of the experience made it possible to easily identify and interpret the analysis elements according to the Activity Theory and, in particular, using the AT checklist for evaluation (Kaptelinin and Nardi, 2006). The Four-X model application has made it possible to quickly identify the mediation of technologies. It also helped to track mutual transformations between operations and actions, and internalization and externalization processes.

The application of Four-X model to our data analysis helped to explain the changes in the interactions with technology and community when MR technology was in use. It was employed with data gathered from our participant observation and that of the teachers, and to videos taken during the experimentation. This helped to verify a key difference between the interactions of the control group (where the TM was a booklet) and the experimental one (where the TM was the MR app). In fact, both in the exploration and in the exchange phases—that involved spontaneous interactions with the mediating artifact (booklet or app) and with the heritage as a physical object (for example a Roman construction)—the experimental group had a very high count of interactions between student and student, student and mediating artefact, and consequently between student and cultural object. In contrast, the control group had just a portion of those interactions.

The Four-X model proved to be very useful in the three phases of design, execution, and analysis of experiences. In particular, it proved to be a useful tool for translating the principles of Activity Theory into teaching practice with technologies and, at the same time, as a tool for analyzing the teaching process.

In summary, the Four-X model favoured:

- The design of the phases of the educational visits by the guides and teachers.
- The analysis of the experimentation with AT checklist.
- The transferability of the trial to England.
- Reflection on pupils' cognitive processes and their interaction with technology.
- The discovery of any tensions (or contradictions) between elements of the activity system (especially in the analysis of videos) (e.g., action instead of internalization during explanation–interference of technology).
- The discovery of different dynamics in the experimental group compared to the control group.

References

Aboulafia, A., Gould, E. and Spyrou, T. (1995). Activity theory vs cognitive science in the study of human-computer interaction. *In*: *The Proceedings of the IRIS (Information Systems Research Seminar in Scandinavia) Conference*. 12, Gjern, Denmark.

Agostini, D. and Petrucco, C. (2018). *Sperimentazione di una App di Realtà Aumentata per comunicare il Patrimonio Culturale:* Hestercombe Gardens Augmented Visit. *In*: *Proceedings Multiconferenza EM&M ITALIA, Progress to Work, Contesti, processi educativi e mediazioni tecnologiche*. Genova, Genova University Press.

Agostini, D. and Piva, M. (2018). *Progetto di Sperimentazione Didattica: Geolocalizziamo la Grande Guerra, Percorsi e Trincee sul Fronte del Monte Grappa e del Fiume Piave*. pp. 109–22. *In*: Masetti, C. (ed.). *Per un Atlante della Grande Guerra, Dalla Mappa al GIS, Labgeo Caraci*. Roma.

Agostini, D. and Petrucco, C. (2020). Tri-AR. An activity theory-based teaching model for design, practice and analysis of educational experiences with mobile technology. pp. 115–22. *In*: Cecchinato, G. and Grion, V. (eds.). *Dalle Teaching Machines al Machine Learning*. Padova University Press, Padova.

Avouris, N., Komis, V., Meletis, M. and Fiotakis, G. (2004). An environment for studying collaborative learning activities. *Journal of Educational Technology & Society*, 7(2): 34–41.

Baloch, H.Z., Aziziah A. and Noorminshah A.I. (2012). Mobile collaborative informal learning design: study of collaborative effectiveness using activity theory. *International Journal of Interactive Mobile Technologies (IJIM)*, 6(3): 34–41.

Barhoumi, C. (2015). The effectiveness of whatsapp mobile learning activities guided by activity theory on students knowledge management. *Contemporary Educational Technology*, 6(3): 221–38.

Bertoncin, M. (2004). *Logiche di terre e acque: le geografie incerte del delta del Po*. Cierre.

Bødker, S. (1996). Applying activity theory to video analysis: how to make sense of video data in human-computer interaction. pp. 147–74. *In*: Nardi, B.A. (ed.). *Context and Consciousness: Activity Theory and Human-Computer Interaction*. Cambridge, MA: MIT Press.

Brown, J.S., Collins, A. and Duguid, P. (1989). Situated cognition and the culture of learning. *Educational Researcher*, 18(1): 32–42.

Burden, K. and Kearney, M. (2015). Conceptualising authentic mobile learning. pp. 27–42. *In*: Churchill, D., Lu, J., Chiu, T.K.F. and Fox, B. (eds.). *Mobile Learning Design: Theories and Application, Lecture Notes in Educational Technology (LNET)*. Springer, Singapore.

Churchill, D., Fox, B. and King, M. (2015). Framework for designing mobile learning environments. pp. 3–25. *In*: Churchill, D., Lu, J., Chiu, T.K.F. and Fox, B. (eds.). *Mobile Learning Design: Theories and Application, Lecture Notes in Educational Technology (LNET)*. Springer, Singapore.

Collins, P., Shukla, S. and Redmiles, D. (2002). Activity theory and system design: a view from the trenches. *Computer-supported Cooperative Work (CSCW)*, 11(1): 55–80.

Cook, J. (2010). Mobile phones as mediating tools within augmented contexts for development. *International Journal of Mobile and Blended Learning (IJMBL)*, 1 July.

Cook, J. and Santos, P. (2015). Three phases of mobile learning state of the art and case of mobile help seeking tool for the health care sector. pp. 315–33. *In*: Churchill, D., Lu, J., Chiu, T.K.F. and Fox, B.

(eds.). *Mobile Learning Design: Theories and Application, Lecture Notes in Educational Technology (LNET)*. Springer, Singapore.

Cowan, P. and Butler, R. (2013). Using activity theory to problematise the role of the teacher during mobile learning. *SAGE Open*, 3(4): 2158244013516155.

Daniels, H. and Warmington, P. (2007). Analysing third generation activity systems: labour-power, subject position and personal transformation. *Journal of Workplace Learning*, 19(6): 377–91.

Daniels, H. (2010). Using and developing activity theory: subject position and discourse in multi-agency settings. pp. 279–94. *In*: Aunio, P., Jahnukainen, M., Kalland, M. and Silvonen, J. (eds.). *Piaget is Dead, Vygotsky is still Alive?* Finnish Educational Research Association, Jyväskylä.

Engeström, Y. (1993). *Developmental Studies of Work as a Testbench of Activity Theory: The Case of Primary Care Medical Practice*.

Engeström, Y., Engeström, R. and Kärkkäinen, M. (1995). Polycontextuality and boundary crossing in expert cognition: learning and problem solving in complex work activities. *Learning and Instruction*, 5(4): 319–36.

Engeström, Y., Miettinen, R. and Punamäki, R.L. (eds.). (1999). *Perspectives on Activity Theory, Learning in Doing: Social, Cognitive and Computational Perspectives*. Cambridge University Press, Cambridge.

Engeström, Y. (1999). Activity theory and individual and social transformation. pp. 19–38. *In*: Engeström, Y., Miettinen, R. and Punamäki, R.L. (eds.). *Perspectives on Activity Theory. Learning in Doing: Social, Cognitive and Computational Perspectives*. Cambridge University Press, Cambridge.

Engeström, Y. (2015). *Learning by Expanding*. Cambridge University Press.

Ergazaki, M., Zogza, V. and Komis, V. (2007). Analyzing students shared activity while modeling a biological process in a computer-supported educational environment. *Journal of Computer Assisted Learning*, 23(2): 158–68.

Falk, J.H. (1991). Analysis of the behavior of family visitors in natural history museums: the national museum of natural history. *The Museum Journal*, 34(1): 44–50.

Falk, J.H. and Dierking, L.D. (2018). *Learning from Museums*. Rowman & Littlefield.

Fjeld, M., Lauche, K., Bichsel, M., Voorhorst, F., Krueger, H. and Rauterberg, M. (2002). Physical and virtual tools: activity theory applied to the design of groupware. *Computer-supported Cooperative Work (CSCW)*, 11(1): 153–80.

Gibson, J.J. (2014). *The Ecological Approach to Visual Perception*. Psychology Press.

Hasan, H. and Kazlauskas, A. (2014). *Activity Theory: Who is Doing What, Why and How*. Faculty of Business – Papers (Archive), January: 9–14.

Holt, G.R. and Morris, A.W. (1993). Activity theory and the analysis of organisations. *Human Organization*, 52(1): 97–109.

Jonassen, D.H. and Rohrer-Murphy, L. (1999). Activity theory as a framework for designing constructivist learning environments. *Educational Technology Research and Development*, 47(1): 61–79.

Jonassen, D.H. (2013). Designing constructivist learning environments. Instructional design theories and models: a new paradigm of instructional theory. pp. 215–39. *In*: Reigeluth, C.M. (ed.). *Instructional-design Theories and Models: A New Paradigm of Instructional Theory*. Routledge.

Kaplan, A. (2017). *The Conduct of Inquiry: Methodology for Behavioural Science*. Routledge.

Kaptelinin, V., Kuutti, K. and Bannon, L. (1995). Activity theory: basic concepts and applications. pp. 189–201. *In*: Blumenthal, B., Gornostaev, J. and Unger, C. (eds.). *Human-Computer Interaction, Lecture Notes in Computer Science*. Springer, Berlin, Heidelberg.

Kaptelinin, V. (1996a). Computer-mediated activity: functional organs in social and developmental contexts. pp. 45–68. *In*: Nardi, B.A. (ed.). *Context and Consciousness: Activity Theory and Human-Computer Interaction*. MIT Press, Cambridge.

Kaptelinin, V. (1996b). Distribution of cognition between minds and artifacts: augmentation of mediation? *AI & SOCIETY*, 10(1): 15–25.

Kaptelinin, V., Nardi, B. and Macaulay, C. (1999). Methods & tools: the activity checklist: a tool for representing the 'space' of context. *Interactions*, 6(4): 27–39.

Kaptelinin, V. and Nardi, B. (2006). *Acting with Technology: Activity Theory and Interaction Design*. MIT Press.

Kaptelinin, V. and Nardi, B. (2012). Activity theory in HCI: fundamentals and reflections. *Synthesis Lectures on Human-centered Informatics*, 5(1): 1–105.

Khoo, K.Y. (2015). Enacting app-based learning activities with viewing and representing skills. Preschool mathematics lessons. pp. 351–72. *In*: Churchill, D., Lu, J., Chiu, T.K.F. and Fox, B. (eds.). *Mobile*

Learning Design: Theories and Application, Lecture Notes in Educational Technology (LNET). Springer, Singapore.

Kuutti, K. (1996). Activity theory as a potential framework for human-computer interaction research. pp. 17–44. *In*: Nardi, B.A. (ed.). *Context and Consciousness: Activity Theory and Human-Computer Interaction.* MIT Press.

Leont'ev, A.N. (1981). *Problems of the Development of the Mind: A. N. Leontyev.* Progress Publishers, Moscow.

Luckin, Rosemary and Danae Stanton Fraser. (2011). Limitless or pointless? An evaluation of augmented reality technology in the school and home. *International Journal of Technology Enhanced Learning,* 3(5): 510–524.

Maslow, A.H. (1966). *The Psychology of Science: A Reconnaissance.* Harper & Row.

Nardi, B.A. (ed.). (1996). *Context and Consciousness: Activity Theory and Human-Computer Interaction.* MIT Press.

Nardi, B.A. (1998). Activity theory and its use within human-computer interaction. *Journal of the Learning Sciences,* 7(2): 257–61.

Norman, D.A. (1999). Affordance, conventions, design. *Interactions,* 6(3): 38–43.

Osborne, R. (2014). *An Ecological Approach to Educational Technology Affordance as a Design Tool for Aligning Pedagogy and Technology.* Doctoral thesis, ORE (Open Research Exeter), University of Exeter, Exeter.

Papadimitriou, I., Tselios, N. and Komis, V. (2007). Analysis of an informal mobile learning activity based on activity theory. pp. 25–28. *In*: Vavoula, G., Kukulska-Hulme, A. and Pachler, N. (eds.). *The Proceedings of the Workshop Research Methods in Informal and Mobile Learning.* WLE Centre, London.

Pask, G. (1976). *Conversation Theory: Applications in Education and Epistemology.* Elsevier, Amsterdam, New York.

Roth, W.M. and Lee, Y.J. (2007). 'Vygotskys neglected legacy': cultural-historical activity theory. *Review of Educational Research,* 77(2): 186–232.

Rozario, R., Ortlieb, E. and Rennie, J. (2015). Interactivity and mobile technologies: an activity theory perspective. pp. 63–82. *In*: Churchill, D., Lu, J., Chiu, T.K.F. and Fox, B. (eds.). *Mobile Learning Design: Theories and Application, Lecture Notes in Educational Technology (LNET).* Springer, Singapore.

Sweller, J. (2011). Cognitive load theory. *Psychology of Learning and Motivation,* 55: 37–76.

Taylor, J., Sharples, M., O'Malley, C., Vavoula, G. and Waycott, J. (2006). Towards a task model for mobile learning: a dialectical approach. *International Journal of Learning Technology,* 2(2-3): 138–58.

Tomkins, S.S. and Messick, S. (1963). Computer simulation of personality. *Frontier of Psychological Theory.* Wiley, New York.

Uden, L. (2006). Activity theory for designing mobile learning. *International Journal of Mobile Learning and Organisation,* 1(1): 81–102.

Ukhtomsky, A.A. (1978). Izbrannye Trudy, *Selected WORKS.* Nauka, Leningrad.

Vygotsky, L.S. (1980). *Mind in Society: The Development of Higher Psychological Processes.* Harvard University Press.

Vygotsky, L.S. (2012). *Thought and Language.* MIT Press.

Wali, E.A. (2008). *Reinterpreting Mobile Learning: An Activity Theoretic Analysis of the Use of Portable Devices in Higher Education.* Institute of Education, University of London, London.

Walker, K. (2010). *Designing for Meaning Making in Museums: Visitor-constructed Trails Using Mobile Digital Technologies.* Doctoral thesis, University of London, UCL Theses.

Waycott, J., Jones, A. and Scanlon, E. (2005). PDAs as lifelong learning tools: an activity theory-based analysis. *Learning, Media and Technology,* 30(2): 107–30.

Zinchenko, V.P. (1996). Developing activity theory. pp. 283–324. *In*: Nardi, B.A. (ed.). *Context and Consciousness: Activity Theory and Human-Computer Interaction.* MIT Press.

CHAPTER 11

Methodologies and Strategies for a Mixed Reality Mobile Learning

Daniele Agostini

IIIIIIIIIIIIIIIIIIIIIIIIIIIIIIIIIIIIIIIIIIIIIIIIIIIIIIIIIIIIIIIIIIIIIIIIIIIIIIIIIIIIIIIIIIIIIIIIIIIIIIIIIIIIIIIIIIIIIIIIIIIIIIIIIIIIIIIIIIIIIIIIIIIIIIIIIIIIIIIIIIII

It does not seem likely that machines will have the effect of dehumanising learning any more than books dehumanise learning. A program for a teaching machine is as personal as a book: it can be laced with humor or be grimly dull, can either be a playful activity or be tediously like a close-order drill.

—Bruner (1977)

Introduction

In the history of the instructional use of technology, it has often been believed that educational technology strengthens the teaching-learning process. However, in-depth analyses focused on hundreds of experiments in the last 30 years (Hattie, 2009; Tamim et al., 2011) continue to show that technology itself does not guarantee a substantial improvement (Rushby and Seabrook, 2008). In particular, the issue, underscored by the Cognitive Load Theory, is that technology seems to have a neutral or average influence and, in some cases, detrimental consequences due to overloading of the cognitive function (Sweller, 2011). Several academic papers point to the fact that the most significant indicator of progress in the use of technologies in education is the choice of suitable methodologies, appropriate to the situation in which they are to be applied (Calvani, 2014; Kirschner et al., 2006).

From this point of view, mobile devices, such as smartphones, are considered more disruptive than conventional devices used in schools (i.e., laptops, interactive whiteboards, etc.) mainly because students in informal everyday situations

Postdoctoral Researcher in Educational Technology and Digital Interpretation, University of Padua.
Email: daniele.agostini@unipd.it

continuously use them. Teaching plans involving mobile devices must take into account the setting up of such a diverse and cross-context learning experience.

Mobile learning (M-learning) started in the 1980s when portable computers (the bleeding-edge of technology at the time) were first brought into the classroom to run educational experiments (Kukulska-Hulme et al., 2009). Only at the end of the 1990s, it really spread due to experimental educational projects aimed at discovering the educational potential of PDAs (personal digital assistant). From the mid-nineties to the present, we can identify definite periods with three different focuses: tools, out-of-class learning, and student mobility ('Big Issues in Mobile Learning: Report of a Workshop by the Kaleidoscope Network of Excellence Mobile Learning' 2006). The first step is characterized by the quest for the right instruments, or rather those that are ideally tailored to the instructional system, the learning and teaching methods, and a special interest in affordances. E-books, learning aids, and digital notebooks were at the forefront of this search, along with data recording and learning objects (Ranieri and Pieri, 2015). The second step focused on out-of-class learning, and detailed testing was carried out on mobile devices that could be used for field trips and museum visits.

Nevertheless, the technology's developmental state still placed substantial limitations on this methodology (Kukulska-Hulme et al., 2009). In the third (and current) step, more consideration is paid to student mobility and, consequently, to learning spaces (real and virtual), and to the relationship between formal and informal learning environments (Winters, 2006). The technology can finally fully support mobile learning: it is available, affordable, and flexible.

The learning process can be context-specific, thanks to the relationship between the learner, the device and the environment. In this stage, M-learning has some very specific affordances that have been made possible by methodological and technical developments. Strigel and Pouezevara (2012) suggest four significant classes of M-learning affordances: accessibility, immediacy, individualization, and intelligence. The first comprehends all the affordances to access learning opportunities, reference materials as well as experts, mentors, and other learners. The second covers on-demand learning, real-time networking, real-time data sharing, and situated learning. Individualization or, better, personalization (the latter emphasizing learning-centeredness) allows people to use their own device and encourages constructive learning. 'Bite-size learning' has a significant role in this approach, since it is a concept that refers to small, highly contextualized bits of information, given at a particular moment when one needs them to be grasped not in a specific order. This approach often matches perfectly with the limited time-frames that informal learning demand. The last class is intelligence, which contains all the technical features related to context awareness, data capture, and multimedia capability.

The third phase of M-learning, even concerning the abbreviation form, comes from e-learning and one can recognize it by looking at those classes. They are partly the same as e-learning, but they also have some unique characteristics. While accessibility and personalization are also under the e-learning umbrella, everything related to concepts described as spontaneous, situated, portable, context-aware, lightweight, informal, personal, and bite-sized are only part of M-learning (Traxler, 2005) (*see* Fig. 15).

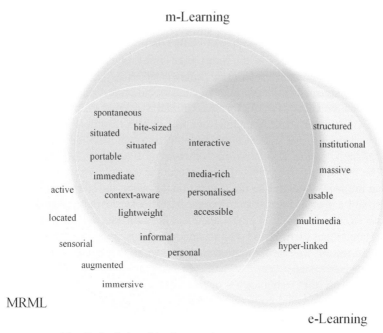

Fig. 15. Evolution of the diagram of Traxler (2005) with MRML.

Teaching strategies that include the use of mobile devices must take into account the context of such a complex learning experience which crosses the borders of formal and informal.

The Transition from Mobile Learning to Mixed Reality Mobile Learning (MRML)

Here mixed reality mobile learning is intended as a continuous advancement and expansion of mobile learning. The main feature of M-learning is that it enables a situated learning experience (Lave and Wenger, 1991) mediated by a particular technology. At its base there is Vygotsky's principle, which asserts that man becomes acquainted with the environment in which he lives by means of instruments and artefacts that expand the proximal development zone (ZPD). ZPD being 'the distance between the actual developmental level as determined by independent problem-solving and the level of potential development as determined through problem-solving under adult guidance, or in collaboration with more capable peers' (Vygotsky, 1980: 86).

Multimedia augmented M-learning was an intermediate phase on the way to the current MRML. This deserves mentioning because of the challenges that were addressed during this transformation: mobile devices, networking, content heterogeneity, delivery, and user requirement issues (Yousafzai et al., 2016). If some of them remain crucial, others have been significantly downsized, in particular, on the technical side. Mobile device issues, like small screens, display resolution, codecs,

OSs heterogeneity, and limited memory have been addressed quite efficiently. Also, networking issues have been significantly diminished, thanks to 4G and 5G networks. We are now at the next step, where the multimedia augmented M-learning become augmented and mixed reality M-learning. Therefore, we need to ask what benefit could come from this transition.

As mobile learning has its roots in e-learning and moves from it, bringing a change of paradigm, likewise the shift from M-learning to MRML brings novelties, such as new tools and new characteristics. In the transition from e-learning to M-learning we moved from attributes, such as personalized, media-rich, usable, hyper-linked, accessible, and connected to spontaneous, situated, portable, informal, and personal (Traxler, 2005). The research on the didactic potential of AR and MR is increasingly focused on concepts, such as situated, portable, context-aware, informal, personal, and bite-sized as developments of the mobile learning third phase (Ranieri and Pieri, 2015). However, new characteristics are emerging: augmented, active, located, immersive, wearable, adaptive, and intelligent are some of them (Fig. 15).

Affordances of Mixed Reality Mobile Learning

The term 'augmented reality' is somehow misleading; in fact, it is not the actual reality, but the perception that we have of it that is augmented (Hugues et al., 2011). Perception is seldom purposeless; it is oriented, at a conscious or unconscious level, to an action (Auvray and Fuchs, 2007). Through AR one can increase the quantity of information perceived, but, most importantly, it can deliver information more effectively. At the same time, one can have a better mastery of actions related to real events (Hugues et al., 2011). Similarly, Vygotsky, Auvray and Fuchs (2007) affirm that using a new tool modifies our relationship with the environment and thus our perception. New tools can modify our 'preceptory space' in a process aimed at achieving 'target action' more efficiently (Bergson, 1911). Being perception oriented, MRML calls for different teaching where one is invited to learn in view of, and through, actions.

AR and MR used for mobile learning produce a situation where, thanks to their peculiar way of incorporating information, the learner could sense a cognitive dissonance between the actual element and the virtual element. This situation induces a powerful motivating tendency, which is intended to minimize dissonance and thus change the existing internal knowledge or combine it with other elements in order to add coherence to one's perception (Munnerley et al., 2012). Another benefit of MR teaching is that it fascinates pupils. In their research, Luckin and Fraser (2011) showed that this technology made learning enjoyable, enhanced motivation and engagement, and made the experience memorable, all of which were critical to meaningful learning.

Other results point out that this technology's proper application encourages students to see themselves as problem solvers and autonomous scholars (Squire, 2010). MRML frees students from a static context and encourages them to personalize their learning context or space via a mobile device. Due to the sensors and networks, this enables not only a situated learning, but also a shared mean, thanks to immediate exchange and editing of content that facilitates an inclusive co-construction of knowledge (Cook, 2010). These same sensors and networks allow learning to take place not only through augmented sight, but also through augmented hearing and haptic feedback. This feature sheds new light on the application of this technology in a variety of contexts: in cases, for example, where a monitor cannot be seen, or where there are people with visual impairments (van der Linden et al., 2012).

Thanks to these resources, students gain a quantitative improvement in their learning and work in terms of performance and speed, but are also able to monitor and organize their behavior. In fact, our understanding of the reality that we reinterpret through a continuous process of sense attribution and the mediation of tools, changes in proportion to the nature of the tool's interaction. In successive research, another criterion has arisen in the field of situated learning: the need for sorting processes (Latour, 1999) and, in particular, for handling one's learning in spatial and temporal dimensions (Munnerley et al., 2012). In summary, the enhancement provided by the tool is quantitative, but also qualitative, and depends on the tool's affordances.

After this excursus of MRML affordances, the most important, probably, is still missing. The foremost is, in fact, its unique capability of surrogating the real experience and making the experience of abstract concepts possible. This feature has been extensively used in many industries for training in the field of transport (Crescenzio et al., 2011), medicine (Barsom et al., 2016), and machine tools (Gavish et al., 2015; Webel et al., 2013). It has also been used in education of course, in particular, STEM (Sırakaya and Sırakaya, 2020) and cultural heritage (Ibañez-Etxeberria et al., 2020), as has been demonstrated in this book.

Thanks to the ability to superimpose interactive three-dimensional objects and imagery, representing concepts and real experiences, the quality of internal models that the students make up in their minds is higher; there is an understanding of the connection between various elements of a complex system (Agostini and Petrucco, 2020). These elements lead to consider the MRML under the experiential learning umbrella.

Approaches and Strategies for a Didactical Use of MRML

AR/MR applications can support new learning paradigms (Chen and Wang, 2008), filling the gap between theory and practice through use of constructive activities. It is for this reason that the choice of setting and teacher's role are so important: the experiences with MRML can undoubtedly be used within a traditional teaching setting in the classroom, for example, but in this way, they would lose much of their great potential (Auld and Johnson, 2015). This is the difference between didactical innovation and a mere technology upgrade. For example, it is critical to discern between tablet learning and M-learning. The former is a sort of e-learning that one can also use on a tablet. It would represent just a technological upgrade to e-learning. The latter has both technological and methodological innovation, and, as already mentioned in this chapter, it is a new kind of e-learning that can be done in mobility.

It is necessary to develop a new curriculum to allow the student to integrate informal learning through MRML technology and conduct experiments outside the school context with informal learning processes. Teachers need to encourage these instances of meaningful learning (Howland et al., 2014), providing students with conceptual means of judging their MRML experiences within the prospect of self-regulated learning and lifelong learning. There are, in fact, experimentations aimed at creating an augmented environment for active learning and that could also be seamlessly used out of school (Land et al., 2013; Miglino et al., 2014; Pérez-Sanagustín et al., 2014).

Experience Dimensions

While designing a new MRML experience, it is important to think about the contexts in which it will take place—may be more than one—and all the elements that will contribute to creating them. Strigel and Pouezevara (2012) researched how mobile learning can be used to increase numeracy skills. They classified 23 projects by using a model that they called 'Variations on Mobile Learning Configurations'. The same model has subsequently been used by Roberts et al. (2015) for research on nearly four thousand tenth grade students in South Africa. This model works in three dimensions with two polarities each: learning context (from formal to informal), kinetic context (from stationary to mobile), and collaborative context (from individual to collaborative).

The model has been adapted to MRML (Fig. 16), adding more specifications for each dimension. We have also chosen to rename 'contexts' in 'dimensions', because we see the sum of these three dimensions in the learning experience context. We will also use, alongside this model, the concept of setting; it could be indoor or outdoor, and in many different environments with different peculiarities. To design MRML experiences is crucial to address the issue of control over environmental

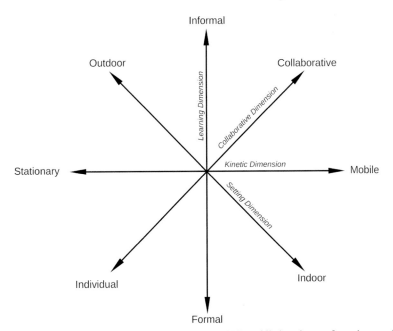

Fig. 16. An adaptation of the Strigel and Pouezevara's (2012) mobile learning configurations model.

variables (CoEV from now on). Each environment, each setting, has its own features that may change more or less frequently, even in the same setting, not to mention in the context of an experience that moves from one setting to another. Environmental variables like space, freedom of movement, freedom of sight, temperature, light, people, and sounds change continuously and need to be taken into due account during the designing of the experience. In fact, devices' sensors are not as flexible and intelligent as human senses, and it is very easy to interfere with or blind some of them. For instance, a GPS sensor cannot work indoors, a microphone cannot recognize a voice in a boisterous environment, and the camera cannot be useful in extreme brightness and contrast conditions.

Learning dimension:

Formal --------- Non-formal --------- Informal
+ CoEV -------------------------------- - CoEV

The learning dimension is defined by two polarities: formal and informal learning. Formal learning is thought with the purpose of teaching and takes place in a specific institution or organization. Informal learning is the result of everyday activities often unrelated to institutions, and it is mostly unintentional. Non-formal learning, as a hybrid between the two, takes place in planned activities which are not necessarily thought of as learning but have an informative or educational element in them (European Centre for the Development of Vocational Training, 2009). As already mentioned in this book, mobile devices, and smartphones in particular, are inherently ubiquitous devices that cross the boundaries between contexts. Designing an M-learning or an MRML experience, one must foresee if how and when this crossing will take place. The more informal the learning, the less CoEV is possible since it is more difficult to foresee where and how the mobile tool would be employed.

Kinetic dimension:

Stationary ----------------------------- Mobile
+ CoEV ----------------------------- - CoEV

The kinetic dimension represents the capability to experience learning on the move; in other words, the level of kinetic freedom offered. It mostly depends on the device used. For example, a smartphone would afford a mobile experience, while a desktop computer would tether the user to a stationary position. There are middle grounds; however, for instance, using a laptop would result in a portable experience, as using a powerful virtual reality kit would end in a luggable experience. The CoEV is inversely proportional to the level of mobility.

Collaborative dimension:

Individual ----------------------------- Collaborative
+ CoEV ----------------------------------- - CoEV

The collaborative dimension indicates the level of interaction with peers required by the proposed experience. It is directly related to the experience's goal because it requires methodological choices in the first stages of the experience design. For instance, a shared product or digital artifact would require collaboration, and following a trail in a scavenger hunt using a device in pairs would also need it. It can be a choice or a logistic necessity. The CoEV become more difficult the more collaborative is the activity, that is, because it needs to take into account the characteristics and the interaction of many users with the technology, instead of just of one of them.

Setting dimension:

Indoor --------------------------------- Outdoor
(Lab) -- (Classroom) -- (Museum) -- (Open-air)
+ CoEV -------------------------------- - CoEV

The setting is the space in which an experience takes place, with all its characteristic features. The most important of them for this classification is the amount and the type of space. Is it possible to employ M-learning in the classroom? It seems so. However, M-learning in a park would be a very different experience, not just for the amount of space, but also the open-air, and plants, the Sun and many other elements. From the technological perspective, some sensors essential in an outdoor setting would be useless indoor, such as the GPS. In contrast, classic AR with visual recognition has always been a challenge outdoors because of the ever-changing lights and contrasts. Needless to say that CoEV is optimal indoor in controlled environments and very difficult outdoor. Wider indoor environments, such as museums, are a middle ground where some environmental variable can be managed.

Every AR/MR experience is developed to take place in a specific setting and context. The combination of learning, kinetic, collaborative and setting dimensions lead to a situation where one can have more or less control over all the environmental variables (CoEV). This means that the teaching/learning methodology, the technology and, generally, the experience proposed changes considerably depending on where one is on the dimensions mentioned above. The idea is to strike a balance between dimensions and CoEV, always depending on the experience's goals.

Project-based Learning

Mixed reality mobile learning can eventually play a leading role in the future in planning constructivistic-type learning experiences, based on individual research and teamwork amongst students (FitzGerald et al., 2013). However, this can only occur in a broader framework that involves a specific pedagogical and methodological approach, designed to promote meaningful learning. This type of learning, according to the constructivist epistemology to which we refer, is only possible if there exist four characteristics: active participation in authentic learning tasks, reflection on and fine-tuning of the construction of personal and social meanings, teamwork, and willingness to learn (Howland et al., 2014).

An approach that contains all of these requirements and which is being, in recent years, increasingly applied effectively is project based learning (PBL) (Strobel and Barneveld, 2009). The main characteristics of the PBL (Thomas, 2000) are as follows: students should learn when they try to solve a real or a realistic problem; they should have effective control over the learning process; teachers should act as facilitators and tutors as regards research and reflection and students should, whenever possible, work in group or pairs. A successful PBL project should, ideally, have the following characteristics (Ertmer and Simons, 2005):

1. A project or realistic problem, tailor-made to suit the capabilities and interests of the students and which requires students to learn specific content or well-defined skills.

2. Groups of three or four students of different abilities and playing interdependent roles, with group incentives and personal responsibility to make each student grow.

3. A multi-faceted assessment system that allows, on the students' side, feedback and revision, objectives, and presentation of results which encourages participation and underlines the social value of the proposal.

4. Active involvement by a group of professional teachers, collaborating and reflecting on the PBL experience and sharing this with other colleagues as well as attending courses on inquiry-based methods of teaching.

This model, which focuses on meaningful content and other 21st century skills (such as problem-solving, critical thinking, communication, collaboration, interdisciplinary teaching, technological know-how, etc.) (Bell, 2010) can best be favored for some key points through integration with MRML:

- Content: contextual presentation of meaningful content.
- Assessment and feedback: immediate feedback and self-assessment.
- Collaboration: cooperative building-up of knowledge thanks to the fact that contents and findings are shared.
- Student Interest: the technological and fun element renders the project more engaging to students.
- Personal Control over one's Learning: personal interaction with the application allows autonomy and personalisation of the learning process and environment.
- Reflection and in-depth study: created and shared digital content can easily be the object of reflection and study thanks to their accessibility and the fact that each stage of the creation process is recorded and visible.

Nevertheless, the single main argument that would recommend the use of PBL with MRML is the MR's capability of surrogating real-life experiences with virtual ones, thus enabling experiential learning, which is at the core of PBL methodology (Efstratia, 2014).

Augmented Ekphrasis Learning

Ekphrasis was defined in the first century BC by Theon as an 'expository speech which vividly brings the subject before our eyes'. Remarkably, one of the earliest examples of ekphrasis we have in literature is the description by Homer in the *Iliad* of Achilles' shield. Before the duel with Hector, Homer describes Hephaestus's shield forged for Achilles in every detail of its mighty appearance and spectacular decoration. This created such a vivid image of the mythological object that it moved artists to depict it in paintings (e.g., Angelo Monticelli, from *Le Costume Ancienou Moderne*, c. 1820; Kathleen Vail) and even to forge it (W.H. Auden; The King of Hanover's Silver-Gilt Shield of Achilles, Philip Rundell for Rundell, Bridge and Rundell, London, 1823, John Flaxman's design, modeled with scenes from the eighteenth-century book of the Iliad) turning a shield that once was the fantasy of a single man into a real object. It happened thanks to the externalization and projection in the reader's mind achieved by the author by means of ekphrasis. Ekphrasis worked in ancient times as a 'backup copy' of important works of art and monuments.

A more modern definition of ekphrasis is 'verbal representation of visual representation' (Heffernan, 1991: 299), while a contemporary, radical one is 'representation in one medium of a real or fictitious text composed in another medium' (Bruhn, 1999: 296). The former definition can well fit with the classic use of ekphrasis widespread in the eighteenth and nineteenth centuries.[49] It is thanks to ekphrastic texts, and accounts of travels such as the Grand Tour, and visitors, that several eighteenth and nineteenth-century landscape gardens have been designed, initially, and then restored in recent times. In MRML experiences, there should be the opportunity of employing what one can call 'augmented ekphrasis', which is the process of representing verbally, textually, or graphically what he sees through the MR device. Such an account is a synthesis of information from the real and the virtual environments, together with information internalized by the student, in the same verbal representation.

The process of externalisation and sharing through ekphrasis should activate the higher-level cognitive processes such as analysing, evaluating and creating, and filling cognitive dissonance (Anderson et al., 2001).

These should be all conditions conducive to meaningful learning. In two different studies (one at Hestercombe, UK and one in Verona, Italy), this strategy has been used during and after visits to cultural heritage using MRML devices. During the augmented visit, students were encouraged to explain what they see, to ask questions, and to give answers. Fulfilling these obligations requires an ekphrastic process not only of the view that the students have in front of them, but of the whole augmented reality with its layers of information, imagery, and models as part of the process of externalisation expected by the Four-X methodology. In both studies, there was a second and a third moment of ekphrasis when students described what they have seen in the visit and represented it through a drawing. The results were encouraging (Agostini and Petrucco, 2020, 2018; Petrucco and Agostini, 2016).

Design-based Learning and Discovery

There is a variety of Project Based Learning where the projects consist of designing and create something. This is called Design-Based Learning (DBL). Its roots are in constructionism (Papert, 1982) and the idea of learning-by-making, that is to create an artefact—tangible or not—as the objective of the activity (Cook, 2010). Often, mixed reality is being used by educators and teachers to create readymade

---

[49] Famous amongst them in England is the '*Ode on a Grecian Urn*' by John Keats, written in 1819.

learning experiences for pupils. Used as it is, passively, it can help remembering, comprehension and interpretation, but it could be a missed opportunity to activate higher thinking processes such as assessment and development. That is why one could consider proposing the creation of a MR experience as the objective of the Design-learning activity.

When one challenge students' expertise, the latter tend to get more involved. Thoughtful discussion, problem-solving, theorising, and drawing hypotheses are all tools that can help foster intelligence. Fully engaged students actively create, improve, and determine their knowledge, rather than passively absorbing it (Van Haren, 2010). Also, DBL is inherently associated with teamwork and collaboration between peers, thus enabling peer-tutoring, development of collaborative skills and ideas contamination. In most cases, employing a DBL methodology with MR technologies has been found to increase students' learning outcomes (Bower et al., 2014; Doppelt et al., 2008).

There are software and platforms which are precisely thought to enable MR DBL starting from K12 age group. They are rooted in Papert's constructivist idea, which gave birth to the Logo programming language and Turtle Graphic,[50] and of which the most up-to-date champions are Scratch 3.0 (https://scratch.mit.edu/) and MIT AppInventor (https://appinventor.mit.edu/). The common characteristic of those platforms is the use of a block-based visual programming language which make coding accessible even to primary school students. CoSpaces Edu and Metaverse Studio are among the most successful MR platforms for DBL. The first is based on a 3D engine and makes possible to create 3D environments, games, storytelling, and simulations. Each 3D object can be programmed thanks to visual blocks. The latter is mostly based on programmable augmented reality with a specific attention to interactive storytelling and scavenger hunts used on mobile devices. It can be programmed using graphical language.

Design-based learning is potent when it uses a virtual environment as the synthesis of many sources of information. For example, an educational activity could be to render a particular environment how it was in the past, or how it will be in the future, on the basis of different sources such as newspapers, paintings, photos, written accounts, trends and statistics. While creating such mixed reality environments, one will realise that the virtual recreations are not just simplifications and synthesis of all the data at its disposal. In fact, one will grow aware that this process sheds light on details that would have otherwise been invisible or overseen. One could say that making the virtual, the real is discovered.

Bricolage: Another Key Concept for Design-Based Learning

Bricoler is a French verb that indicates that activity of manual labour made at home. It can be done as a distraction, a hobby, or to save the money of a professional worker. While in a professional context the results are often seen as sloppy works, it is not necessarily so. Mounier in his *Traité du caractère* praised the *bricolage*

---

[50] Logo is an easy-to-use programming language created in 1967 by Wally Feurzeig, Cynthia Solomon, and Seymour Papert, explicitly thought for children. The Turtle Graphic was added by Papert and it is a triangular cursor on the screen, capable of drawing vectorial graphic.

attitude in 1946 as revealing an '*aptitude for games, the resourcefulness, the ability to get out of complex difficulties or to take advantage of means of fortune, the ability to make plans, sometimes the taste to manufacture, rearrange*' (Mounier, 1946: 640). Subsequently, Claude Lévi-Strauss (1966) elaborated the concept of the *bricoleur* as the '*savage mind*' who uses pre-existing things, '*the means at hands*', in new ways, in contrast with the engineer, the '*scientific mind*' who designs and create from scratch new tools and systems. In 1966, Jaques Derrida (1993) criticised this idea, which would make the *bricoleur*'s divergent thinking something inferior to the engineer's scientific thinking. He maintained that, in the first place, it is not possible to be the '*absolute origin of his own discourse*', more, '*the engineer is a myth produced by the bricoleur*' (Derrida, 1993: 6). In this perspective, the *bricoleur* just wants to be effective and have a job done. He or she has no particular interest in the tidiness or stability of a tool or a system (Mambrol, 2016). It is the same approach we can see nowadays looking at the Internet. Through tools like 'how-to' and 'DIY (Do It Yourself)' webpages and videos,[51] *bricolage* knows an unprecedented success. There are three main features of the bricolage process which recommends it for DBL:

- The bricoleur has an initial idea when he/she starts his work but is capable of reaching a compromise with the real context and what is actually needed between the materials at hand and the effectiveness of the solution. The bricoleur is a problem solver, or, at least, a problem fixer.

- Has a variety of tools, which are not project-specific, but ready to be used if needed just when the idea comes to bricoleur's mind. They serve both for construction and destruction. His/her tools and knowledge might not always be the perfect fitting for the situation, but he/she will adapt them to get the job done. He/she is able to find new relations and configurations between the available materials and items.

- The bricoleur relies on his experience continuously and looks backwards to decide what could be done for the present situation, how to use apparently incompatible materials, and how to create order and give sense to the situations in front of him/her. Thus, it is crucial to appreciate that the bricoleur continues to gather knowledge, skills and competences and is far from being naïve in front of situations (Blankenship, 2020; Vallgårda and Fernaeus, 2015).

Finally, thanks to the implementation of a bricolage approach in DBL practice, new perceptions of the potential of technologies could arise, together with new ways of usage, and new content expressions. In this sense, bricolage is not only suited as a design activity, but also as a way to enhance creativity, and create new knowledge thanks to co-constructive activities. The same kind of approach can be used to create MR experiences with DBL tools, such as CoSpaces Edu, Metaverse Studio, and AppInventor. In fact, to test technologies, solve problems, and finalize

---

[51] Youtube (youtube.com) is the most famous example of this tendency. It features countless how-to channels that cover every possible subject (engines, computers, electronics, plumbing, woodwork, medicine, etc.). Another very well-known website is IFixIt (ifixit.com) which explains how to fix more than ten thousand devices from more than one hundred thousand issues. The social network TikTok (https://www.tiktok.com) has been the last to follow this lead.

projects, one may need a way to prototype an app rapidly and easily. This is often the default requirements for using technology in schools because of constraints in schools' programmes and deadlines.

Bring Your Own Device

Bring your own device (BYOD) is a well-tested approach that has been able to gather consensus in its educational applications (Afreen, 2014; Song, 2014). As the name suggests, it consists of relying on students' personal devices instead of distributing institutional ones for learning and teaching purposes. Students should bring their own devices into the formal educational context. It may be adopted to avoid all the expenses involved in buying a set of devices to be used at school or in other educational institution. Its main advantage is nonetheless methodological and not an economical one. In fact, the BYOD approach enhances seamless learning possibilities, allowing students to continue their learning processes at home with the same device, in the same virtual environment, and precisely at the point where they left it. They are also facilitated in bringing back to school their works, materials, and projects started or improved at home.

BYOD also favors motivation, responsibility and, more importantly, the adaptive process with the software tools and contents that the teacher proposes in the classroom, thanks to the use of a well-known hardware device that is also emotionally linked with the student. The learning is, therefore, more personalized. Those affordances should improve the learning process and enhance both the quantity and quality of students' work with the proposed tools and tasks (Afreen, 2014; Song, 2014). The reverse of the medal are the rules that a teacher has to put in place to manage all these personal devices working in the classroom. Students could be tempted to use them for other purposes than learning. Often, these devices have messaging apps and free access to the Internet, which could act as distractors. Luckily, nowadays, one has ample choices when it comes to applying restrictions to mobile devices and tablet. Parent's and teacher's control can be granular, and the school should configure its Internet connection to allow only permitted resources and websites (Cheng et al., 2016; Eslahi et al., 2014; Svensk, 2013).

If that can work neatly with M-learning, to employ BYOD in MRML is not easy. The problem is that MRML needs very specific sensors to be available in the device, such as gyroscope and GPS. Also, the interface, the screen size, and brightness, the resolution of the device camera, the quantity of RAM and of storage, the OS, and thus the compatibility with specific MR frameworks are all crucial factors in MRML experiences. One cannot depend on the fact that all students would have compatible devices. On the contrary, experience teaches that many students would have issues or find it impossible to use MRML software. With the democratization of MR technologies, these problems are not as blocking as they were only four years ago, but are nonetheless taken into account when designing a MRML experience.

Immersive Learning

In the MR field, immersion is the sensation of being physically present in virtual reality. The immersion can be more or less complete, depending on the number of human senses involved and their digital reproduction quality. A virtual reality that is hardly distinguishable from the actual experience of the 'real' takes the name of 'simulated reality'. Simulated realities are the next step of immersion; still, the immersion is not limited to senses' involvement in a computer-based virtual reality. The engagement, or the sense of presence, depends on other factors as well, and VR headsets and computer technology are not the only way to achieve immersion. One can create artificial realities in virtual spaces as well as artificial realities in real spaces to deliver the same or a better level of immersion, albeit at a greater cost. For instance, speaking of visual arts, 18th-century English landscape gardens were created after paintings and descriptions of landscapes from the Grand Tour, capriccios, italianates, and so forth. The Georgian landscape garden itself was an artificial reality made to allow a person to become immersed in these kinds of paintings and atmospheres, into an English Arcadia, with the addition of some exotic oriental elements.

The panorama, patented by the Scottish portraitist Robert Barker, in 1787, is another example of immersion in real space. It is a technique very similar to the one in use for virtual panoramas. The image is displayed in a 360-degree view on a circular canvas that surrounds the viewer. As the modern panoramic pictures require a virtual spherical space on which to be located or, sometimes, a physical semi-cubic or toric space in the case of virtual caves,[52] the panorama requires a circular building made specifically for it. Spectators need to be on a central platform half the height of canvas; an object could pop-out from the canvas to provide an immersive foreground. The light must be provided from above and concealed at the same time to merge with the image by means of a canopy seamlessly.

The pantoscope, known in Italy as *mondo nuovo* (Italian for 'new world'), in England as a peep box, or raree-show, and in Germany as *Guckkasten*, was an instrument known since the 15th century but mainly used in the 17th, 18th and 19th centuries. Leon Battista Alberti created the earliest specimens in the 1430s. Generally, it consisted of a box with one or several holes, with or without lenses, allowing one to look inside. Inside, it was possible to see drawings on paper, usually landscapes with monuments or large public events, with effects of transparency, often a night-day effect, and sometimes, animated figures, such as little puppets. The light was provided from one or more candles, the brightness of which was managed by an ingenious aperture system. However, size and weight apart, it was somewhat similar to a modern headset (akin to Google Cardboard): the observers looked with one or both eyes in a dark box to see a luminous screen and immerse themselves in a scene.

Finally, technologies and instruments are not the only ways to achieve immersion. Imagination, our natural, non-technological device with which we are all endowed, can also be used to attain the same goal. The technique, in this case, resides in the correct ways of storytelling and engagement that activate it. Good stories, books, plays, films, and music are capable of immersion, thanks to their capacity to activate the imagination and, at the same time, focus the attention of the audience while excluding most of the surrounding environment. In this respect, they are lesser forms

---

[52] Virtual Caves are rooms where the virtual reality is projected on the walls, usually, on four sides of a cube. Thanks to special glasses, one is able to move in the world and interact with 3D objects.

of hypnosis.[53] Finally, dreams and dream-like situations, such as daydreams, are other situations where the imagination attains immersion without external agencies' help.

All the above-mentioned immersive tools and situations have great educational potential. The immersion is one of the real super powers of the technologies in the virtuality continuum. In particular, mixed and virtual reality are capable of surrogating experiences due to their immersive affordances: the ability to plunge users into a virtual experience with some of their senses. Most of the experiential learning qualities are, therefore, also immersive learning ones, with some advantages and some disadvantages. The main disadvantage is that immersive learning, most of the times, requires a device, like a smartphone or a headset, as technological mediator. Many researches show that this technology is still not perfect and far from transparent. In fact, students or users that have too much interest in the device and the technology, tend to be distracted by the device and its collateral functions, and, consequently, lose concentration during the visit. Other issues range from ergonomics to user interface design, from software glitches to motion sickness. Because of these factors, users may get frustrated and the learning process hindered (Agostini and Petrucco, 2018; Schott and Marshall, 2021).

On the other hand, immersive learning has the advantage of seamless augmentation with contextual information, comprehensive vision, sense of place, safe environment, virtually infinite repetitions of the experience by just 'starting it again', and all the unique features already expressed as general for MRML. It encourages a transition towards a student-centered approach where students are more in charge of their learning process. In addition, there is the appeal and the engagement factor on users, typical of new, game-like, technologies.

Imagination Importance in Learning and Mixed Reality

'I am enough of an artist to draw freely upon my imagination. Imagination is more important than knowledge. For knowledge is limited, whereas imagination encircles the world' said Albert Einstein, as quoted in *What Life Means to Einstein: An Interview*, by George Sylvester Viereck' in *The Saturday Evening Post*, the 26 October 1929 (Viereck, 1929: 117).

In contrast to the lesser animals of this world, humans are endowed with a powerful imagination. Although recent studies suggest that chimpanzees and gorillas can pretend that an object is something different and that rats can try to calculate how to get a reward on the basis of previous experiences (Ólafsdóttir et al., 2015), the animal kingdom comes nowhere near the complexity of human imagination. Humans can simulate extremely complex scenes and use the imagination to solve problems, plan, invent, and understand, starting from the basis of their experience and knowledge but going far beyond them. When we see a phenomenon, our imagination is immediately at work to interpret and explain it, as well as to consider other comparable phenomena (Mithen, 2001; Riegler, 2001). Without that ability, research and knowledge building would not be possible. This is the sense of the Einstein's statement quoted above. An

---

[53] Hypnosis being 'a state of focused attention' reduced peripheral awareness, and better capacity to respond to suggestion (Elkins et al., 2015).

excellent example for supporting this hypothesis is science fiction, which emerged in the 17th and 18th centuries only to go mainstream in the 19th and 20th centuries.

Through this genre of literature, various writers, scholars, philosophers, and scientists have the opportunity to use their imaginations to create, develop, spread their ideas, and speculate about the future of society and technology, as well as about many things yet to be discovered, such as the existence of other worlds, alien civilizations, and celestial bodies.[54] It might also create interest in specific scientific development and influence people's expectations on it (Kriz et al., 2010; Lee, 2019; Raven, 2017). Science fiction literature is one example of the use of imagination to create alternative worlds and realities that we could call 'fantasy'. Fantasy is a continuous work of imagination to create alternative realities was at the centre of English enlightenment taste. The debate about alternative realities was continued in the 20th century by J.R.R. Tolkien, who, in his essay *On Faery-stories* (1947), called them 'secondary worlds' and 'sub-creation'—the real world being the primary 'creation'. Through these stories, humankind has two powers. The first is that of the 'sub-creator', who can make visions of fantasy effective by the exercise of will. The second one is the 'escape', which creates the possibility of escaping from reality and finding refuge in fantasy.

It is not by chance that these powers are at the core of the modern IT and entertainment industry. They have been recently enhanced by new technologies that provide the ability to actually design and edit 'secondary worlds' or, as we call them now, 'virtual worlds'. Software programs, such as Second Life,[55] OpenSimulator, and Minecraft have democratized sub-creation, allowing people with reasonable computer expertise to create virtual worlds from scratch easily: shaping sky, terrain, and environment or deciding on vegetation, buildings, objects, and inhabitants. Once created, a virtual world can be open to other people; it can be shared. People can enter it, meet other visitors, share experiences, and contribute to the world. This technology's capabilities have quickly been recognized, and these worlds have been used for entertainment, training, and educational purposes. It is interesting how these virtual, secondary worlds change the primary reality and peoples' behavior within it. There are a huge number of experiences, for example, in the field of safety at the workplace and training or the use of machines, not to mention driving and flight simulators. Several skills, trained in the secondary world, translate with some accommodation into skills in the primary world (skill transfer process). Of course, the most effective method is to blend virtual simulation with real-world training (Beckem and Watkins, 2012; Korteling et al., 2017; Sitzmann, 2011).

---

[54] *See* Francis Godwin's *The Man in the Moone* (1638), Margaret Cavendish's *The Blazing World* (1666), Bernard le Bovier de Fontenelle's *Conversations on the Plurality of Worlds* (1686), Samuel Madden's *Memoirs of the Twentieth Century* (1733), Voltaire's *Micromégas* (1752), Louis-Sébastien Mercier's *The Year 2440* (1771), and all Jules Verne's novels (starting from 1851 with *A Voyage in a Balloon*).

[55] Second Life, created in 2003, is one of the most famous multi-user virtual environment (MUVE) where people can run a parallel existence. They have jobs, earn money, have friends, and go 'out' for cultural activities. Real companies and industries have their shops and their representatives in the Second Live World. On the other hand, Minecraft is a so-called 'Sandbox Game', which can also be multi-user, where one can shape the world freely, not having any particular goal. It has widely adopted from education institution for the ease of use and its capabilities.

Imagination lets us swiftly adapt to any scenario, when necessary, augmenting and accommodating our perceptions accordingly.[56] In fact, this is what distinguishes reality from actuality, for the first is always a mainly subjective experience mediated by perceptions, while the second is the physical event as a camera might record it (Derrida, 2003). Drawing a parallel with the reality-virtuality continuum, here is proposed an imagination continuum where, on one pole, one finds 'actuality' and, on the other, 'fantasy'. Perception would, in that case, take the place of an actuality with a mild amount of imagination in it, resulting in an augmented—or interpreted— actuality (Fig. 17). This might prove handy for classifying activities and experiences, which imply the activation of imaginative processes.

The power to share imagination and fantasy is today greater than ever due to MR apps and headsets that support these technologies with their ubiquitous character.

Fig. 17. Imagination continuum.

Augmented Storytelling

This chapter has already highlighted the importance of elements, such as experience, ekphrasis, immersion, and imagination. Storytelling is a single key approach that opens preferential ways to all these elements. Plus, it is ideal to be embedded in DBL contexts. It can be classical, oral or written; it can be digital and multimedia-based; or it can be augmented, based on mixed reality and mobile technologies. Classical storytelling has been employed from the beginning of human civilization to teach and pass on information on the world, human nature, and culture (Coulter et al., 2007). From its very beginning, the oral story has been accompanied by some sort of media backup. In fact, it is possible that cavemen used a mixture of oral and drawn expressions to communicate stories: technically, a cross-modality information transfer (Nóbrega et al., 2018). The human brain has a 'storytelling mode' that allows retaining much information from a story from other kinds of texts or expressions. This seems to be possible thanks to the integrated employment of different intelligence dimensions (Gardner, 2000).

---

[56] There is a joke in Italy that sometimes adults play on children, or even among themselves. The game starts with the adult pretending he or she has a rubber band in hands, moving them as if he or she actually had one and encouraging the other person to do the same or to follow closely his or her movements. Suddenly, he or she mimes the movement of aiming and snapping the band against the other person. If not already aware of this joke, the other person will close the eyes and protect the face before realizing that there is no rubber band to be thrown. This is an example of how our imagination can augment our perception of reality. This unconscious mechanism is observable in many situations, like enjoying a mime artist show, but it is a crucial device for the survival of humans as their brains are continuously trying to simulate, predict, and interpret what is going to happen.

In the centuries, storytelling has been done with the help of pictures, capable of 'speaking' and teaching, and to be 'read', even for illiterates. This is the case, for instance, of Christian biblical and evangelical episodes. In history, they have been 'read' by the people mostly through pictures in the churches than on written bibles (Kielian, 2019; Saint-Martin, 2000). Therefore, digital storytelling was a natural evolution with up-to-date media support, namely multi-media support, capable of increasing the range of modalities, the engagement with the content, and the interaction with the story. Digital storytelling has been vastly employed, tested, and assessed in the educational field and always been regarded as one of the critical approaches for the developing of 21st-century skills (Lambert and Hessler, 2018; Petrucco, 2009; Robin, 2008, 2006).

Augmented storytelling is a subgenre of digital storytelling that relies on the capabilities of augmented and mixed reality technologies to tell the story and to make it more engaging and interactive. It is increasingly used in journalism (Pavlik and Bridges, 2013) and education. Concerning educational experiences, there are several in museums (Bimber et al., 2003; Katifori et al., 2014; Recupero et al., 2019), in schools (Berreth et al., 2020; Billinghurst and Duenser, 2012; Juan et al., 2008; Macauda, 2018; Yilmaz and Goktas, 2017) and in outdoor environments (Dow et al., 2005; Guimarães et al., 2015; Liestøl, 2014; Petrucco and Agostini, 2016).

In museums, augmented storytelling is appreciated for its ability to immerse the visitor in the story, showing ancient environments, items, and monuments as it were, besides allowing to visualize their changes in the centuries. Museums make use of both fixed installations and mobile devices (BYOD or institutional) with MR software. Lately, the latter tendency is preferred due to the enrichment it represents for the visit experience. Due to mobile augmented storytelling in museums, the visit and the trails are personalized, adaptive, interactive, and engaging. This is due to experimentations, such as the CHESS (cultural heritage experiences through socio-personal interactions and storytelling) project (Katifori et al., 2014), and the 'Ara as it was' project at the Ara Pacis monument in Rome (Recupero et al., 2019).

As for the augmented storytelling studies at school, they are mainly based on smartphones and tablets, and occasionally on laptops or head-mounted displays. They prove that, employed in the story creation process, it improves narrative skill, length, and creativity of stories, if compared with traditional storytelling media (Yilmaz and Goktas, 2017). Augmented storytelling enhances creativity and out-of-the-box thinking (Berreth et al., 2020), allowing students to find interest in topics that they could not experience in the real world. Multimodality and interactivity are keys to engagement and long-term memory integration processes. When using such technology, students work harder to understand the content and its meaning. Additionally, students that usually find traditional text-based learning difficult to retain seem to especially benefit from this new media (Billinghurst and Duenser, 2012). This fact opens the perspective of augmented storytelling as an inclusive approach to teaching and learning.

In outdoor environments, as for MR technology in general, it is more challenging to use augmented storytelling due to the limited control over environmental variables. Nevertheless, there are brilliant examples of its employment which have found a new, explorative way of doing storytelling. Software, such as FreshAiR (maybe discontinued), ARIS, Huntzz, and Metaverse Studio—but also MIT AppInventor for those good at block programming—facilitate true mobile and augmented storytelling experiences to take place. They feature maps, GPS tracking, Bluetooth beacons' support, 360-degrees imagery and videos, images and 3D models overlay, items inventory, characters personalization, and various triggers. Some of these apps have collaborative modalities, too (Crawford Pokress, 2015; Dunleavy, 2014; Holden, 2015), which allow the teacher to create student-centered activities, employ peer-tutoring, and single student teacher guidance.

Compared with standard digital storytelling, augmented storytelling adds location and context awareness and opens new possibilities, such as collaborative inquiry approach (Markouzis and Fessakis, 2015) and real-world learning experiences. Thanks to device sensors or other measurement instruments, students can draw connections between what they are learning and their environment. The augmented storytelling platforms and apps also work as a scaffolding for their mobile, seamless learning experiences, helping them complete tasks, and attribute meanings to the experience (Kamarainen et al., 2013).

A last consideration is about the research made during two school visits to the Landscape Garden of Hestercombe, the same experimentation reported in Agostini and Petrucco (2018). Analyzing students' reports and drawings on the visit, it is evident that the stopovers where augmented storytelling was employed were the most appreciated and the most remembered of the whole visit. Photos and MR are compelling together, but they would be hugely diminished without an underlying story of the people involved, which is one of the reasons it was decided to keep the human guide in the experiences instead of trying to replace him/her with technology. Storytelling is not only what keeps a series of photos together, but it is also the reason why they have been taken in the first place.

The Mirage of the 'New'

Mirages are optical effects caused by the refraction of light in a medium, usually the air, of which the density varies, usually because of the different temperature. It would be fascinating to consider them as a natural form of virtuality, of an optically modified perception of the landscape. It is not known how and when people took this idea from the natural world to use it in their crafts, but it is known that these natural virtualities have profoundly influenced cultures in stories and legends and arts.

To distinguish 'perceived reality' from virtuality is never easy. Oases in the desert, flying ships, flying islands, immense coasts and islands, impossible sunrises, and even mysterious fading cities have been only a few of the examples of these phenomena, incorporated in the literature throughout the ages. The mirage optical effect is not dissimilar to the new augmented and mixed reality technology. If the word 'new' is suitable for the devices in use nowadays, it is not applicable to the idea of augmented and mixed reality. The final effect recreated by means of these devices has been pursued for centuries but is not yet as perfected as the imagination would like. Even the newest incarnation of these ideas, the one that gave the name to them, is 26-years old.

In 1992, Thomas P. Caudell and David W. Mizell, researchers at Boeing Computer Services, Research, and Technology, created a headset with an integrated heads-up display (HUD) and sensors to help engineers in repairing the Boing 747 aeroplane. In 1975, 17 years earlier, Virginians David A. Bosserman and Charles F. Freeman patented a device called a 'toric reflector'. It consisted of a headset that puts a semi-transparent screen in front of one eye, projecting information as distant virtual images which are superimposed on the real world (U.S. Patent No. 4,026,641). This technology seems to be at the base of products, such as modern HUDs and even Google Glasses. Looking back a little further, most rangefinders and viewfinders on consumer and professional cameras, since the second decade of the 20th century, sport a sort of augmented reality system that allows the photographer to better compose, focus, and expose a picture. They superimpose an informative layer on the view. One can find an ancestor of it in the 'drawing frame', in use since the 17th century, which helped painters in framing the landscape and that, with the 'grid' gadget worked as a guide to the eye to maintain the right proportions and distances while drawing (Martinet and Chatel, 2001: 61–62).

In the 18th century, the 'Claude Glass',[57] also known as 'black mirror', was a convex hand mirror tinged with colors, which were usually dark. Tourists and painters used it for its effect of framing the landscape, softening the lines, and emphasizing tonal variations. Some of its variations included having a transparent colored glass instead of a mirror (Kinsley, 2008). It is captivating to notice how both the Claude Glass and smartphones bring the traveller to forego the real, natural view of the landscape for a mirage of a mediated version of it rendered by a device that changes, improves, or re-interprets it. These technologies have always been controversial. Thinking about how Instagram and other apps work, allowing anyone to use 'filters', one can think that some critic may refer to them as 'one of the most pestilent inventions for falsifying nature and degrading art which was ever put into an artist's hand', except that it is a John Ruskin quote against the 'black convex mirror', which was so effectively promoted by Thomas Gray, Thomas West, and William Gilpin. This seems to suggest that not only the ideas, but even the fears and the criticisms, are legacy of the past[58] (Willim, 2013).

Humphry Repton, one of the prominent English landscape designers, active in the second half of the 18th and beginning of the 19th century, had his original idea about the augmented representation of reality. In his 'red books', which he often made when he was asked to landscape a garden, he drew detailed maps of the estate but also views of the garden before and after the proposed modifications. The technique he used consisted of drawing a page with the new landscape on it and covering part of it with paper flaps on which he drew the existing landscape. The result was a transition effect, leading to an actual *dis*-covery of the imagined landscape. The transition mechanism and the type of content are very similar to the

---

[57] It was named after Claude Lorraine, 17th-century landscape painter as it was supposed to help the painters to achieve similar results.

[58] For current criticism example, read 'Instagram is Debasing Real Photography', Kate Bevan https://www.theguardian.com/technology/2012/jul/19/instagram-debasing-real-photography.

ones used in many AR apps, while the medium differs. Also, all these tools were mobile or, at least, portable.

This brief overview demonstrates how, deepening the research, one began to be aware of these parallel ideas, techniques, and ultimately, of tastes between the contemporary new media technologies and applications and the 18th century. These ideas seem to have been propagated up to the present day. This can be seen as an expansion of what Manovich (2002) called 'the fractal structure of new medias'. Like a fractal, a media object has the same similar structure on different scales[59] and, as just demonstrated, also on different time frames throughout the centuries. It is as if there was a recursive self-similarity in these ideas and technologies that should enable one to interpret better and organize reality.[60] This seems to be confirmed by historical studies that found that self-similarities are 'footprints' of iterative processes (Farmer et al., 2015). They tend to emerge in systems that are continuously transformed by recursive operations, meaning that the result of each prior transformation becomes the starting point of the subsequent one (Mandelbrot, 1982).

Nonetheless, this does not mean that a medium, or an idea of media, would be the same over and over. Variability and flexibility would be the keys due to new technology. There is an abundance of content and relative ease in its creation. The quantity of photos, videos, three-dimensional models, and general information at our disposal is massive; therefore, the focus of new media technology is to help in creating content, as well as in storing, organizing, and providing efficient access to it. The use of AR and MR technology goes in this direction.

Teachers and educational practitioners must be aware of this background of history, ideas, and materials that lies unseen behind the sparkling 'new' technologies that they use. It is important to know that there are other ways and other 'unplugged' methods used extensively in the past, which could serve as useful alternatives and inspiration nowadays. Finally, it is somehow encouraging to know that even criticisms of the adoption of such technologies have been present in the past, in almost identical terms to the ones used nowadays. Shall we be able to use MRML methodology and technology in such an effective way to address most of these concerns?

Final Thoughts

Observation of institutional and educational contexts highlights that the most practised teaching method is the 'traditional' one, i.e., lecturing. It has several limitations, particularly in the face of new educational challenges presented by new generations of students, as the growing phenomenon of learning disabilities. Such generations used to be presented as Digital Natives (Prensky, 2001) and as 'Millennials' and 'Generation Z' (Dimock, 2019) and their peculiarity are to have learning styles that are very different from former generations.

---

[59] Fractal mathematic is used nowadays in the recreation of virtual photorealistic landscapes. This technique has been tested since 1980 (Carpenter, 1980).

[60] The same fractal-like structure is inherent in theoretical tools: for example, in Activity Theory. AT systems, in fact, can be embedded as part of the other activity systems.

The reason is that every generation has a preferred type of medium with which it interacts since early childhood. If for the previous generation it was television, after 1982 it became computer and then the Internet (Dede, 2005; Oblinger, 2003). In the last decades, computers, the Internet, and mobile technologies have been increasingly pervasive in new generations' lives, thus modelling learning styles and cognitive patterns. Lately, concepts such as 'millennials' or 'digital natives' have been reconsidered by many researchers in order to clarify that they are not a 'generation with an innate knowledge of how to use new technologies' (Kirschner and De Bruyckere, 2017). Because of all this evidence, one may consider the possibility that new generations would need a different way of teaching. If, in school contexts, one continues to use a communication channel that is not students' preferred one, the communication will not be effective. There is also the possibility that they are listening to more than one channel as their preferred mode, and so, to utilize only one channel, say, the auditive one, could bring them to be distracted from what's happening on other channels.

The usual way of lecturing would have another disadvantage: it wants interactivity. New generations are used to a multi-modal interactive approach. Traditional lecture and writing should also be used; they are essential along with the capacity of focussing and increasing the attention span. The suggestion is that, probably, educators should use as the primary means of communication channels and modalities that are cognitively the pupils preferred—this, in order to facilitate the learning process. The learning contexts represent a second challenge. If, historically, schools, universities and libraries were the physical places were to access the knowledge, today that is no more the case. Information is available everywhere at any time thanks to the Internet, computers and mobile devices.

The kind of society in which we live is often called 'Information Society' (Castells, 2009). In this scenario, educational institutions need to provide students with the instruments to verify the information and to recognise authoritative, trustworthy sources. In fact, on the Internet, often, one would find both a point of view and its contrary.[61] To address this issue, researchers have come up with a set of skills needed from a person in the Twenty-first Century society, especially for young people that need to work and face very different professional requirements than in the past. They called them '21st Century Skills' (Trilling and Fadel, 2009). This situation has been remarked upon by institutions, and several frameworks have been created with the aim of spreading those skills. The European Union started drafting the 'Strategy of Lisbon' (in 2001) and several competence framework programs followed.

Having this as the background, it is crucial to think about what educational institutions can do about everything that students learn outside, in non-formal and informal contexts, especially when they learn things connected with school curricula. Teachers need to be aware of it and to valorize all that information. Students should be able to bring it back into the class and share it. It would be up to the teacher to

---

[61] Sometimes the information is wilfully made to be wrong, to misinform the Internet reader. In many countries, at this moment, it is ongoing a wide debate on the so called 'fake news' and several of the greatest hi-tech players are committed to oppose this phenomenon.

mediate, connect, and help to create shared meanings and, ultimately, authoritative knowledge. Informal and formal learning contexts should not be parallel, never-touching lines, but intertwined. Methodologies such Flipped Classroom (Fulton, 2012), WebQuest (Dodge, 1995), Project-based Learning (Bell, 2010) and Design Based Learning (Van Haren, 2010) are employed to answer these challenges.

The use of new mobile MR technologies as mediators of teaching and learning is a right way forward in this sense and should act as a bridge between different learning contexts.

References

Afreen, R. (2014). Bring your own device (BYOD) in higher education: opportunities and challenges. *International Journal of Emerging Trends & Technology in Computer Science (IJETTCS)*, 3(1): 4.

Agostini, D. and Petrucco, C. (2018). *Sperimentazione di una App di Realtà Aumentata per comunicare il Patrimonio Culturale: l'Hestercombe Gardens Augmented Visit*. pp. 2–9. In: *Progress to Work, Contesti, processi educativi e mediazioni tecnologiche, Proceedings della Multiconferenza EM&M ITALIA 2017*. Genova University Press, Genova.

Agostini, D. and Petrucco, C. (2020). Drawings as a tool for assessment of cultural heritage understanding: reports on e-learning. *Media and Education Meetings*, 8 (January): 128–33.

Anderson, L.W., Krathwohl, D.R., Airasian, P.W., Cruikshank, K.A., Mayer, R.E., Pintrich, P.R., Raths, J. and Wittrock, M.C. (eds.). (2001). A taxonomy for learning, teaching, and assessing. *A Revision of Bloom's Taxonomy of Educational Objectives*. Abridged edition, Longman.

Auld, G. and Johnson, N.F. (2015). Teaching the 'other': curriculum 'outcomes' and digital technology in the out-of-school lives of young people. pp. 163–81. In: Bulfin, S., Johnson, N.F. and Bigum, C. (eds.). *Critical Perspectives on Technology and Education. Palgrave Macmillan's Digital Education and Learning*. New York

Auvray, M. and Fuchs, P. (2007). *Perception, Immersion et Interactions Sensorimotrice en Environnement Virtuel*. *Intellectica*, 45(1): 23–35.

Barsom, E.Z., Graafland, M. and Schijven, M.P. (2016). Systematic review on the effectiveness of augmented reality applications in medical training. *Surgical Endoscopy*, 30(10): 4174–83.

Beckem, J.M. and Watkins, M. (2012). Bringing life to learning: immersive experiential learning simulations for online and blended courses. *Journal of Asynchronous Learning Networks*, 16(5): 61–70.

Bell, S. (2010). Project-based learning for the 21st century: skills for the future. *The Clearing House: A Journal of Educational Strategies, Issues and Ideas*, 83(2): 39–43.

Bergson, H. (1911). *Matter and Memory*. George Allen & Unwin Ltd., London.

Berreth, T., Polyak, E. and FitzGerald, P. (2020). Story-go-round: augmented reality storytelling in the multidisciplinary classroom. pp. 1–2. In: *ACM SIGGRAPH 2020 Educator's Forum*. Association for Computing Machinery, New York.

Billinghurst, M. and Duenser, A. (2012). Augmented reality in the classroom. *Computer*, 45(7): 56–63.

Bimber, O., Encarnação, L.M. and Schmalstieg, D. (2003). The virtual showcase as a new platform for augmented reality digital storytelling. pp. 87–95. In: *Proceedings of the Workshop on Virtual Environments EGVE '03*. Association for Computing Machinery, New York.

Blankenship, B. (2020). Bricolage and student learning. *Student Success*, 11(2): 122–26.

Bower, M., Howe, C., McCredie, N., Robinson, A. and Grover, D. (2014). Augmented reality in education—cases, places and potentials. *Educational Media International*, 51(1): 1–15.

Bruhn, S. (1999). Piano poems and orchestral recitations. Instrumental music interprets a literary text. pp. 277–99. In: Bernhart, W., Scher, S.P. and Wolf, W. (eds.). *Word and Music Studies Defining the Field: Proceedings of the First International Conference on Word and Music Studies at Graz*. Editions Rodopi, Amsterdam, Atlanta.

Bruner, J.S. (1977). *The Process of Education*. Harvard University Press.

Calvani, A. (2014). *Come fare una lezione efficace, Carrocci editore*. Roma.

Carpenter, L.C. (1980). Computer rendering of fractal curves and surfaces. pp. 109. In: *Proceedings of the 7th Annual Conference on Computer Graphics and Interactive Techniques SIGGRAPH '80*. Association for Computing Machinery, New York.

Castells, M. (2009). An introduction to the information age. pp. 152–64. *In*: Thornham, S., Bassett, C. and Marris, P. (eds.). *Media Studies: A Reader*. New York University Press.

Caudell, T. and Mizell, D.W. (1992). Augmented reality: an application of heads-up display technology to manual manufacturing processes. *In*: *The Proceedings of the Twenty-Fifth Hawaii International Conference on System Sciences*.

Chen, R. and Wang, X. (2008). An empirical study on tangible augmented reality learning space for design skill transfer. *Tsinghua Science and Technology*, 13(S1): 13–18.

Cheng, G., Guan, Y. and Chau, J. (2016). An empirical study towards understanding user acceptance of bring your own device (BYOD) in higher education. *Australasian Journal of Educational Technology*, 32(4).

Cook, J. (2010). Mobile phones as mediating tools within augmented contexts for development. Article, *International Journal of Mobile and Blended Learning (IJMBL)*, 1 July.

Coulter, C., Michael, C. and Poynor, L. (2007). Storytelling as pedagogy: an unexpected outcome of narrative inquiry. *Curriculum Inquiry*, 37(2): 103–22.

Crawford Pokress, S. (2015). MIT app inventor: democratising personal mobile computing. pp. 67–83. *In*: Holden, C., Dikkers, S., Martin, J., Litts, B. et al. (eds.). *Mobile Media Learning. Innovation and Inspiration*. ETC Press, Pittsburgh.

Crescenzio, F., De, M., Fantini, F., Persiani, L., Di Stefano, P., Azzari and Salti, S. (2011). Augmented reality for aircraft maintenance training and operations support. *IEEE Computer Graphics and Applications*, 31(1): 96–101.

Dede, C. (2005). Planning for neomillennial learning styles. *Educause Quarterly*, 28(1): 7–12.

Derrida, J. (1993). Structure, sign, and play in the discourse of the human sciences. *In*: Natoli, J. and Hutcheon, L. (eds.). *A Postmodern Reader*. SUNY Press.

Derrida, J. (2003). The deconstruction of actuality. An interview with Jacques Derrida. pp. 245–72. *In*: Culler, J.D. (ed.). *Deconstruction: Critical Concepts in Literary and Cultural Studies*. Taylor & Francis.

Dimock, M. (2019). Defining generations: where millennials end and generation Z begins. *Pew Research Center*, 17(1).

Dodge, B. (1995). WebQuests: a technique for internet-based learning. *Distance Educator*, 1(2): 10–13.

Doppelt, Y., Mehalik, M.M., Schunn, C.D., Silk, E. and Krysinski, D. (2008). Engagement and achievements: a case study of design-based learning in a science context. *Journal of Technology Education*, 19(2): 22–39.

Dow, S., Lee, J., Oezbek, C., MacIntyre, B., Bolter, J.D. and Gandy, M. (2005). Exploring spatial narratives and mixed reality experiences in Oakland cemetery. pp. 51–60. *In*: *The Proceedings of the 2005 ACM SIGCHI International Conference on Advances in Computer Entertainment Technology ACE '05*. Association for Computing Machinery, New York.

Dunleavy, M. (2014). Design principles for augmented reality learning. *TechTrends*, 58(1): 28–34.

Efstratia, D. (2014). Experiential education through project based learning. pp. 1256–60. *In*: *Procedia— Social and Behavioral Sciences 152 (October), ERPA International Congress on Education*, 6–8 June 2014, Istanbul, Turkey.

Elkins, G.R., Barabasz, A.F., Council, J.R. and Spiegel, D. (2015). Advancing research and practice: the revised APA division 30 definition of hypnosis. *American Journal of Clinical Hypnosis*, 57(4): 378–85.

Engeström, Y.R., Miettinen, R. and Punamäki, L. (eds.). (1999). Perspectives on activity theory. *Learning in Doing: Social, Cognitive and Computational Perspectives*. Cambridge University Press, Cambridge.

Ertmer, P. and Simons, K. (2005). Scaffolding teachers efforts to implement problem-based learning. *International Journal of Learning*, 12(January).

Eslahi, M., Naseri, M.V., Hashim, H., Tahir, N.M. and Saad, E.H.M. (2014). BYOD: current state and security challenges. pp. 189–92. *In*: *IEEE Symposium on Computer Applications and Industrial Electronics (ISCAIE)*.

European Centre for the Development of Vocational Training. (2009). *European Guidelines for Validating Non-formal and Informal Learning*. Information Series, Office for Official Publications of the European Union, Luxembourg.

Farmer, S., Henderson, J. and Robinson, P. (2015). Commentary traditions and the evolution of premodern religious and philosophical systems: a cross-cultural model. *Comparative Civilizations Review*, 72(72).

FitzGerald, E., Ferguson, R., Adams, A., Gaved, M., Mor, Y. and Thomas, R. (2013). Augmented reality and mobile learning: the state of the art. *International Journal of Mobile and Blended Learning*, 1 October 2013.

Fulton, K. (2012). Upside down and inside out: flip your classroom to improve student learning. *Learning & Leading with Technology*, 39(8): 12–17.

Gardner, H. (2000). *The Disciplined Mind*. Penguin Books, Harmondsworth.

Gavish, N., Gutiérrez, T., Webel, S., Rodríguez, J., Peveri, M., Bockholt, U. and Tecchia, F. (2015). Evaluating virtual reality and augmented reality training for industrial maintenance and assembly tasks. *Interactive Learning Environments*, 23(6): 778–98.

Guimarães, F., Figueiredo, M. and Rodrigues, J. (2015). Augmented reality and storytelling in heritage application in public gardens: Caloust Gulbenkian foundation garden. *Digital Heritage*, 1: 317–20.

Hattie, J. (2009). *Visible Learning: A Synthesis of Over 800 Meta-analyses Relating to Achievement*. Routledge, Milton Park.

Heffernan, J.A.W. (1991). Ekphrasis and representation. *New Literary History*, 22(2): 297–316.

Holden, C. (2015). ARIS: augmented reality for interactive storytelling. pp. 67–83. *In*: Holden, C., Dikkers, S., Martin, J., Litts, B. et al. (eds.). *Mobile Media Learning. Innovation and Inspiration*. ETC Press, Pittsburgh.

Howland, J.L., Jonassen, D.H. and Marra, R.M. (2014). *Meaningful Learning with Technology*. Pearson.

Hugues, O., Fuchs, P. and Nannipieri, O. (2011). New augmented reality taxonomy: technologies and features of augmented environment. pp. 47–63. *In*: Furht, B. (ed.). *Handbook of Augmented Reality*. Springer, New York.

Ibañez-Etxeberria, A., Gómez-Carrasco, C.J., Fontal, O. and García-Ceballos, S. (2020). Virtual environments and augmented reality applied to heritage education: an evaluative study. *Applied Sciences*, 10(7): 2352.

Juan, C., Canu, R. and Giménez, M. (2008). Augmented reality interactive storytelling systems using tangible cubes for edutainment. pp. 233–35. *In*: *Eighth IEEE International Conference on Advanced Learning Technologies*.

Kamarainen, A.M., Metcalf, S., Grotzer, T., Browne, A., Mazzuca, D., Tutwiler, M.S. and Dede, C. (2013). EcoMOBILE: integrating augmented reality and probeware with environmental education field trips. *Computers & Education*, 68(October): 545–56.

Katifori, A., Karvounis, M., Kourtis, V., Kyriakidi, M., Roussou, M., Tsangaris, M., Vayanou, M. et al. (2014). CHESS: personalized storytelling experiences in museum. pp. 232–35. *In*: Mitchell, A., Fernández-Vara, C. and Thue, D. (eds.). *Interactive Storytelling. Lecture Notes in Computer Science*. Springer, Cham.

Kielian, A. (2019). Education through art: the use of images in catholic religious education. pp. 335–44. *In*: Buchanan, M.T. and Gellel, A.M. (eds.). *Global Perspectives on Catholic Religious Education in Schools*; vol. II: *Learning and Leading in a Pluralist World*. Springer, Singapore.

Kinsley, Ë. (2008). *Women Writing the Home Tour, 1682–1812*. Routledge, London.

Kirschner, P.A., Sweller, J. and Clark, R.E. (2006). Why minimal guidance during instruction does not work: an analysis of the failure of constructivist, discovery, problem-based, experiential, and inquiry-based teaching. *Educational Psychologist*, 41(2): 75–86.

Kirschner, P.A. and De Bruyckere, P. (2017). The myths of the digital native and the multitasker. *Teaching and Teacher Education*, 67(October): 135–42.

Korteling, H.J.E., Helsdingen, A.S. and Sluimer, R.R. (2017). An empirical evaluation of transfer-of-training of two flight simulation games. *Simulation & Gaming*, 48(1): 8–35.

Kriz, S., Ferro, T.D., Damera, P. and Porter, J.R. (2010). Fictional robots as a data source in HRI research: exploring the link between science fiction and interactional expectations. pp. 458–63. *In*: *19th International Symposium in Robot and Human Interactive Communication*.

Kukulska-Hulme, A., Sharples, M., Milrad, M., Arnedillo-Sanchez, I. and Vavoula, G. (2009). Innovation in mobile learning: a European perspective. *International Journal of Mobile and Blended Learning (IJMBL)*, 1(1): 1335.

Lambert, J. and Hessler, B. (2018). *Digital Storytelling: Capturing Lives, Creating Community*. Routledge.

Land, S.M., Smith, B.K. and Zimmerman, H. (2013). Mobile technologies as mindtools for augmenting observations and reflections in everyday informal environments. pp. 214–28. *In*: *Learning, Problem Solving, and Mindtools: Essays in Honor of D.H. Jonassen*. Taylor and Francis.

Latour, B. (1999). *Pandora's Hope: Essays on the Reality of Science Studies*. Harvard university press.

Lave, J. and Wenger, E. (1991). *Situated Learning: Legitimate Peripheral Participation*. Cambridge University Press.

Lee, C.A.L. (2019). What science fiction can demonstrate about novelty in the context of discovery and scientific creativity. *Foundations of Science*, 24(4): 705–25.

Lévi-Strauss, C. (1966). The savage mind. *Nature of Human Society*. University of Chicago Press, Chicago.

Liestøl, G. (2014). Along the appian way. Storytelling and memory across time and space in mobile augmented reality. pp. 248–57. *In*: Ioannides, M., Magnenat-Thalmann, N., Fink, E., Žarnić, R., Yen, A.Y. and Quak, E. (eds.). *Digital Heritage, Progress in Cultural Heritage: Documentation, Preservation, and Protection, Lecture Notes in Computer Science*. Springer, Cham.

Linden, J., Braun, T., Rogers, Y., Oshodi, M., Spiers, A., McGoran, D., Cronin, R. and O' Dowd, P. (2012). Haptic lotus: a theatre experience for blind and sighted audiences. pp. 1471–72. *In*: *CHI '12 Extended Abstracts on Human Factors in Computing Systems*. Association for Computing Machinery CHI EA '12, New York.

Luckin, R. and Stanton Fraser, D. (2011). Limitless or pointless? An evaluation of augmented reality technology in the school and home. *International Journal of Technology Enhanced Learning*, 3(5): 510–24.

Macauda, A. (2018). Augmented reality environments for teaching innovation. *Research on Education and Media*, 10(2): 17–25.

Mambrol, N. (2016). Claude Levi Strauss' concept of bricolage. *Literary Theory and Criticism*, 21 March 2016.

Mandelbrot, B.B. (1982). *The Fractal Geometry of Nature*. W.H. Freeman, San Francisco.

Manovich, L. (2002). *The Language of New Media*. MIT Press, Leonardo, Cambridge.

Markouzis, D. and Fessakis, G. (2015). Interactive storytelling and mobile augmented reality applications for learning and entertainment: a rapid prototyping perspective. pp. 4–8. *In*: *International Conference on Interactive Mobile Communication Technologies and Learning*.

Martinet, M.M. and Chatel, L. (2001). *Jardin et Paysage en Grande-Bretagne au XVIIIe Siècle, Collection CNED-Didier Concours, Didier Érudition*. Paris.

Miglino, O., Di Ferdinando, A., Di Fuccio, R., Rega, A. and Ricci, C. (2014). Bridging digital and physical educational games using RFID/NFC technologies. *Journal of E-Learning and Knowledge Society*, 10(3).

Mithen, S.J. (2001). The evolution of imagination: an archaeological perspective. *Substance*, 30(1): 28–54.

Mounier, E. (1946). *Traité du caractère, Esprit*. Éditions du Seuil, Paris.

Munnerley, D., Bacon, M., Wilson, A., Steele, J., Hedberg, J. and FitzGerald, R. (2012). Confronting an augmented reality. *Research in Learning Technology*, 20(August).

Nóbrega, V.A., Miyagawa, S. and Lesure, C.L. (2018). Cross-modality information transfer: a hypothesis about the relationship among prehistoric cave paintings, symbolic thinking, and the emergence of language. *Frontiers*, February.

Oblinger, D. (2003). Boomers, Gen-Xers & Millennials. Understanding the new students. *Educause Review*, 500(4): 37–47.

Ólafsdóttir, H.F., Barry, C., Saleem, A.B., Hassabis, D. and Spiers, H.J. (2015). Hippocampal place cells construct reward-related sequences through unexplored space. *E-Life*, 4(June): e06063.

Papert, S. (1982). Mindstorms: children, computers, and powerful ideas. *Harvester Studies in Cognitive Science*. 14, Harvester Press, Brighton.

Pavlik, J.V. and Bridges, F. (2013). The emergence of augmented reality (AR) as a storytelling medium in journalism. *Journalism & Communication Monographs*, 15(1): 4–59.

Pérez-Sanagustín, M., Hernández-Leo, D., Santos, P., Delgado Kloos, C. and Blat, J. (2014). Augmenting reality and formality of informal and non-formal settings to enhance blended learning. *IEEE Transactions on Learning Technologies*, 7(2): 118–31.

Petrucco, C. (2009). *Apprendere con il digital storytelling*. *Italian Journal of Educational Technology*, 17(1): 4–4.

Petrucco, C. and Agostini, D. (2016). Teaching cultural heritage using mobile augmented reality. *Journal of E-Learning and Knowledge Society*, 12(3).

Prensky, M. (2001). Digital natives, digital immigrants, Part 1. *On the Horizon*, 9(5): 1–6.

Ranieri, M. and Pieri, M. (2015). Mobile learning. *Dimensioni teoriche, modelli didattici, scenari applicativi, Unicopli*. Milano.

Raven, P.G. (2017). Telling tomorrows: science fiction as an energy futures research tool. *Energy Research & Social Science, Narratives and Storytelling in Energy and Climate Change Research*, 31(September): 164–69.

Recupero, A., Talamo, A., Triberti, S. and Modesti, C. (2019). Bridging museum mission to visitors experience: activity, meanings, interactions, technology. *Frontiers in Psychology*, 10.

Riegler, A. (2001). The role of anticipation in cognition. *AIP Conference Proceedings*, 573(1): 534–41.

Roberts, N., Spencer-Smith, G., Vanska, R. and Eskelinen, S. (2015). From challenging assumptions to measuring effect: researching the nokia mobile mathematics service in South Africa. *South African Journal of Education*, 35(2): 1045–1045.

Robin, B.R. (2006). The educational uses of digital storytelling. pp. 709–16. *In: Association for the Advancement of Computing in Education (AACE)*.

Robin, B.R. (2008). Digital storytelling: a powerful technology tool for the 21st century classroom. *Theory into Practice*, 47(3): 220–28.

Rushby, N. and Seabrook, J. (2008). Understanding the past—illuminating the future. *British Journal of Educational Technology*, 39(2): 198–233.

Saint-Martin, I. (2000). Catechism in images. *Archives de Sciences Sociales Des Religions*, 111(3): 7–7.

Schott, C. and Marshall, S. (2021). Virtual reality for experiential education: a user experience exploration. *Australasian Journal of Educational Technology*, 37(1): 96–110.

Sharples, Mike (ed.). (2006). *Big Issues in Mobile Learning: Report of a Workshop by the Kaleidoscope Network of Excellence Mobile Learning Initiative*. University of Nottingham, Nottingham.

Sitzmann, T. (2011). A meta-analytic examination of the instructional effectiveness of computer-based simulation games. *Personnel Psychology*, 64(2): 489–528.

Sırakaya, M. and Sırakaya, D.A. (2020). Augmented reality in STEM education: a systematic review. *Interactive Learning Environments*, 0(0): 1–14.

Song, Y. (2014). 'Bring your own device (BYOD)' for seamless science inquiry in a primary school. *Computers & Education*, 74(May): 50–60.

Squire, K.D. (2010). From information to experience: place-based augmented reality games as a model for learning in a globally networked society. *Teachers College Record*, 112(10): 2565–2602.

Strigel, C. and Pouezevara, S. (2012). *Mobile Learning and Numeracy: Filling Gaps and Expanding Opportunities for Early Grade Learning*. GIZ, Berlin.

Strobel, J. and van Barneveld, A. (2009). When is PBL more effective? A meta-synthesis of meta-analyses comparing PBL to conventional classrooms. *Interdisciplinary Journal of Problem-based Learning*, 3(1).

Svensk, K. (2013). *Mobile Device Security: Exploring the Possibilities and Limitations with Bring Your Own Device (BYOD)*.

Sweller, J. (2011). Cognitive load theory. *Psychology of Learning and Motivation*, 55: 37–76.

Tamim, R.M., Bernard, R.M., Borokhovski, E., Abrami, P.C. and Schmid, R.F. (2011). What forty years of research says about the impact of technology on learning: a second-order meta-analysis and validation study. *Review of Educational Research*, 81(1): 4–28.

Thomas, J.W. (2000). *A Review of Research on Project-based Learning*. Autodesk Foundation, San Rafael.

Tolkien, J.R.R. (1947). On fairy-stories. pp. 38–89. *In*: Lewis, C.S. (ed.). *Essays Presented to Charles Williams*. Oxford University Press, London.

Traxler, J. (2005). Defining mobile learning. pp. 261–66. *In: IADIS International Conference Mobile Learning*.

Trilling, B. and Fadel, C. (2009). *21st Century Skills: Learning for Life in Our Times*. John Wiley & Sons.

Vallgårda, A. and Fernaeus, Y. (2015). Interaction design as a bricolage practice. pp. 173–80. *In: Proceedings of the Ninth International Conference on Tangible, Embedded, and Embodied Interaction TEI '15*. Association for Computing Machinery, New York.

van der Linden, J., Braun, T., Rogers, Y., Oshodi, M., Spiers, A., McGoran, D., Cronin, R. and O'Dowd, P. (2012). Haptic lotus: a theatre experience for blind and sighted audiences. pp. 1471–1472. *In: CHI '12 Extended Abstracts on Human Factors in Computing Systems*.

Van Haren, R. (2010). Engaging learner diversity through learning by design. *E-Learning and Digital Media*, 7(3): 258–71.

Viereck, G.S. (1929). What life means to Einstein. *The Saturday Evening Post*, 26 October 1929, Curtis Publishing Co.

Vygotsky, L.S. (1980). *Mind in Society: The Development of Higher Psychological Processes*. Harvard University Press.

Webel, S., Bockholt, U., Engelke, T., Gavish, N., Olbrich, M. and Preusche, C. (2013). An augmented reality training platform for assembly and maintenance skills. *Robotics and Autonomous Systems, Models and Technologies for Multi-modal Skill Training*, 61(4): 398–403.

Willim, R. (2013). Enhancement or distortion? From the claude glass to instagram. pp. 353–59. *In*: Reader, S. (ed.). *Projections 09, Editorial Collective*. Lund University, Lund.

Winters, N. (2006). What is mobile learning? pp. 5–9. *In*: Sharples, M. (eds.). *Big Issues in Mobile Learning: Report of a Workshop by the Kaleidoscope Network of Excellence Mobile Learning Initiative*. University of Nottingham, Nottingham.

Yilmaz, R.M. and Goktas, Y. (2017). Using augmented reality technology in storytelling activities: examining elementary students narrative skill and creativity. *Virtual Reality*, 21(2): 75–89.

Yousafzai, A., Chang, V., Gani, A. and Noor, R.M. (2016). Multimedia augmented m-learning: issues, trends and open challenges. *International Journal of Information Management*, 36(5): 784–92.

Index